RECENT DEVELOPMENTS IN THEORETICAL STUDIES OF PROTEINS

Advanced Series in Physical Chemistry

Published

Vol. 1: Physical Chemistry of Solids: Basic Principles of Symmetry and Stability of Crystalline Solids
H. F. Franzen

Vol. 2: Modern Electronic Structure Theory
ed. D. R. Yarkony

Vol. 3: Progress and Problems in Atmospheric Chemistry
ed. J. R. Barker

Vol. 4: Molecular Dynamics and Spectroscopy by Stimulated Emission Pumping
eds. H.-L. Dai and R. W. Field

Vol. 5: Laser Spectroscopy and Photochemistry on Metal Surfaces
eds. H.-L. Dai and W. Ho

Vol. 6: The Chemical Dynamics and Kinetics of Small Radicals
eds. K. Liu and A. Wagner

Advanced Series in Physical Chemistry – Vol. 7

RECENT DEVELOPMENTS IN THEORETICAL STUDIES OF PROTEINS

Editor

Ron Elber

Department of Physical Chemistry
Hebrew University
Israel

World Scientific
Singapore • New Jersey • London • Hong Kong

Published by

World Scientific Publishing Co. Pte. Ltd.

P O Box 128, Farrer Road, Singapore 912805

USA office: Suite 1B, 1060 Main Street, River Edge, NJ 07661

UK office: 57 Shelton Street, Covent Garden, London WC2H 9HE

British Library Cataloguing-in-Publication Data
A catalogue record for this book is available from the British Library.

RECENT DEVELOPMENTS IN THEORETICAL STUDIES OF PROTEINS

ISBN 981-02-2196-7

Printed in Singapore.

ADVANCED SERIES IN PHYSICAL CHEMISTRY

INTRODUCTION

Many of us who are involved in teaching a special-topic graduate course may have the experience that it is difficult to find suitable references, especially reference materials put together in a suitable text format. Presently, several excellent book series exist and they have served the scientific community well in reviewing new developments in physical chemistry and chemical physics. However, these existing series publish mostly monographs consisting of review chapters of unrelated subjects. The modern development of theoretical and experimental research has become highly specialized. Even in a small subfield, experimental or theoretical, few reviewers are capable of giving an in-depth review with good balance in various new developments. A thorough and more useful review should consist of chapters written by specialists covering all aspects of the field. This book series is established with these needs in mind. That is, the goal of this series is to publish selected graduate texts and stand-alone review monographs with specific themes, focusing on modern topics and new developments in experimental and theoretical physical chemistry. In review chapters, the authors are encouraged to provide a section on future developments and needs. We hope that the texts and review monographs of this series will be more useful to new researchers about to enter the field. In order to serve a wider graduate student body, the publisher is committed to making available the monographs of the series in a paperbound version as well as the normal hardcover copy.

Cheuk-Yiu Ng

PREFACE

Modeling, simulations and theoretical studies of biological molecules have expanded tremendously in the last few years. Some of the growth was attributed to the rapid advances in computer technology. Faster and more economical computers made it possible to simulate larger systems for longer periods of time at lower costs. The impact on simulations of biological macromolecules such as proteins was thus profound.

However, this was only part of the story. Equally significant has been the explosion in the number of computational tools available today to investigators in the field. This is especially striking considering that the field was overwhelmed, just a few years ago, by essentially one methodology — the Molecular Dynamics (MD) approach. The strength of MD is its simplicity and generality. The prime weakness of the MD approach is the restriction on the time scale and the limited conformation space that one can sample.

A few recent uses of the MD approach are described by K. Kuczera. The studies of dynamics and thermodynamics of the globins are of special interest. This is due to the wealth of experimental data available, which makes detailed atomic studies worthwhile. Furthermore, since many processes in the globins occur rapidly (e.g. nitric oxide diffusion and recombination), the limited time scale of Molecular Dynamic simulations is less of a problem.

Due to the specificity of proteins and their complexity, it was difficult to come up with an analytical theory of proteins. It is desirable of course to have methods of sufficient generality to address more than one or a few special cases.

Difficulty is not impossibility. Motivated by a problem (protein folding) of utmost importance in biochemistry and biophysics, researchers from different fields entered the area of theoretical biophysics. Armed with a set of tools new to biophysics of proteins, researchers understood a number of beautiful and general principles of protein design. The "polymer physicists"' approach to proteins is discussed by Garel, Orland, and Thirumalai.

The title of the chapter "Analytical theories of protein folding" clearly indicates that this is one contribution that is independent of the revolution in computational power.

Another development motivated by the protein folding problem is the work by Luthey-Schulten, Goldstein and Wolynes. This investigation aims at the design of an effective potential that will fold a protein to its correct structure at maximum speed. The principle of the design of such a potential came from the "physicists"' approach to proteins and spin glass models. In fact, Peter Wolynes should be credited for many early studies and ideas on the relation between spin glasses and proteins.

Even if an exact potential energy surface is provided, the problem is not yet solved. It is still a non-trivial problem; how to find the global (free) energy minimum among all of the possible alternative conformations. Fortunately, a number of original algorithms appeared in the field of "global optimization", suggesting that the problem previously believed to be too difficult to solve may be more tractable after all. Straub describes the philosophy and the applications of the new class of optimization techniques.

Low energy minima provide information on probable structures. However, they cannot tell us about kinetics. To bridge the gap between static information and dynamics it is necessary to compute properties related to transitions between alternative minima. Possible approaches to the problem are described in Chapter 2.

Finally we are concerned with electrostatic and ion channels. Ion channels belong to a specific type of proteins that transport ions through membranes. While electric field is important almost everywhere in molecular biology, its importance cannot be overemphasized in channel, in which a charged entity (the ion) is passed through a low dielectric medium (the membrane) using a special biological machinery (the channel). The discussion on electrostatic properties in channels and beyond, in dynamics and in statics of biomolecules, is provided in the last chapter by Eisenberg.

Ron Elber

CONTENTS

CHAPTER 1

DYNAMICS AND THERMODYNAMICS OF GLOBINS

KRZYSZTOF KUCZERA

Departments of Chemistry and Biochemistry,
University of Kansas,
2010 Malott Hall, Lawrence, KS 66045
USA

Contents

1. Introduction

Molecular dynamics simulations of macromolecules were launched over 15 years ago[1] and have since then become an increasingly used and powerful tool in understanding basic physico-chemical properties, structure, dynamics and functions of these complex and interesting systems. In this chapter I review our current progress in this understanding, illustrated with results obtained for proteins of the globin family — mainly myoglobin and hemoglobin.

The power of computer simulations of molecular dynamics lies in the enormous detail of information they provide by probing motions of individual atoms on a subpicosecond time scale. Analysis of this information leads to important insights into the microscopic underpinnings of observed molecular properties. Although approximate, simulation results are currently of sufficient accuracy to be complementary to experimental studies. In this complementary relationship, simulation methods are tested by comparison with available experimental data, used to interpret the data in terms of atomic-level events, and finally to predict novel phenomena and suggest new experiments. A growing number of studies employ the joint experimental-simulation approach to macromolecular systems.[2–6]

A molecular dynamics simulation consists of solving Newton's equations of motion for a system of particles on a computer, and analyzing the resulting trajectory — the collection of snapshots of the particles' positions and velocities at consecutive points in time. To solve Newton's equations we need two important pieces of information about our system: the initial conditions (initial positions and velocities of particles) and the forces acting on the particles in any given configuration. The forces are usually derived from semi-empirical analytical potential energy functions. The trajectory is analyzed to yield quantities of interest — including time evolution and averages for structural and energetic properties. Finally, the results are interpreted in terms of observable quantities and compared with available experimental data. After a short introduction to the properties of myoglobin and hemoglobin, the remaining parts of the Introduction deal with different aspects of simulations in detail. For excellent reviews of molecular dynamics simulations see Ref. 7, for biomolecular applications — Ref. 8, and for free energy simulations — Refs. 9–11.

1.1. *Myoglobin and Hemoglobin Structure and Function*

Myoglobin (Mb) and hemoglobin (Hb) are an "oxygen carrying team" in vertebrate organisms.[12] Hemoglobin binds oxygen in the lungs and transports it through the bloodstream to other parts of the organism, while myoglobin stores oxygen in muscle cells. Both Mb and Hb bind the oxygen to an iron atom in prosthetic heme groups. The heme is covalently bound to the protein and is tightly surrounded by the polypeptide chain.

Myoglobin is a monomer, consisting of a single polypeptide chain and one heme group. The protein has a compact globular structure, with eight alpha-helical fragments labeled A to H. In the case of sperm whale Mb, for which most of the simulations were performed, the protein has 153 amino acid residues, and is roughly spherical in shape, with a radius of ca. 15 Å. The structure of sperm whale myoglobin is shown in Fig. 1.

Hemoglobin is a tetramer, consisting of two α and two β subunits. The subunits differ slightly in size and sequence, but have spatial structures very similar to each other, and to Mb. Normal human hemoglobin (HbA) consists of 576 amino acid residues and four heme groups. It is roughly spherical in shape, with a radius of ca. 30 Å. The structure of human hemoglobin is shown in Fig. 2.

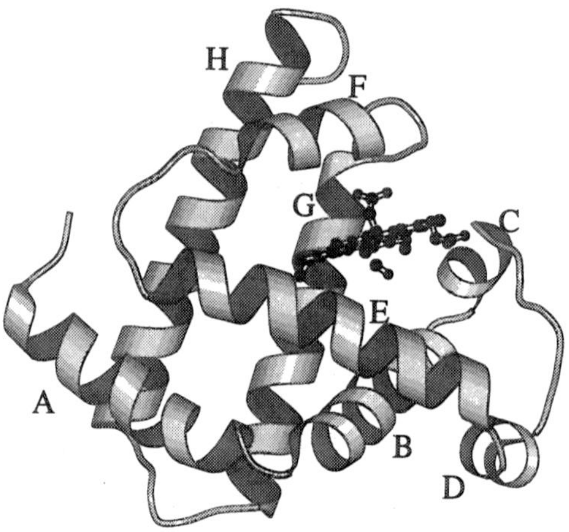

Fig. 1. Three-dimensional structure of myoglobin. Protein backbone shown as helical ribbons, labeled A–H, connected by loops; heme and CO ligand shown in ball-and-stick representation. Structure of sperm whale MbCo[18] taken from PDB file 1 mbc.[114,115] Drawing created using program MOLSCRIPT.[116]

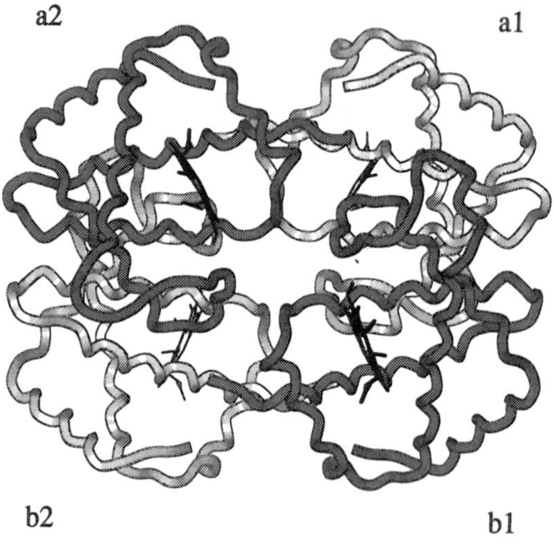

Fig. 2. Three-dimensional structure of hemoglobin. Protein backbone traced out for the four subunits, α (upper right), β (lower right), α_2 (upper left), and β_2 (lower left). Structure of human deoxy-Hb[15] taken from PDB file 3hhb.[114,115] The $\alpha_2\beta_2$ coordinates were generated by rotating $\alpha_1\beta_1$ by 180° around the y (vertical) axis. Drawing created using program MOLSCRIPT.[116]

Although they are not enzymes, myoglobin and hemoglobin participate in chemical reactions — ligand binding and dissociation at the heme. The possible ligands include oxygen (O_2), carbon monoxide (CO), and nitric oxide (NO). What makes myoglobin and hemoglobin especially interesting systems for molecular dynamics simulations is that motion plays an important role in their biological function. Two types of motion are relevant, involving equilibrium fluctuations and a nonequilibrium structural transition.

From the refined X-ray structures of both liganded and unliganded Mb[13,14] and Hb,[15,16] it can be seen that there is no evident pathway for ligand travel between the heme binding site and the external environment. Thus, structural fluctuations must accompany ligand diffusion through the protein. Crystal structures of liganded and unliganded heme systems — protoheme, Mb or Hb,[17] indicate that a structural change occurs upon ligand dissociation. In the liganded or oxy state, the heme iron atom has six ligands — four are pyrrole nitrogens of the heme, one is a nitrogen atom ($N_{\epsilon 2}$) of a histidine sidechain (the proximal histidine) which covalently links the heme to the protein, and the sixth is a small, mobile molecule — O_2, CO, NO etc. The liganded heme is planar, with the Fe atom in the mean heme plane[17] (see Fig. 3a and b). After dissociation of the sixth ligand, the heme assumes the unliganded or deoxy structure; the Fe atom moves out of the mean heme plane by about 0.5 Å, while the heme itself becomes domed[17] (see Fig. 3c and d).

In myoglobin this shift results in a small-scale tertiary structural change involving mostly a shift of the F helix (containing the proximal histidine residue) relative to the heme. The backbone rms deviation between liganded and unliganded structures is only about 0.3 Å.[18] In hemoglobin the tertiary structural change of the subunits is of similar scale as in myoglobin; however, there is a relatively large quaternary structural change, consisting of a rotation of the two halves of the molecule by about 15° and a relative translation by about 1 Å.[12] This structural transition is intimately involved in the regulation of Hb affinity for oxygen, and also plays an important role in Hb sickling, which will be described in Sec. 4.

Clearly, knowledge of the dynamics of structural fluctuations, ligand diffusion and the structural transition at the atomic level is important if we wish to understand the mechanism of the biological function of heme proteins.

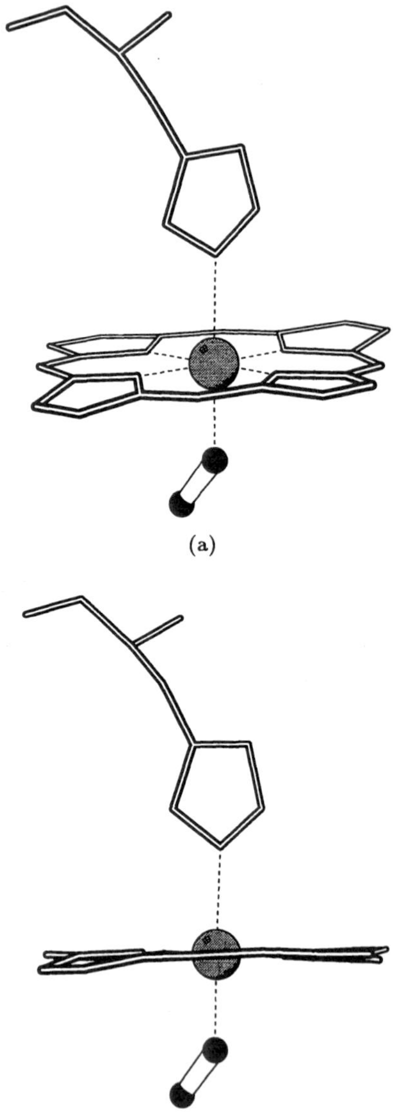

(a)

(b)

Fig. 3. Comparison of liganded and unliganded heme structures. (a) and (b) — two views of the planar six-liganded heme, including the proximal histidine (His 93) and CO, from sperm whale MbCO[18] (PDB file 1 mbc[114,115]).

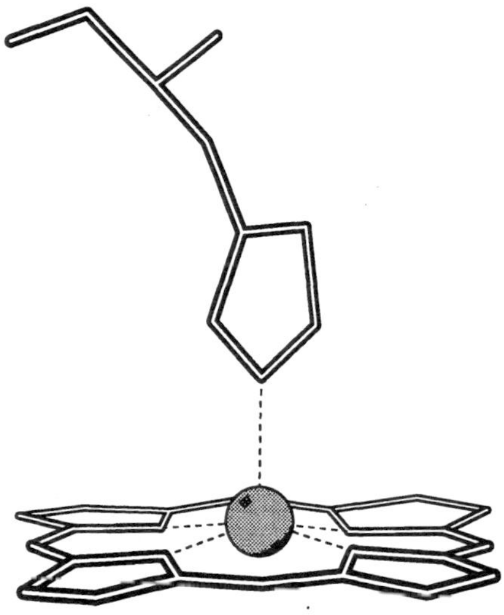

(c)

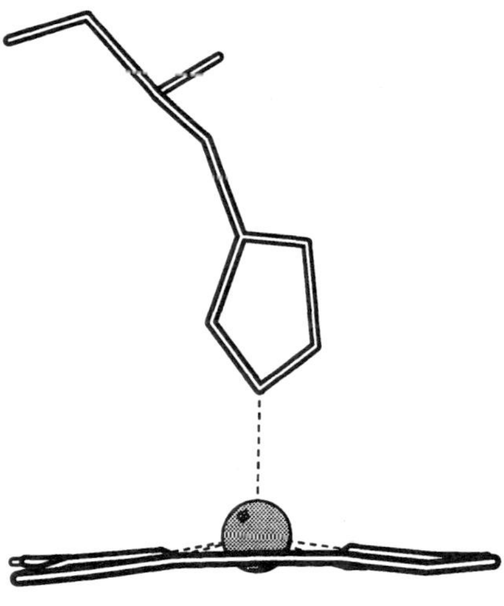

(d)

Fig. 3. (c) and (d) — two views of the domed five-liganded heme, including the proximal histidine (His 93), from sperm whale deoxy-Mb[14] (PDB file 1 mbd[114,115]). For clarity, only heavy atoms of heme porphyrin skeleton and His residue are shown. The central Fe atom is displayed as a sphere of 0.5 Å radius; bonds between Fe and ligands shown as dashed lines. Drawings created using program MOLSCRIPT.[116]

1.2. *Potential Energy Functions*

In classical mechanics the answers to all questions concerning molecular structure, interactions and dynamics lie in the potential energy function. The potential energy determines the relative stabilities of different molecular structures. The dynamics of the system is determined by forces acting on the atoms, which are the first derivatives of the potential with respect to atomic coordinates. The second derivatives of the potential energy with respect to atomic coordinates are the harmonic force constants, which enable the calculation of molecular vibrational spectra.

At present the only viable approach for macromolecular systems is to use empirical energy functions, which can usually be expressed in a simple form, such as:

$$\begin{aligned} U = & \frac{1}{2} \sum_{\text{bonds}} k_b(b - b_0)^2 + \frac{1}{2} \sum_{\text{angles}} k_\theta(\theta - \theta_0)^2 \\ & + \frac{1}{2} \sum_{\text{impropers}} k_\omega(\omega - \omega_0)^2 \\ & + \frac{1}{2} \sum_{\text{dihedrals}} k_\phi[1 + \cos(n\phi + \delta)] \\ & + \sum_{\text{atom pairs } ij} \left[\frac{q_i q_j}{R_{ij}} - \frac{A_{ij}}{R_{ij}^6} + \frac{B_{ij}}{R_{ij}^{12}} \right], \end{aligned} \tag{1}$$

which is employed in CHARMM[19] and other macromolecular simulation programs. In this equation b, θ, ω, ϕ and R_{ij} are the actual bond lengths, valence bond angles, improper and proper dihedral angles and atom-atom distances for a given configuration of the system, respectively; $k_b, b_0, k_\theta, \theta_0, k_\omega, \omega_0, k_\phi, n, \delta, q_i, A_{ij}$ and B_{ij} are parameters. One way of looking at potential energy functions of this type is to treat them simply as mathematical models leading to certain predictions. However, both the method of derivation and the form of the function gives the parameters physical meaning.

The first three terms in Eq. (1) describe energies of deformation from equilibrium of chemical bond length (b), bond angles (θ) and improper dihedrals (ω) and angle deformations from equilibrium values. These distortions are usually small under normal conditions, allowing the energy to be described in the form of a sum of harmonic (Hooke's law) terms. The presence of these terms assures that throughout the simulation the molecule preserves its basic chemical structure.

The fourth term in Eq. (1) describes the energy of torsion around chemical bonds — it is periodic and is modeled as a cosine term (or sum of several cosines with different periodicities). The presence of the $\cos(n\phi)$ term shows that the molecule will sample n distinct potential energy minima as the complete rotation around the chemical bond is performed.

The last line of Eq. (1) describes the nonbonded interactions between pairs of atoms, and has two contributions: electrostatic interactions between point charges and a Lennard–Jones 6–12 potential which models the interplay of weak attraction between neutral atoms at moderate distances and strong repulsion on close contact.

The force constants k_b, k_θ, k_ϕ and k_ω can be determined from vibrational spectroscopy, equilibrium values for bond lengths and angles b_0, θ_0 and ω_0 — from crystallography and microwave spectroscopy of model compounds. Dihedral angle torsional potentials can be obtained from infrared and microwave spectroscopic data. Nonbonded interaction parameters q, A and B can be determined by a wide variety of methods, both experimental (e.g. crystallography, thermodynamic measurements, gas phase scattering) and theoretical (quantum mechanical calculations). Generally, parameter values are determined in studies of small molecule models such as amino acids or their fragments. The assumption that the same parameters can be used to describe macromolecules is verified by comparison of calculated and observed properties.

The representation of the potential energy of a macromolecular system by Eq. (1) involves a number of approximations, some of which are discussed in Sec. 5. For a more complete discussion of potentials, see Ref. 8.

For a peptide or protein system, the potential energy would thus involve the terms describing deformations of all the bonds, angles and dihedrals from their minimum values, as well as nonbonded interactions between atoms which are not chemically bonded, but are brought close together due to the three-dimensional folding of the molecule (tertiary structure). To simulate a solvated protein, we add additional terms in the potential, describing bond and angle deformations of the water molecules as well as water-water and water-protein nonbonded interactions.

It is truly amazing that such a simple functional form of the potential can describe the wide range of interactions occurring in macromolecules —

formation of salt links, hydrogen bonds, solvation, complementary fitting of hydrophobic surfaces. The simple form is also misleading. For a protein of about 100 amino acids there might be 50,000–100,000 individual terms in the potential Eq. (1). This is a complicated multidimensional function, the basic properties of which are not yet fully understood.

1.3. *Solving Newton's Equations of Motion*

Consider a molecular system composed of N atoms. Newton's equations of motion can be expressed as

$$m_i \mathbf{a}_i = m_i \frac{d^2\mathbf{r}}{dt^2} = \mathbf{F}_i \quad i = 1, \ldots, N \tag{2}$$

where m_i is the mass, $\mathbf{a}_i$ the acceleration, $\mathbf{r}_i = (x_i, y_i, z_i)$ the position, and $\mathbf{F}_i$ the force acting on atom i, and t is the time. Given a potential U describing the potential energy as a function of coordinates (e.g. Eq. (1)), the forces can be expressed as partial derivatives of U with respect to the atomic positions (i.e. components of the gradient):

$$\mathbf{F}_i = -\boldsymbol{\nabla}_i U = -\left(\frac{\partial U}{\partial x_i}, \frac{\partial U}{\partial y_i}, \frac{\partial U}{\partial z_i}\right) \quad i = 1, \ldots, N\,. \tag{3}$$

A widely used and popular method of numerically solving Newton's equations of motion (Eq. (2)) with the forces (Eq. (3)) is the Verlet algorithm,[20] which can be derived from two Taylor expansions:

$$\mathbf{r}_i(t \pm \Delta t) = \mathbf{r}_i(t) \pm \frac{d\mathbf{r}_i}{dt}\Delta t + \frac{1}{2}\frac{d^2\mathbf{r}_i}{dt^2}(\Delta t)^2 + \cdots\,. \tag{4}$$

After substituting $\mathbf{v} = \frac{d\mathbf{r}}{dt}$ and $\mathbf{a}_i = \frac{\mathbf{F}_i}{m_i}$ adding the two equations, Eq. (4) leads to an equation for propagation of the positions accurate to terms of the fourth power in Δt:

$$\mathbf{r}_i(t + \Delta t) = 2\mathbf{r}_i(t) - \mathbf{r}_i(t - \Delta t) + \frac{\mathbf{F}_i}{m_i}(\Delta t)^2 + O[(\Delta t)^4] \tag{5}$$

while subtracting them gives the equation for propagating the velocities accurate to terms of the third-power in Δt:

$$\mathbf{v}_i(t + \Delta t) = \frac{\mathbf{r}_i(t + \Delta t) - \mathbf{r}_i(t - \Delta t)}{2\Delta t} + O[(\Delta t)^3]\,. \tag{6}$$

The above equations provide a numerical solution for Newton's equations of motion by generating positions $\mathbf{r}_i$, and velocities $\mathbf{v}_i$ for all atoms $i = 1, \ldots, N$ at times $t+\Delta t$, $t+2\Delta t, \ldots$. The Verlet algorithm has very interesting properties similar to those of the exact solutions.[21] The algorithm has been used to simulate dynamics of a variety of systems, from simple liquids to macromolecules. For biological macromolecules, with or without inclusion of explicit solvent water, a time step Δt of 1–2 fs is usually used.[8] Current biomolecular simulations are done for systems of 10^3–10^4 atoms, and trajectories of 100 ps to 1 ns. Calculating a 1 ns trajectory with a 1 fs time step involves 10^6 evaluations of energy and gradient.

1.4. *Initial Conditions*

To start a simulation, we need initial coordinates and velocities for all atoms. The starting structure is usually obtained from X-ray crystallography or NMR experiments.[8,22] First, energy minimization is performed to ensure consistency of the structure with the potential energy function employed, and to bring the coordinates close to a local minimum, where forces acting on atoms are small. The next stage is heating, consisting of a series of consecutive random velocity assignments from Maxwell–Boltzmann distributions corresponding to increasing temperatures, interspersed with periods of unperturbed dynamical evolution. The purpose of this phase is to bring the system close to the desired final temperature and evenly distribute kinetic energy throughout the macromolecule. Heating is followed by an equilibration phase, when the system is allowed to evolve, with temperature monitored and adjusted as necessary to the desired target value. The equilibration is usually assumed to be complete when the temperature of the system does not change in a systematic way. In MD simulations, the temperature T is defined through the kinetic energy K:

$$K = \frac{1}{2}\sum_{i=1}^{N} m_i v_i^2 = f\, k_B T\,, \tag{7}$$

where k_B is the Boltzmann constant and f is the number of degrees of freedom ($f = 3N - 6$ usually, since we eliminate the overall translations and rotations of the system). The final stage of a simulation is the generation of a trajectory which will be used to analyze the properties of interest. The trajectory is simply a set of coordinates (and possibly velocities) of the system — "snapshots" of its time evolution.

1.5. *Analyzing Simulations: Averages*

Experimental measurements are performed not on individual molecules, but on macroscopic samples. As a result, observed quantities are averages, both over the molecules in the sample and over the time scale of the measurement. For systems at equilibrium, statistical mechanics provides a connection between such averages and properties of microscopic states, defined by specific positions and velocities (or momenta) of the individual atoms. In the canonical ensemble, in which the number of molecules N, volume V and temperature T are kept constant, the probability distribution for points in the phase space (p, q) is[23]:

$$\rho(p,q) = \frac{e^{-H(p,q)/k_BT}}{\int dpdq \; e^{-H(p,q)/k_BT}} \tag{8}$$

and the ensemble average of any quantity $X(p, q)$ is:

$$\langle X \rangle = \int dpdq \; X(p,q)\rho(p,q), \tag{9}$$

where p and q are the momenta and coordinates of all the N particles (atoms) in the system, $H(p, q)$ is the Hamiltonian, which gives the energy of the system as a function of p and q, and k_B is the Boltzmann constant. Usually $H(p,q) = K(p)+U(q)$, where K is the kinetic, and U the [potential energy.] In calculating ensemble averages we are considering an infinite number of copies of our system prepared with different initial conditions at a fixed point in time. Ensemble averages may be directly evaluated from Monte Carlo simulations of many-particle systems.[24]

A significant amount of information can be derived from molecular dynamics simulations. Since we are solving Newton's equations of motion, the resulting dynamics represents the true classical time evolution of our system, yielding forms, amplitudes, time scales and energetics for different microscopic events, such as atomic fluctuations, side chain rotations, helix-helix reorientations, etc. which occur within the time frame of the simulation.[8] In order to make molecular dynamics simulation results comparable to experimental observations, we also have to evaluate averages. Since in MD we are usually dealing with a single copy of a system evolving over some time interval, we can directly evaluate time (or trajectory) averages, defined as

$$\bar{X}_t = \frac{1}{t}\int_0^t X(\tau)d\tau \approx \frac{1}{N_t}\sum_{k=1}^{N_t} X(t_k), \tag{10}$$

where $X(\tau) = X(p(\tau), q(\tau))$ is the value of quantity X at time τ and the final equation shows how in practice the time average is estimated as the arithmetic average over the N_t trajectory points $t_k, k = 1, \ldots, N_t(t = N_t \Delta t)$. Thus, in molecular dynamics the time evolution of a single copy of the system is determined and X is averaged over the microstates sampled in the trajectory.

The *ergodic hypothesis* states that these two averages should be equivalent in the long time limit[25]:

$$\langle X \rangle = \bar{X} = \lim_{t\to\infty} \bar{X}_t . \tag{11}$$

We will assume that this is true and in further discussions use the term "average" to denote both time and ensemble averages at equilibrium. Real computer simulations can only be propagated for finite time intervals, providing approximate estimates of true averages. If the finite set of configurations and momenta sampled during a simulation is able to correctly reproduce the properties of the system, it is called a representative sample. To check whether a given sample is representative, the usual way is to compare calculated and observed properties of the simulated system, monitor averages for systematic change, and compare results from separate trajectories started from different initial conditions. That configuration samples obtained by MD simulations are usually representative follows from the method of preparing the trajectory (see Sec. 1.4) and from Eq. (3). Since the forces acting on the atoms are driving them towards regions of lower potential energy U, we expect that after neglecting the initial heating and equilibration phase, the system will sample regions of relatively low U. These are the physically most meaningful configurations, which have the largest weights in the ensemble average (Eqs. (9) and (27)).

1.6. *Equilibrium vs. Nonequilibrium Dynamics*

Molecular dynamics simulations may be used to study two general classes of motion — equilibrium fluctuations, and nonequilibrium or relaxation type dynamics. The first type of motion may be generated by the procedure described in Sec. 1.4, where the system was carefully prepared in a conformation close to the experimentally observed (and presumably biologically active) one, with a random thermal distribution of velocities. The simulated dynamics is then expected to model thermal fluctuations at equilibrium of the experimental system.

To understand the difference between equilibrium and nonequilibrium dynamics, let us consider the standard example of irreversible gas expansion.[25,26] The gas initially occupies one part of the volume of a container divided by a partition. After the partition is removed, the gas irreversibly expands to fill the whole available volume. Analogously, in simulations of nonequilibrium dynamics, we prepare a system in a nonequilibrium state and follow its time evolution as it relaxes towards a new equilibrium. This nonequilibrium initial state may have unequal distributions of kinetic[27] or potential energy.[5,6,28,29]

Observations of relaxation in a macroscopic sample will provide results averaged over the constituent molecules. The individual molecules will generally exhibit somewhat different time scales and pathways of relaxation, due to differences in their initial conditions. In order to adequately simulate observed relaxation phenomena, it is not sufficient to calculate a single long trajectory, since the system will tend to reach its new equilibrium after some characteristic time. Rather, it is necessary to generate a number of independent trajectories, starting with different initial conditions, and obtain from them the average time evolution. Clearly, variations in relaxation time scales and pathways between the different trajectories also constitute valuable information about properties of the studied system.

2. Dynamics: Equilibrium Fluctuations of Myoglobin

A number of molecular dynamics simulations of different forms of myoglobin under different conditions have been reported: met-Mb in vacuum at 300 K,[30] MbCO in vacuum at 325 K and 80 K.[29] MbCO under different environmental conditions at 100 and 300 K,[31] solvated holomyoglobin[32] and solvated apomyoglobin.[33] These studies unfold an interesting picture of equilibrium fluctuations in myoglobin, which provide important insights into the general nature of protein structural flexibility and its relation to biological function. The questions studied included a characterization of the protein potential energy surface, average structure and fluctuations, nature of short and long term motion, and influence of environment and temperature on structure and dynamics.

2.1. *Number of Minima*

One of the more important results of molecular dynamics simulations was the characterization of the potential energy surface over which the motion of macromolecules occurs. This problem was undertaken by Elber and Karplus,[34] who performed energy minimizations for a series of structures from a 300 ps simulation of myoglobin at 300 K in vacuum.[30] They calculated the differences between pairs of trajectory structures separated by different time intervals. The differences were measured as the root mean square atomic position deviations (RMSD) after superposition. For two molecular structures a and b, the RMSD is defined as:

$$\mathrm{RMSD}_{ab} = \sqrt{\frac{1}{N}\sum_{i=1}^{N}\left[(\mathbf{r}_i)_a - (\mathbf{r}_i)_b\right]^2} \tag{12}$$

where $(\mathbf{r}_i)_a$ and $(\mathbf{r}_i)_b$ are the positions of atom i in structures a and b, respectively. To ensure that the RMSD measures only internal structure differences, a and b are first superimposed by rigid-body translations and rotations.

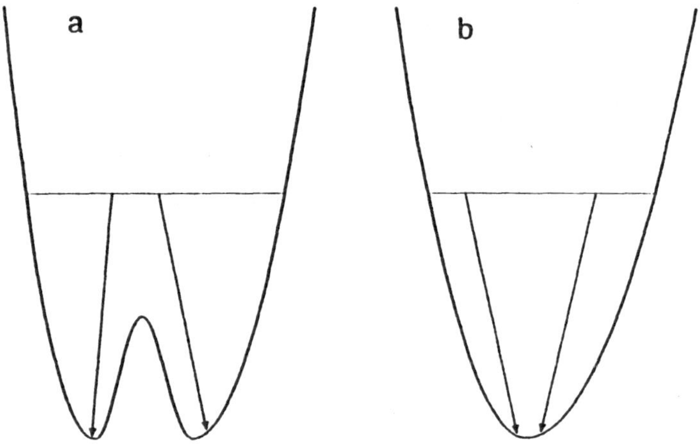

Fig. 4. Criterion for determining whether two trajectory structures correspond to the same potential energy minimum. (a) Divergence; trajectory structures correspond to different local minima. (b) Convergence; trajectory structures correspond to the same local minimum. (Analogous to Fig. 1 from Ref. 34.)

Elber and Karplus[34] found that for pairs of trajectory points separated in time by less than 0.2 ps, the RMSD between the minimized structures was lower than the RMSD between the corresponding starting structures. For pairs of trajectory points separated in time by more than 0.2 ps the reverse was true — the RMSD between the minimized structures was greater than for the corresponding starting coordinates.

As can be seen from the graphical illustration in Fig. 4, the logical conclusion is that at 300 K the protein resides within a single potential energy minimum for about 0.2 ps on average, with transitions between different minima occurring on longer time scales.[34] The energies of the sampled potential energy minima were within 20 K (40 cal/mol) of each other. Thus, they would all have significant populations at 300 K.[34]

2.2. *Structural Characterization*

One of the basic characteristics of a molecular dynamics trajectory is the RMSD of the average coordinates from the experimental structure used to start the simulation. In vacuum simulations RMSD values between 100 ps trajectory averages and the starting X-ray structures were in the range of 1.0 to 1.9 Å.[29–31] In simulations which include sufficient amount of solvent water, the RMSD from the X-ray structures was generally smaller than in vacuum, and was in the 1.0 Å range.[31] Generally, rms deviations from the starting structures are about twice the magnitude of the corresponding trajectory root mean square atomic fluctuations (see Sec. 2.3). Thus, the trajectories explore essentially the same region of conformational space as that occupied by experimentally determined structures. A measure of the explored conformational space can be obtained from Fig. 5 in which the C_α traces of several structures from the simulation[29] are presented.

There are several possible reasons for the systematic deviations between experimental and MD average structures. The most important is probably differences in the environment. Crystal proteins are highly solvated[35] and are in contact with crystal neighbors. In vacuum simulations both these environmental components are missing, and this must influence the structure and dynamics even if corrective measures such as using a distance-dependent dielectric "constant"[29] or scaling down of atomic charges lying close to the protein surface[30] are used. Based on comparison of vacuum and fully solvated trajectories of Mb generated using the same simulation protocol, solvation appears to decrease RMSD from the X-ray structures

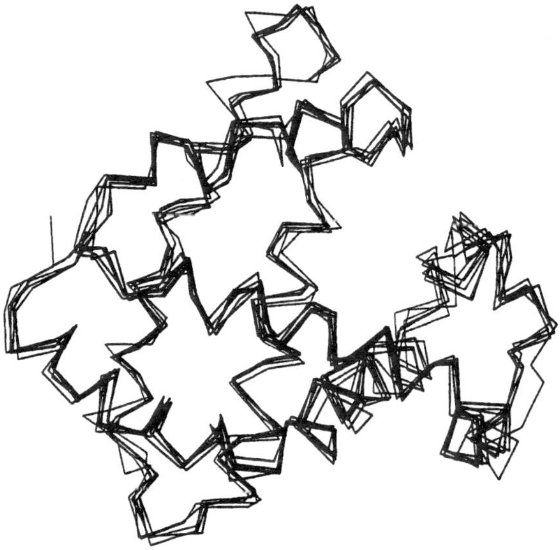

Fig. 5. C_α traces of five trajectory structures sampled every 30 ps from a vacuum simulation of MbCO[29] and the structure of sperm whale MbCO.[18]

by about 40%.[31] Differences between MD simulations in water and X-ray structures may be explained in part by lack of crystal neighbors in the trajectories and may depend on the details on crystal packing. A simulation of a partially hydrated "cluster" of three MbCO molecules yielded average RMSD from X-ray structures very similar to those found for a single MbCO.[31]

Systematic errors in the simulations are another major factor contributing to the deviations. The main effects in this class are insufficient sampling of the conformational space, the approximate nature of the potential energy functions used and further approximations introduced in evaluating the nonbonded interaction part of the potential, mainly the treatment of the long-range electrostatic interactions. The influence of truncation schemes on protein dynamics has been examined extensively by Bernard Brooks and coworkers.[31,36]

An important parameter used to characterize macromolecular size is the radius of gyration. Vacuum simulations give radii of gyration for myoglobin that are 1–3% lower than the experimental value, while solvated simulations tend to deviate less than 1%.

2.3. *Atomic Fluctuations at Room Temperature*

Another important characteristic of a trajectory is the mean square atomic position fluctuations (MSF), defined as

$$\langle \Delta r_i^2 \rangle = \frac{1}{N_t} \sum_{k=1}^{N_t} (\mathbf{r}_i(t_k) - \langle \mathbf{r}_i \rangle)^2 \tag{13}$$

with

$$\langle \mathbf{r}_i \rangle = \frac{1}{N_t} \sum_{k=1}^{N_t} \mathbf{r}_i(t_k) \tag{14}$$

where the angled brackets denote trajectory (time) averages, $\mathbf{r}_i(t_k)$ is the position of atom i at step k and N_t is the total number of integration steps. To simplify considerations, the atomic position MSF are usually further averaged over classes of atoms, such as whole residues, backbone, sidechains, etc.

Average values of backbone atom mean square fluctuations found in myoglobin molecular dynamics trajectories of ca. 100 ps length at room temperature were 0.3–0.4 Å^2 in vacuum.[29–31] This is in excellent agreement with experimental estimates of 0.4 Å^2 (Ref. 37) obtained from X-ray temperature factors ("isotropic B factors"), calculated according to the formula[38]:

$$\langle \Delta r^2 \rangle = \frac{3B}{8\pi^2} . \tag{15}$$

In the simulations mean square fluctuations for sidechain atoms were approximately two times higher than in the backbone[29]; residues in loops exhibited higher mobility than those in α-helices, in agreement with experimental results.[37] Thus, the overall character of the atomic fluctuations is reproduced correctly.

A comparison of the fluctuations as a function of residue number shows significant deviations between vacuum simulations and X-ray results, with the largest differences occurring in the known areas of crystal intermolecular contacts.[29] Essentially, several loops and the polypeptide chain termini exhibit significantly larger fluctuations in the crystal. There are several possible reasons for these discrepancies — the system environment, systematic errors in the simulations and the presence of rigid-body contributions to protein dynamics in the crystal. At room temperature, the influence of static disorder on the observed X-ray B factors appears to be small.[37] In

MD simulations the center of mass linear and angular momentum of the system are constrained to be zero. Thus, we do not have overall rotations or translations in a simulation of a protein in vacuum; some motions of protein relative to the solvent can occur in simulations including water.

The basic shape of the B factor vs residue number dependence for the 1217 heavy atoms of MbCO may be obtained using a ten-parameter model of rigid body motions based on the TLS method of Schoemaker and Trueblood[39] without considering any internal motions at all.[29] The myoglobin loops, located on the protein surface and far from the center of mass naturally exhibit the largest mobilities in this model. In view of this result, it is difficult to decompose the atomic fluctuations from the crystallographic B factors into an "intrinsic" part due to internal motions, and an "external" one from rigid-body translations/rotations. Thus, loops are both regions of highest "intrinsic mobility", as found in MD simulations, and also regions of expected largest librational mobility.

Several studies of myoglobin dynamics in water have been published.[31-33] For simulations following a similar protocol, hydration tended to increase the mean square fluctuations by about 10% relative to vacuum.[31] Again, in simulations of 100 ps length at room temperature, very good agreement between calculated and observed atomic fluctuations was found.[31,32] The atomic fluctuations as function of residue number showed closer agreement to crystallographic results as compared to vacuum simulation results.[32] This suggests that crystal packing may play a less important role in determining atomic mobility, while the presence of solvent and possibility of rigid-body motions are more significant.

2.4. *Temperature Dependence of Atomic Fluctuations: The Glass Transition*

Computer simulations of biological molecules at temperatures outside the physiological range can yield useful information about their properties. Calculations at elevated temperatures are usually employed to efficiently sample conformational space.[40] Simulations at low temperatures provide insight into basic physical properties.[29,31,41] An additional stimulus for low-temperature simulations is the wealth of available experimental data, including X-ray crystallography,[42] Mössbauer spectroscopy[43,44] and neutron scattering.[45] This makes possible both the validation of simulation results by comparison with experimental data and explanation of the observed phenomena using the detailed microscopic picture of the system that emerges from the simulations.

For a harmonic oscillator we expect a linear increase of mean square position fluctuations $\langle \Delta x^2 \rangle$ with temperature T. This result can easily be derived in statistical mechanics[23]: for a system with potential energy $U = \frac{1}{2} f \cdot x^2$, the canonical ensemble average of U is $\frac{1}{2} k_B \cdot T$, where k_B is the Boltzmann constant. Thus

$$\frac{1}{2} k_B T = \langle U \rangle = \frac{1}{2} f \langle x^2 \rangle \quad \text{or} \quad \langle x^2 \rangle = \frac{k_B T}{f}. \tag{16}$$

The experimental data can be summarized as follows. In all methods, a monotonic increase of atomic fluctuations with increasing temperature is found. The X-ray results, which provide position fluctuations for all non-hydrogen atoms, include a contribution from static disorder[37] and exhibit a rather weak temperature dependence, with the backbone atom mean square

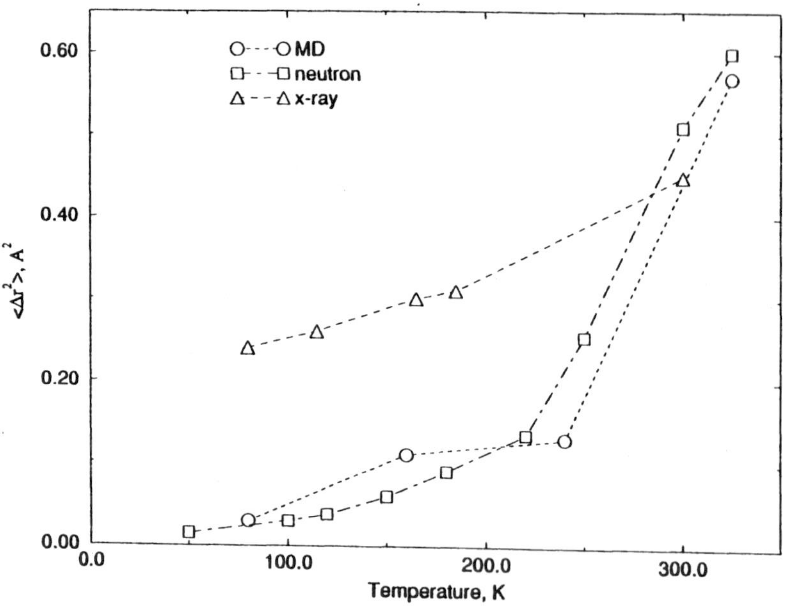

Fig. 6. Temperature dependence of atomic mean square fluctuations of myoglobin. Fluctuations averaged over all non-exchangeable hydrogen atoms from 50 ps molecular dynamics trajectories of MbCO in vacuum (circles) and from incoherent neutron scattering (squares),[29,41,48] compared with average backbone atom fluctuations derived from X-ray crystallographic B-factors (triangles).[42]

fluctuations $\langle \Delta r^2 \rangle$ changing from about 0.2 Å^2 at 80 K to about 0.5 Å^2 at 300 K.[42] Mössbauer spectroscopy measures position fluctuations of the iron atom which are coupled to motions of the other protein atoms[43,44] while incoherent neutron scattering measures fluctuations averaged over all nonexchangeable hydrogen atoms.[41,45] Both of these methods are insensitive to static disorder, and both give a similar picture: at low temperatures Mb behaves like a harmonic solid and $\langle \Delta r^2 \rangle$ increases linearly with temperature, while at higher temperatures (above 210 K for Mössbauer spectroscopy and above 180 K for neutron scattering), the mean square atomic fluctuations become anomalously large.[44,45] The conclusion is that new forms of motion become activated at about 200 K — the protein undergoes a dynamical transition, similar to the melting observed in glasses.[46,47] These new modes of motion are suggested to be torsional transitions.[45]

Results from molecular dynamics simulations of Mb on 100 ps time scales both in vacuum and in different solvation states[31,48] are qualitatively similar to the Mössbauer spectroscopy and neutron scattering data, exhibiting a harmonic behavior at low temperatures and faster-than-harmonic increase of $\langle \Delta r^2 \rangle$ with temperature above 200 K. There is also quantitative agreement: simulated and experimental fluctuation amplitudes are quite similar over the studied temperature range of 40–350 K. A comparison of temperature dependence of atomic fluctuations from several experimental studies and MD simulations is given in Fig. 6.

The X-ray determined atomic fluctuations at room temperature are close to those found in neutron scattering, Mössbauer spectroscopy, and MD simulations. However, the low temperature X-ray results are significantly higher than those found in the other methods. This indicates that the low temperature X-ray B factors are significantly influenced by the presence of static structural disorder in crystals, as has been suggested.[37] This conclusion is supported by results of MD simulations in which low temperature MbCO trajectories were generated by scaling down velocities at different points in a room temperature trajectory, leading to trapping of the molecule in distinct, non-overlapping regions of conformational space.[29]

2.5. *Time Evolution of Atomic Fluctuations*

Time scales of atomic position fluctuations in myoglobin have been analyzed.[29] This was done by dividing the MD trajectory into consecutive blocks of equal length, calculating fluctuations within each block and averaging over blocks. Varying the length of the blocks enables the determination of the development of the fluctuations in time. The results for backbone atoms of MbCO in vacuum at 325 K (120 ps simulation) and 80 K (50 ps) are presented in Fig. 7.

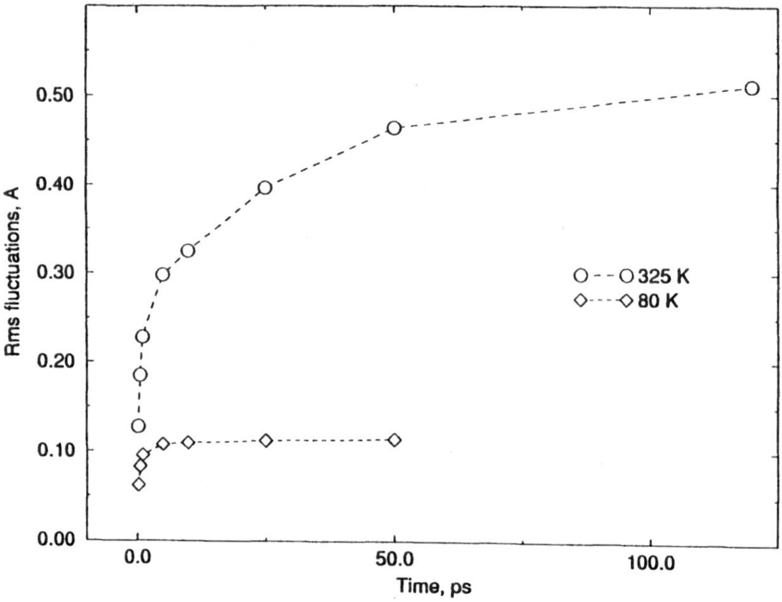

Fig. 7. Development of average backbone atom root mean square fluctuations with time in vacuum simulations of MbCO at 80 K and 325 K. (Analogous to Fig. 6a from Ref. 29.)

At 325 K, the root mean square atomic fluctuations (RMSF) reach about 25% of their final value in 0.2 ps and about 60% in 10 ps; there is a continued slow increase beyond this time, indicating that convergence has not occurred in the 120 ps simulation. At 80 K, 50% of the final value of RMSF is reached within 0.2 ps, and in 10 ps the trajectory has nearly converged.[29]

The ratios of the fluctuations at the two temperatures as a function of time is shown in Fig. 8. At 0.2 ps, the root mean square fluctuation ratio is very close to the harmonic value of 2, and deviations from the harmonic ratio increase with increasing time scale. This is in agreement with the results of Elber and Karplus[34] that myoglobin resides in a single potential energy minimum for times of about 0.2 ps and samples different minima on

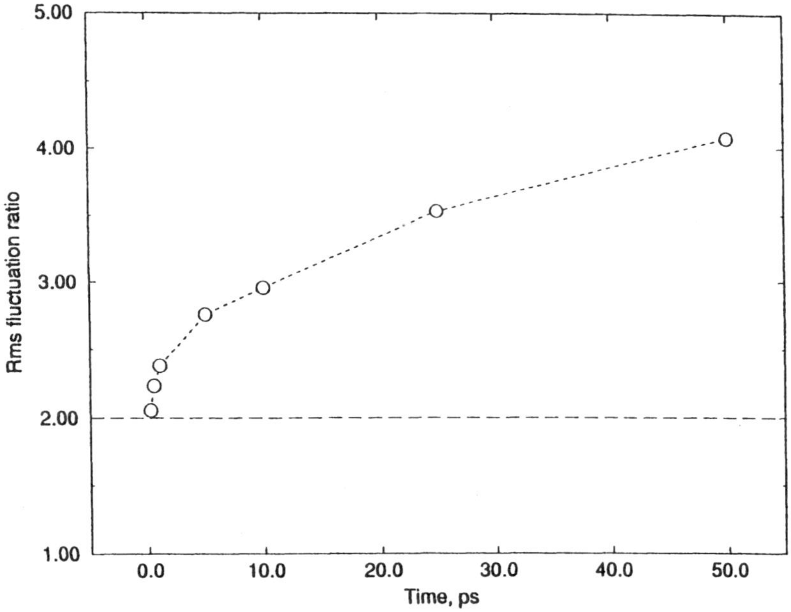

Fig. 8. Anharmonocity in MbCO atomic fluctuations. Ratio of average backbone root mean square at 325 and 80 K as a function of time scale. Dashed line shows value of 2 expected for harmonic motion. (Analogous to Fig. 7 from Ref. 29.)

longer time scales (see Sec. 2.1). For the 0.2 ps averages at 325 K and 80 K, fluctuations as a function of residue number are quite uniform. At room temperature nonuniformity develops with increasing time scale, suggesting that such effects as increased mobility of loops relative to helices are due to the anharmonic components of the dynamics, i.e. the transitions between substates. At 80 K fluctuations remain approximately uniform throughout the simulations, indicating that the protein resides in a single minimum.

A confirmation of the fact that MD simulations predict correct time scales and amplitudes of atomic motions in myoglobin, is the fact that trajectories can be used to successfully simulate incoherent neutron scattering structure factors and correlation functions over a range of temperatures.[41,48,49] The experimental and simulated correlation functions show a decay with a time scale of 0.2 ps at low temperatures, and the appearance of an additional slower (tens of picoseconds) decay above the glass transition temperature.[45] In view of the simulation results, the 0.2 ps decay is related to vibrations within a single potential energy well, while the slower relaxation is due to transitions between wells.

2.6. *Character of Myoglobin Motion*

The overall picture of protein equilibrium fluctuations emerging from the myoglobin simulations is consistent with the substate model of Frauenfelder.[50] At room temperature the macromolecules perform vibrational-type motion within a single potential energy well (or substate) on a time scale of 0.2 ps. This motion is harmonic and uniform in amplitude along the polypeptide chain. Dynamics on longer time scales involves transitions between different minima (substates). This type of motion is strongly anharmonic, and exhibits marked differentiation along the polypeptide chain, with loop regions moving with larger amplitudes than α-helices. The molecular events associated with the anharmonic motional component appear to be dihedral angle transitions, involving mostly sidechains in loops and surface residues.

Interestingly, a number of sidechains buried in the protein interior exhibit large dihedral angle fluctuations[29]; most of these are close to the known internal cavities of myoglobin.[51] Simulations of ligand escape from myoglobin suggest that the ligand visits several of these cavities before exiting to the outside environment.[52] The mobile internal residues appear to function as "gates" between the different cavities and between cavities and the protein surface[52] (see Sec. 3.2). Thus, there is a direct connection between the simulated myoglobin structural fluctuations and the biological function of the protein.

3. Dynamics: Nonequilibrium Relaxation

When high resolution X-ray structures of six-liganded (further denoted as "liganded") and five-liganded ("unliganded") myoglobin were determined, it became apparent that:

(a) the greatest change upon ligand dissociation occurs in the heme, with the Fe atom moving from a position in the mean heme plane in the liganded form to about 0.5 Å out of the plane; some heme doming also occurs, i.e. the heme pyrrole nitrogens also become displaced from the mean plane defined by all 24 heme skeleton heavy atoms (see Fig. 3).

(b) the protein tertiary structure changes upon ligand dissociation; the backbone rms deviation between the two structures is about 0.3 Å and there is a general shift of the proximal part of the protein away from the heme.[13,14,18]

(c) no evident pathway for ligand transport from the environment to the heme iron was found.

Measurements of ligand X rebinding after photodissociation from myoglobin and other heme protein systems[50,53,54] have provided the following phenomenological description of the kinetics:

$$MbX \rightleftharpoons (Mb + X)_{\text{pocket}} \rightleftharpoons (Mb + X)_{\text{protein}} \rightleftharpoons (Mb + X)_{\text{solution}} \,. \quad (17)$$

Here the subscript specifies the location of the ligand X, which directly after dissociation resides in the heme pocket and consequently diffuses through the protein to the outside solvent. Each of the designated species may involve a number of substates.

At low temperatures (below 200 K), the geminate rebinding of the CO ligand is nonexponential (geminate rebinding occurs when the dissociated ligand does not leave the protein matrix prior to recombination). This has been explained by the presence of a distribution of barrier heights associated with the different substates that equilibrate slowly relative to the rebinding, i.e. the presence of an inhomogeneous protein population.[53,55] As the temperature is raised, the geminate recombination of CO becomes exponential and can be described by a single barrier[54]; this behavior has been attributed to the presence of a homogeneous protein population on the rebinding time scale.

CO rebinding is exponential at room temperature, and occurs at a time scale of $\simeq$ 100 ns.[54] In contrast, NO geminate recombination at room temperature is nonexponential, with 80% ligands rebound within 300 ps.[3] Computer simulations indicate that a number of phenomena may occur in myoglobin on the 1–100 ps time scale of NO rebinding, including structural fluctuations,[3,5,6,28] protein relaxation induced by ligand photodissociation,[3] heme vibrational cooling,[27] heme/protein relaxation[3,28,56] and ligand diffusion.[5,6] Some or all of these effects could thus influence the rebinding process.

The barrier to ligand rebinding is complex, a combination of electronic[57,58] and steric[59–61] factors. The electronic component arises from the crossing of potential energy surfaces due to the changes in spin state at the iron atom, from a quintet in the unliganded to a singlet in the liganded form.[61]

3.1. *Protein Structural Change: Simulating A "Proteinquake"*

Ligand photodissociation brings a heme protein to a nonequilibrium state — the heme geometry and protein tertiary structure are those of the liganded form, while the potential energy corresponds to an unliganded system. The system starts relaxing towards the new equilibrium, a process called a "functionally important motion" (FIM) or "proteinquake" by Ansari *et al.*[50]

The first simulation of such an event was for CO photodissociation from protoheme and an isolated hemoglobin α chain in vacuum.[28] Starting with coordinates and velocities from liganded trajectories of these systems, a sudden transition to the unliganded state was generated by replacing the Fe–C bond with a weak repulsive interaction and altering the heme parameters. From several independent trajectories the authors found a fast initial displacement of the Fe atom on a 50–150 fs time scale in both protoheme and α chain. The average out-of-plane displacement of the iron atom during 1–10 ps after photodissociation initiation was about 0.1 Å less in the α chain than in the free protoheme.[28]

Molecular dynamics simulations have also been used to study vibrational cooling in laser excited deoxymyoglobin and ferrocytochrome-c.[27] Here the simulated process was the energy transport from a hot heme to the polypeptide part of the system, rather than nonequilibrium structural dynamics. For both deoxymyoglobin and ferrocytochrome-c eight independent trajectories at 300 K were generated, in which either 54 kcal/mol or 81 kcal/mol of kinetic energy was added to the 24 heavy atoms of the heme skeleton by velocity scaling (these energies correspond to absorption of photons of wavelengths 530 and 353 nm, respectively, as used in experimental photodissociation studies). Equilibration between potential and kinetic energy in the heme occurred within ca. 100 fs. Averages over the independent trajectories showed that the decay of the heme excess vibrational temperature was nonexponential, with about 50% loss occurring within 1–4 ps and the remainder in 20–40 ps.

A more detailed picture of the heme and protein dynamics following photodissociation of CO from myoglobin has been presented in Refs. 3 and 56. Four independent 100 ps dissociative simulations started with coordinates and velocities taken at 50, 60, 80 and 90 ps from an equilibrium simulation of liganded MbCO at 325K in vacuum.[29] A transient repulsion between the CO carbon and heme iron was introduced to simulate the motion of the system on the repulsive hypersurface of the electronic excited state generated by photon absorption, while the remaining CO...heme nonbonded interactions were set to zero and unliganded heme parameters were introduced. At times between 100 and 150 fs, the potential was switched to the standard form for the unliganded heme and free CO. This procedure mimicked the surface crossing from the excited state to the ground state by a sudden approximation approach. The ground state trajectories were then propagated for 100 ps.[3,56] A similar procedure was followed for the model system heme-imidazole-CO (HICO), for which a 70 ps liganded trajectory and a set of six 5 ps long dissociative trajectories starting at 5 ps intervals between 40 and 65 ps were generated.

In the free heme, an initial ultrafast Fe out-of-plane motion on a 50 fs average time scale was found, analogously to Ref. 28. Structural relaxation was complete within 1 ps after the start of the dissociative trajectories, with Fe out-of-plane displacement equal to that of an energy-minimized deoxyheme. In the Mb simulations, the initial ultrafast relaxation occurred on the same average time scale of 50 fs as in the free heme. This process accounted for about 80% of Fe the out-of-plane displacement found in the 100 ps trajectories and involves adjustments of the bonds and angles at the iron atom to the unliganded potential. Additional slower relaxations were also found in the simulations, on time scales of order of 1 ps and 10 ps. These involved adjustments of F helix, which is covalently linked to the heme through the proximal histidine sidechain, and more general changes in protein structure, such as relative translations and rotations of helices.[3] The average Fe atom out-of-plane displacement during the last 5 ps of the four 100 ps dissociative trajectories was still over 0.1 Å less than in the energy-minimized free heme-imidazole system, indicating that the structural relaxation of the protein was not complete.

The Fe out-of-plane displacement $z(t)$ can be considered as the reaction coordinate for the structural transition from the liganded to the unliganded state. The behavior of this coordinate at times longer than 50 fs in the dissociative simulations was analyzed in detail.[56] The average over four trajectories $\langle z(t) \rangle$ exhibited nonexponential relaxation that could be modeled as biexponential or power law functions in the 100 ps time interval. The individual trajectories showed rather different $z(t)$ evolutions, suggesting the presence of an inhomogeneous component in the relaxation process.[56]

The nonexponential relaxation of $\langle z(t) \rangle$ provides further evidence for the existence of a hierarchy of relaxation times in myoglobin. This is in accord with expectations for complex systems such as disordered crystals, glasses and macromolecules.[3,62–65] Excellent agreement was found between the simulated $\langle z(t) \rangle$ curve and time dependence of the experimental frequency shift $\Delta\nu(t)$ of heme band III, which is associated with heme doming.[66] This indicates that the band III shift is a linear function of the Fe out-of-plane motion in the 100 ps time scale, and validates the simulation results. The simulation results suggest a possible explanation of the nonexponential NO geminate recombination to heme proteins at room temperature. Namely, the heme and protein relaxation occurring in times of 1–100 ps could modulate the barrier to rebinding, which has been related to the Fe out-of-plane displacement.[64] Since this relaxation occurs on the same time scale as the NO recombination process, the nonexponentiality would thus be due to the presence of a time-dependent barrier. The relaxation would not be expected to influence CO rebinding which takes place on a much longer time scale ($\simeq$ 100 ns).[54] Recent simulations have suggested that nonexponential ligand rebinding may also be explained by ligand diffusion through the protein[5,6] (see below).

3.2. *Ligand Escape Pathways*

Considerable effort has been devoted to investigating the pathways of ligand diffusion through heme proteins between the binding site and the external medium. Two extreme models of ligand diffusion have been presented, one of escape along a well-defined, essentially unique path, the other of motion through a large number of alternative channels. These models have been called the "hole" and "sponge" models, respectively, by Elber and coworkers.[67]

In pioneering work, Case and Karplus[60] defined the "hole" model by suggesting the possibility of ligand escape along a relatively direct path obtained from analysis of the myoglobin crystal structure. This path involves mainly the rotation of the His 64 (the distal histidine) and Val 68 sidechains, opening an exit from the heme pocket to the external environment.[60] The free energy profile for ligand motion along this path was later studied.[68,69] When xenon atoms bound to internal cavities in myoglobin crystals were found,[51] the question of alternate ligand escape paths emerged. The possibility of several escape paths was suggested by simulations of motion

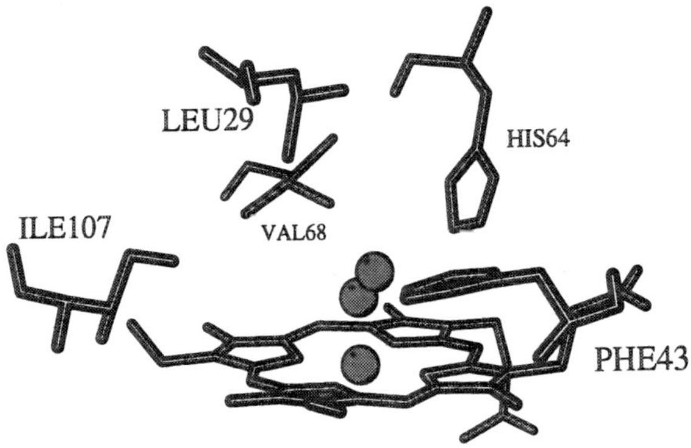

Fig. 9. Residues surrounding myoglobin heme pocket. Only heavy atoms displayed for clarity. The heme Fe and the CO ligand atoms displayed as spheres. Structure of sperm whale MbCO[18] taken from PDB file 1mbc.[114,115] Drawing created using program MOLSCRIPT.[116]

of a probe particle through rigid[70] and flexible[4] myoglobin. Residues surrounding the ligand binding site in sperm whale myoglobin are shown in Fig. 9.

Further support for the existence of multiple paths was provided by Elber and Karplus,[52] who employed their Locally Enhanced Sampling (LES) method to investigate CO diffusion through myoglobin. In the LES approach multiple noninteracting copies of the ligand are simulated moving in the field of the protein. This enables simultaneous generation of multiple ligand trajectories with only a small computational overhead compared to a single trajectory.[52] The LES simulations detected more possible paths than previous studies.[4,60,70] This was partly due to the fact that LES was designed to efficiently sample conformations in a many-particle system by imparting a high effective temperature to the ligand. Thus, LES may be used in exact calculations of time-independent properties, such as free energy differences[71] or global energy minimization,[72] but yields only approximate results for time evolution.[73]

An improved version of Locally Enhanced Sampling called cLES has been introduced[73,74] which includes a binary collision correction, leading to a significant improvement in description of dynamical properties. The cLES method was used to simulate Ar atom diffusion through solvated myoglobin.[73] Four 200 ps cLES trajectories with 10 ligand copies each were compared with four 700 ps standard single ligand trajectories. The cLES protocol yielded essentially the same diffusion constant for Ar in the heme pocket as the single particle simulations. Nine events of Ar atom escape were detected in the 40 ligand trajectories (4 simulations of 10 copies each), while no escapes were found in the single particle simulations.[73] The cLES Ar atom escaped along two pathways — the first leading through the AB/G and EF/G cavities, and the second directly from the heme cavity to the exterior, next to His 64. The small size of the Ar atom may facilitate its passage through the narrow gate, associated with relatively small His 64 sidechain fluctuations. Escape pathways passing near His 64 were detected in previous simulations of ligand diffusion, which used test particles to model the ligand.[4,60,70]

The cLES method gives results much closer to single ligand trajectories than does LES, but the effective ligand temperature in cLES is still elevated.[73] Thus, there exists the possibility of exploring configurations not accessible in single copy simulations. Ligand diffusion from the heme pocket to several internal cavities of myoglobin has been found in several standard simulations[5,6] (see Sec 3.3). The diffusion of Xe atoms from the exterior to internal cavities of Mb has been inferred from crystal structures.[51] So far, no ligand escape event from myoglobin was detected in standard, single copy, MD simulations up to the nanosecond time scale (4 simulations of 700 ps for solvated Mb[73] and 16 simulations of 500 ps in vacuum[5]). Thus, the location of the ligand exit and the nature of diffusion routes still awaits conclusive experimental and theoretical characterization.

Computer simulations have also been used to study ligand diffusion through a different globin, the protein lupine leghemoglobin.[75–77] Leghemoglobin has a similar overall fold to myoglobin, but exhibits faster ligand diffusion. Initial simulations using the approximate LES protocol indicated that CO diffusion should be faster in leghemoglobin than in myoglobin,[52,75] in accord with experimental differences between myoglobins and leghemoglobins. Furthermore, the LES studies indicated that the "hole" model was appropriate for leghemoglobin, i.e. that a single dominant ligand exit path exists in this protein.[75] The LES estimate of the ligand escape reaction coordinate was refined using the self-penalty walk approach,[76] and the free energy profile along this path was calculated.[77] The simulations predict that the ligand escape occurs in two steps. In the first (activated) step, the ligand hops from the heme pocket to the protein interior after overcoming a barrier ca. 3 kcal/mol; the second step involves free diffusion to the protein surface. An interesting observation was that the diffusive process involved significant protein motion.[77]

3.3. *Ligand Rebinding*

Spectroscopic studies of ligand rebinding are a crucial source of information about the dynamics of heme proteins. Simulations of rebinding can thus be thought of as a "holy grail" of computational studies of these systems, enabling a direct comparison of calculated and experimental quantities, as well as providing a microscopic model for a phenomenon of crucial importance in understanding protein dynamics. There is significant difficulty in constructing accurate potentials for the rebinding process — the dissociative excited state of the heme generated by the initial light absorption has to be described, as well as the ground state potentials of the liganded and unliganded hemes, the transition between which involves a change of spin state (the "electronic barrier"). In the classical simulations, curve crossing between the potentials of different electronic states has to be considered, at which points the Born–Oppenheimer approximation breaks down and the concept of a molecular potential energy function loses meaning. In spite of these formidable difficulties, two simulations of ligand rebinding to myoglobin have been performed so far — with[5] and without[6] an electronic barrier.

3.3.1. *NO Recombination with Myoglobin without an Electronic Barrier*

The authors[6] started from a 200 ps simulation of MbNO surrounded by 351 water molecules, from which 100 independent trajectories of the dissociated system were generated. The photodissociation was simulated by changing the potential energy of the heme-ligand system to the unliganded potential. This was in turn replaced by a "rebinding" potential — a combination of the liganded and unliganded potentials created using geometry-dependent switching functions. The "rebinding" potential changed from the liganded form at short Fe...NO distances (and NO orientations favorable for rebinding) to the unliganded form at large Fe...NO distances. In 89 out of the 100 trajectories, rebinding occurred within 5 ps, and in 7 more — within 15 ps. Thus the simulations predicted that 96% of ligands would rebind within 15 ps. In all the 96 trajectories in which rebinding was found, the ligand remained in the heme pocket directly neighboring the Fe binding site; in the 4 trajectories in which rebinding did not occur within 15 ps the NO molecule left the heme pocket.[6]

The kinetic progress curve of the rebinding process could be described as consisting of a 0.4 ps lag phase, in which the ligands moved away from the heme (the simulated photodissociation event), followed by a roughly exponential decay with a time constant of 2 ps. This was markedly different from the experimental rebinding curve, which could be represented by several nonexponential functions, including a double exponential with time constants of 28 and 280 ps and almost equal amplitudes.[3] The authors attribute the significantly faster simulated geminate recombination rate to the absence of the electronic barrier; this observation has been confirmed by others.[5]

Based on their simulations, the authors interpreted the observed nonexponential NO rebinding to Mb as caused by residence of ligands in different sites within the protein. The fast 28 ps component would be due to ligands residing in the heme pocket and the slower component to those occupying other, more remote sites. In contrast to Refs. 3 and 56, no significant role of protein structural relaxation in ligand rebinding was inferred. A further discussion is presented in Sec. 3.3.3.

3.3.2. *NO Recombination with Myoglobin with an Electronic Barrier*

Molecular dynamics simulations using two coupled electronic surfaces were employed to investigate the geminate recombination of NO to mutants of sperm whale myoglobin.[5] For each of the mutants Phe^{29}, Leu^{29} (native), Val^{29}, and Ala^{29}, 100 independent dissociative trajectories of 10 ps length were started from a 10 ps equilibration simulation of liganded MbNO in vacuum. The molecules were originally placed on the repulsive potential energy surface of the electronic excited state. Crossing to the ground state potential curve then occurred at random, according to the Landau–Zener formula. The ground state potential was a combination of the "pure" liganded and unliganded form potentials with geometry dependent coefficients. Inclusion of the excited electronic state and the simulated potential surface crossing generated an effective barrier for ligand recombination (2.3 kcal/mol at optimal ligand orientation[5]). Parameters for the potential surfaces were taken from previous simulations[3] and experimental data.

The simulation results were in very good agreement with experimental data. For the wild type Leu^{29} 72% of the ligands remained unbound in the simulation, vs 82% in the experiment. Especially significant was the correct reproduction of the ordering of the recombination yields for the four mutants on the picosecond time scale: $Phe^{29} > Leu^{29} > Val^{29} > Ala^{29}$. The variation of rebinding rate with residue type could be explained in terms of differential NO diffusion. Residue 29 is located at the back of the heme pocket; placing a bulky residue at this position restricts the volume available for ligand diffusion in the pocket and accelerates recombination. This point was illustrated by counting the number of collisions of the NO ligand with residues lining the heme pocket — the bulkier the substituent at position 29, the more collisions with NO it experiences, on the average.[5]

Another piece of experimental evidence correctly predicted by the simulations was the fact that on the nanosecond time scale, recombination in the mutant Phe^{29} is slower than in Ala^{29}.[78] To study this problem, 36 trajectories of 500 ps each were calculated (18 for Phe^{29} and 18 for Ala^{29}) by extending those trajectories in which rebinding did not occur during the initial 10 ps. Seven of the Phe trajectories rebound, mostly in the first 10 ps of the extended simulations, and could be considered part of the fast rebinding phase. In contrast, thirteen of the Ala trajectories exhibited rebinding, spread over the whole 500 ps simulation time. The long term rebinding of NO was associated with the ligand diffusion out of the heme pocket. In the case of Ala^{29}, essentially free diffusion occurs between the heme pocket and the cavity at the B/G helix contact, while in Phe^{29} access to the B/G cavity is blocked. Ligands leaving the heme pocket tend to occupy two "waiting sites" — in the EF/heme corner or close to the CD loop; their re-entry into the heme pocket is blocked by the bulky phenylalanine ring. These effects can explain why on the nanosecond time scale rebinding is slower for Phe^{29} (Ref. 5).

No long term Fe out-of-plane relaxation beyond a few picoseconds was found in the simulations for all the studied mutants. The NO rebinding curves could be explained purely in terms of ligand diffusion, with little role of heme/protein relaxation,[5] in contrast to the interpretation proposed in Ref. 3. (See discussion below.)

3.3.3. *NO Recombination to Myoglobin, Ligand Diffusion and Protein Relaxation*

There appears to be conflicting experimental and computational evidence for the relative role of ligand diffusion and heme relaxation in the NO geminate recombination process.[3,5,6,56] An attempt to clarify the issue is presented below.

In the simulations,[3,56] relaxation of the iron atom out-of-plane displacement coordinate and protein structure on several different time scales were found, including a long term component that was not complete in 100 ps. These results are in accord with a large number of experimental observations. First, it is clear that there is a structural transition upon ligand dissociation in heme proteins, consisting of a Fe atom out-of-plane displacement and a tertiary structure change of the protein (see Sec. 1).[13,14,17,18] Second, there is a large body of experimental evidence that relaxation in complex systems is nonexponential, covering a wide range of time scales.[3,62–64] Third, the average Fe atom out-of-plane displacement $\langle z(t) \rangle$ could be overlaid on the time evolution of the heme band III shift $\Delta\nu(t)$, which is a known heme structure marker.[66] Numerous spectroscopic results indicate that the structural transition in myoglobin has a long term component (10 ps and longer).[66,79–81] A large number of experimental results have been interpreted by assuming that the iron out-of-plane displacement is the reaction coordinate for the heme protein structural transition, and that the barrier to ligand rebinding can be related to this displacement.[55,64,82,83] Finally, while CO geminate rebinding to myoglobin at room temperature is exponential with a time scale of $\simeq$ 100 ns, NO rebinding is nonexponential and takes place in 10–100 ps.[3] Based on this combination of simulation and experimental results, it was proposed that a possible mechanism for nonexponential NO rebinding was one in which the population of Mb molecules is homogeneous (i.e. occupying a single substate or family of substates), with a time-dependent barrier for ligand recombination, modulated by the nonexponential Fe out-of-plane relaxation.

The main advantage of the work[3] is that ligand photodissociation was modeled in a simple way, with the long term motion of the ligand on the pure unliganded potential energy surface. This approach provides the simplest way to simulate the behavior of myoglobin after photodissociation, when ligand rebinding does not occur. The extent of the relaxation was studied by comparing protein and model system results; the Fe atom displacement was investigated in great detail, with the ultrafast 50 fs component removed and the remainder smoothed out to remove high-frequency oscillations. A relative disadvantage of the study[3] is the fact that only four independent trajectories were considered. However, care was taken to avoid correlation in the starting points and qualitatively similar behavior was found in all four calculated trajectories. It has also been pointed out that the parameters used in Ref. 3 lead to a heme displacement that is somewhat too large, and that this could lead to exaggerated distortion of the heme/protein structure.[6] The average Fe out-of-plane displacements over the final 5 ps in the four dissociative trajectories were 0.58–0.62 Å,[3] which is only slightly larger than the equilibrium value for deoxy-Mb from X-ray crystallography of 0.55 Å.[13] It is doubtful that such a small change should lead to the significant qualitative differences between the results of Ref. 3 on the one hand and Refs. 5 and 6 on the other, but this possibility cannot be completely excluded.

In the two rebinding simulation studies,[5,6] two phases of geminate recombination were found — fast recombination, in which the photodissociated ligand does not leave the heme pocket, and a slow phase, in which the ligand leaves the heme pocket and eventually re-enters it. Experimental results clearly indicate that the free ligand can diffuse through the protein. This is borne out by the existence of the equilibrium liganded and unliganded forms of Mb.[13,14,18] Crystal structures of xenon atoms bound in several internal cavities of myoglobin indicate the possibility of complex diffusion pathways and multiple binding sites for small ligands within the protein.[51] Simulations provide a detailed microscopic description of the diffusion pathways and time scales, improving our general understanding of this important phenomenon.[5,52] Simulations were able to correctly predict the ordering of the picosecond rebinding rates of four Mb mutants at the 29th position, as well as the reversal of the ordering for Phe^{29} and Val^{29} on the nanosecond time scale.[5] Since residue 29 is not in direct contact with the heme, it is expected, and simulation results confirm, that the heme behavior will be essentially the same for all the studied 29 mutants. Additionally, rebinding experiments with cobalt in place of the heme iron yielded kinetic results similar to iron myoglobin.[84] The fact that cobalt is not displaced out of the heme plane after ligand dissociation suggests that the iron out-of-plane motion does not significantly influence rebinding. In all, the essential features of nonexponential geminate recombination of NO to myoglobin could be explained completely by ligand diffusion between multiple binding sites in the protein, with little or no role for heme/protein relaxation.

Several reasons for the reduced role of protein relaxation have been given. One possibility is that if the transition state for the heme structural change is similar to the unliganded state, the protein relaxation should have little influence on the rebinding rate.[6,59] An estimate of the distribution of the Fe out-of-plane displacement values in the transition state gave results intermediate between those in the unliganded and liganded forms,[6] so this argument is inconclusive. Also, the potential energy surface from which the authors[6] extract properties of the transition state is very approximate. A common finding in the two rebinding simulations is that no significant long term Fe atom out-of-plane displacement was detected in 100 ps[6] and 500 ps[5] trajectories in which no rebinding occurred. Thus, any protein structural transitions occurring on this time scale did not appear to significantly influence the z coordinate. (However, it should be noted that visual inspection of Fig. 5 from Ref. 6 indicates some systematic variation of the Fe displacement during the first 10–20 ps; these results are difficult to quantify because of the small scale of the figure and presence of large-amplitude high-frequency oscillations.) This lack of long term (10–100 ps and longer) relaxation of the heme is in contradiction to numerous experimental results.[66,80,81]

The main advantage of the rebinding simulations[5,6] is that they attempt the direct calculation of observable quantities — the free ligand (or unliganded heme) concentrations. The simulations use large numbers of trajectories (ca. 100), enabling better averaging than was possible in Refs. 3 and 56. The drawback of the rebinding studies is the apparent lack of long term heme relaxation. One possible explanation of this result is that at the current level of theory rather approximate ground-state potential energy surfaces have to be employed, using a combination of pure liganded and unliganded potentials with conformation dependent switching functions.[6,67] It is not clear whether sufficiently realistic representations of the relatively complex heme systems can be obtained in this way. Some aspects of the ligand rebinding may have to be described quantum mechanically.[5] Finally, in both rebinding simulations[5,6] there is no electronic barrier for rebinding in the ground state, so eventual protein relaxation can only modulate the energies of the liganded and unliganded states, and not the activation barrier between them. It is the modulation of the barrier that is the suggested cause for correlation between Fe out-of-plane displacement and rebinding rates.[3,64,66]

In conclusion, computer simulations indicate that several types of phenomena occur in heme proteins at room temperature on the 1–100 ps time scale after ligand photodissociation — structural fluctuations,[3,5,6,28] heme vibrational cooling,[27] heme/protein relaxation,[3,28,56] ligand diffusion and geminate recombination.[5,6] The rebinding simulations[5,6] clearly suggest an important role of ligand diffusion in NO recombination to myoglobin. The different simulations give mixed results concerning the amplitude of the heme/protein structural relaxation and its importance in the rebinding process.[3,5,6,56] In view of the strong experimental evidence for relaxation processes in heme proteins on time scales of 1–100 ps and longer, the relative roles of ligand diffusion and structural relaxation are in need of further, more detailed study.

4. Thermodynamics of Hemoglobin

4.1. *Free Energy in Thermodynamics and Kinetics*

The usual thermodynamic description of a macroscopic molecular system starts with an introduction of the internal energy E, which is defined as the sum of kinetic and potential energies of all the molecules excluding the motion of the center of mass and interactions with external fields (Ref. 26, Chap. 6). The internal energy is a function of the state of the system. The first law of thermodynamics explains how E changes due to exchange of energy between the system and its surroundings such as in compression, heating, phase transitions or chemical reactions. The second law of thermodynamics states that the total entropy in the system and its surroundings increases in spontaneous, irreversible processes and reaches a maximum at equilibrium.

In practical situations, whether in the scientific laboratory, industrial plant or everyday life, most systems are either under conditions of constant temperature and volume ($T, V = \text{const}$) or constant temperature and pressure ($T, P = \text{const}$). Under these circumstances the second law of thermodynamics can be expressed as "the free energy of a system decreases in spontaneous, irreversible processes and reaches a minimum at equilibrium". The free energy under consideration is the Helmholtz free energy A for $T, V = \text{const}$:

$$\Delta A = \Delta E - T\Delta S \tag{18}$$

and the Gibbs free energy G for $T, P = \text{const}$:

$$\Delta G = \Delta H - T\Delta S \,, \tag{19}$$

where T is the temperature, $H = E + PV$ is the enthalpy, P is the pressure, and V is the volume of the system. For condensed phase systems the volume changes in most processes are small, $\Delta E \approx \Delta H$ and thus $\Delta G \approx \Delta A$.

Due to the property that the free energy reaches a minimum in the state of equilibrium, A and G are called the thermodynamic potentials at $T, V = \text{const}$ and $P, T = \text{const}$, respectively. G is a fundamental quantity in chemistry, because there is a simple relation between the reaction Gibbs free energy ΔG and the equilibrium constant $K : K = e^{-\Delta G/k_B T}$, where k_B is the Boltzmann constant. For a solute which can exist in two forms A and B in dilute solution (e.g. 1,2-dichloroethylene in *cis* and *trans* conformers, the the native and denatured forms of a protein, the HCN and HNC isomers):

$$\begin{aligned} &A \rightleftharpoons B \text{ (dilute solution)}\,, \\ &K = \frac{[B]}{[A]} = e^{-\Delta G/k_B T}\,, \end{aligned} \tag{20}$$

where $\Delta G = G^o(B) - G^o(A)$, $[A]$ and $[B]$ are the equilibrium concentrations, $G^o(A)$ and $G^o(B)$ are the standard Gibbs free energies of formation of species A and B, respectively. Equation (20) shows that the species (isomer, conformer, ...) with lower Gibbs free energy G will have the higher concentration.

For two solutes C and D which can associate to form a dimer $C : D$,

$$\begin{aligned} &C + D \rightleftharpoons C : D \text{ (dilute solution)}\,, \\ &K_{\text{assoc}} = \frac{[C : D]}{[C][D]} = e^{-\Delta G_{\text{assoc}}/k_B T}\,, \end{aligned} \tag{21}$$

where $\Delta G_{\text{assoc}} = G^o(C : D) - G^o(C) - G^o(D)$, and $[C]$, $[D]$, and $[C : D]$ denote concentrations of free C, free D, and the $C : D$ complex respectively. The molecules will tend to aggregate more strongly (high concentrations of $C : D$) for large negative values of the free energy of association ΔG_{assoc}.

The Gibbs free energy determines not only the equilibrium properties of molecular systems, but the kinetics of change as well. The standard transition state theory expression gives the rate of a chemical reaction k in terms of the activation Gibbs free energy $\Delta G^{\#}$, which is the difference between the free energies of the transition state and the reactants[26]:

$$k = \frac{k_B T}{h} e^{-\Delta G^{\#}/k_B T},$$

where h is the Planck constant.

4.2. *Statistical Mechanics of Free Energy Simulations*

The connection between G and observable quantities, such as equilibrium constants and rate constants, makes calculations of free energy especially valuable. It has long been the goal of computer simulations to provide not only energies and enthalpies, but also free energies and entropies of molecular states. In recent years computational methods have been proposed which make possible calculation of free energies from computer simulations of molecular systems. A number of methods have been developed, enabling evaluation of free energy changes for both conformational processes and chemical change. The attention here will be focussed on the latter type of events; most of the methods described can be used with only slight modification for conformational processes as well. For reviews of methods and applications see Refs. 8, 10 and 11.

The calculation of free energy starts with the standard formula of statistical mechanics giving the connection between the Helmholtz free energy A and the partition function Q for the canonical ensemble, in which the number of molecules N, volume V, and temperature T are kept constant[23]:

$$A(N,V,T) = -k_B T \ln Q(N,V,T), \tag{22}$$

$$Q(N,V,T) = \frac{1}{h^{3N} N!} \int dp dq e^{-H(p,q)/k_B T}, \tag{23}$$

where p and q are the momenta and coordinates of all the N particles (atoms) in the system and $H(p,q)$ is the Hamiltonian. The factor $N!$ is needed only if the particles are indistinguishable. The probability distribution for points in the phase space (p,q) is given by Eq. (8) and the ensemble average of any quantity $X(p,q)$ by Eq. (9).

In most cases the energy can be partitioned into a kinetic part K, depending only on momenta and a potential part U, depending only on coordinates: $H(p,q) = K(p) + U(q)$. The free energy can then be expressed as[8]:

$$A = C(N,V,T) - k_B T \ln Z, \tag{24}$$

where $C(N,V,T)$ is a function of the number of particles N, volume V, and temperature T, obtained by evaluating the part of the integral involving the kinetic energy $K(q)$, and

$$Z = \int dq e^{-U(q)/k_B T} \tag{25}$$

is the classical configuration integral. The probability distribution for points in the configuration space q is then easy to evaluate:

$$\rho(q) = \frac{\int dp e^{-[K(p)+U(q)]/k_B T}}{\int dp dq e^{-[K(p)+U(q)]/k_B T}} = \frac{1}{Z} e^{-U(q)/k_B T} \tag{26}$$

and the ensemble average of any quantity $Y(q)$ which depends on coordinates only is:

$$\langle Y \rangle = \int dp Y(q) \rho(q) = \frac{1}{Z} \int dq Y(q) e^{-U(q)/k_B T}\,. \tag{27}$$

Evaluations of ensemble averages of simple mechanical quantities such as the coordinates ($Y = q$) or potential energy ($Y = U(q)$) from molecular dynamics (MD) or Monte Carlo (MC) simulations have become routine.[7,8,24] That is because MD and MC methods generate samples of configurations in the physically important regions of low potential energy $U(q)$, i.e. with relatively high Boltzmann factors $e^{-U/k_B T}$, and these same configurations contribute most to the integral in Eq. (27). It is important to note that for such calculations the knowledge of the value of Z (or Q) is not necessary. In MD simulations we use the ergodic hypothesis to equate the ensemble average with the trajectory average, effectively estimating the latter as a simple arithmetic mean over trajectory configurations (see Sec. 1.5). Thanks to special methods of sampling configurations, Monte Carlo estimates of the ensemble average may also be calculated as the arithmetic mean over the sample, without knowledge of Z or Q.[85]

For a direct evaluation of A (Eq. (24)), the configurational integral Z of the system has to be determined. This has proved to be intractable for condensed phase systems.[24] A simple illustration of the problems involved can be seen by considering the following formula[24]:

$$\frac{\int dq e^{U(q)/k_B T} e^{-U(q)/k_B T}}{\int dq e^{-U(q)/k_B T}} = \frac{V^N}{Z}$$

which allows the expression of the configurational integral through an ensemble average:

$$Z = V^N \langle e^{U(q)/k_B T} \rangle^{-1}\,.$$

This average turns out to be extremely difficult to calculate through standard MD or MC approaches, since the sampled configurations represent relatively low potential energy $U(q)$ and high Boltzmann factors $e^{-U(q)/k_BT}$, for which the averaged function $e^{U(q)/k_BT}$ is small.[24]

4.3. *Thermodynamic Perturbation*

To circumvent the need for calculating partition functions, free energy simulations methods concentrate on calculating the differences between free energies of thermodynamic states for molecular systems. This does not place serious restrictions on the range of problems under study, because most practical applications involve differences of thermodynamic properties between two states (see Sec. 4.1). Moreover, since free energies G and A are functions of state, their changes can be evaluated along any path connecting the two states under consideration; the path can be chosen so as to make calculations easiest. Free energy differences are calculated in the following way. Consider two systems with potential energies U_0 and U_1. Neglecting differences in the kinetic energy, we can use Eq. (24) and write:

$$\Delta A - A_1 - A_0 = -k_BT \ln \frac{Z_1}{Z_0} \tag{28}$$

where Z_0 and Z_1 are the configuration integrals corresponding to potentials U_0 and U_1, respectively. We can express the ratio of partition functions using a mathematical identity

$$\frac{Z_1}{Z_0} = \frac{1}{Z_0}\int dq e^{-\beta U_1(q)} = \frac{1}{Z_0}\int dq e^{-\beta U_0(q)} e^{-\beta[U_1(q)-U_0(q)]} = \langle e^{-\beta[U_1-U_0]}\rangle_0$$

where the last expression can be recognized as the average of the function $e^{-\beta[U_1-U_0]}$ over the canonical ensemble for state 0 (i.e. the initial state, with potential U_0), leading to the so-called "perturbation formula":

$$\Delta A = -k_BT \ln \langle e^{-\beta[U_1-U_0]}\rangle_0 . \tag{29}$$

This formula provides the following practical recipe for calculating ΔA: Generate a sample of configurations q for state 0 (by MD or MC), for each q evaluate $e^{-\beta[U_1(q)-U_0(q)]}$ and average the results over the sample, finally substitute the average into Eq. (29).

The formula in Eq. (29) involves averaging over all possible configurations, or, equivalently over infinitely long trajectories (Eq. (11)). In simulations we always deal with limited samples and finite trajectories. With sufficiently careful preparation, involving choice of starting structure, energy minimization, heating and equilibration, a representative sample of configurations of the initial state (0) can usually be generated by MC or MD simulations (see Sec. 1.5). In order to obtain an accurate estimate of the average from Eq. (29), these same configurations also have to be representative for the final state (Eq. (1)). In order to ensure this condition, in practical applications the perturbation from the initial to the final state is not performed in a single step, as in Eq. (29), but is broken up into several stages, with introduction of one or more intermediate states. This is called multistep or multiwindow thermodynamic perturbation.

A coupling parameter λ is introduced to construct a hybrid potential energy function $U(\lambda)$, which mixes the properties of the two molecular states in such a way that varying λ from 0 to 1 changes the potential energy function from U_0 to U_1.[9] The simplest way to do this is the linear coupling approach:

$$U(\lambda) = (1-\lambda)U_0 + \lambda U_1 = U_0 + \lambda(U_1 - U_0) = U_0 + \lambda\Delta U\,, \tag{30}$$

$$\Delta U = U_1 - U_0\,, \tag{31}$$

where ΔU is called the perturbation potential. Simulations are performed for a series of values of the coupling parameter λ falling between 0 and 1. Each of such a simulation is called a perturbation step or window.

To calculate the free energy difference between states 0 and 1, we use a generalization of Eq. (29):

$$\Delta A(\lambda' \to \lambda'') = -k_B T \ln \langle e^{-\beta[U(\lambda'')-U(\lambda')]}\rangle_{\lambda'}\,, \tag{32}$$

where the angled brackets signify a statistical ensemble average for a system with potential energy $U(\lambda')$, which we approximate by the time average over the simulation (window) corresponding to λ'. The total free energy change is calculated by accumulating the differences between windows so as to span the whole interval between 0 and 1:

$$\Delta A = \sum_i \Delta A_i\,. \tag{33}$$

It should be noted that we only need to have simulations for the starting values (λ') of the coupling parameter in Eq. (32), while the end points (λ'') are arbitrary. Thanks to this property, simulations for the initial ($\lambda =$

0) and final ($\lambda = 1$) states are not necessary to calculate ΔA. Another consequence is the possibility of using the double-wide sampling technique, in which one evaluates the free energy changes half a step forward and half a step backward in λ.[86] This procedure reduces the size of the perturbation and leads to more accurate free energy differences.

It has been observed that a necessary condition for accuracy is that the free energy change from Eq. (32) does not exceed 1–3 times k_BT.[10] This is not strictly true, as can be seen by considering two potentials differing by an arbitrary constant: $\Delta U = U_1 - U_0 = C$. In such a case, the probability distributions for the two states (Eq. (22)) will be the same; the dynamics of the two systems, determined by forces which are partial derivatives of the potential (Eq. (3)), will also be identical. It is easy to see that the exact result of $\Delta A = C$ will be obtained from an arbitrary sample of configurations (including a sample of size 1, which is certainly not representative), independent of the magnitude of C. Thus, if the initial and final surfaces are approximately parallel, we can expect to obtain good estimates of the free energy change even for large values of ΔA and relatively small samples of configurations.

4.4. *Thermodynamic Integration*

Thermodynamic integration (TI) is another method of free energy calculation based on the coupling parameter approach. In TI we start by defining the free energy change as a function of the coupling parameter λ. For a separable Hamiltonian $H(p,q) = K(p) + U(q)$, assuming coupling between potential energies only, we get:

$$A(\lambda) = -k_BT \ln Q_\lambda = C(N,V,T) - k_BT \ln Z_\lambda \,, \tag{34}$$

where Q_λ is the canonical partition function (Eq. (23)) and Z_λ is the configurational integral (Eq. (25)) for the system with potential $U(\lambda)$. Differentiation of Eq. (34) with respect to λ leads to the main TI formula:

$$\frac{\partial A(\lambda)}{\partial \lambda} = \langle \frac{\partial U(\lambda)}{\partial \lambda} \rangle_\lambda \tag{35}$$

which states that the derivative of the free energy difference with respect to the coupling parameter λ is equal to the average of the corresponding derivative of the hybrid potential over the canonical ensemble for the system with potential $U(\lambda)$. For the linear coupling scheme the TI formula becomes especially simple:

$$\frac{\partial U(\lambda)}{\partial \lambda} = \Delta U = U_1 - U_0\,, \tag{36}$$

i.e. the derivative is independent of λ. The desired free energy difference between the final and initial states is obtained by integration of the derivative

$$\Delta A = A_1 - A_0 = \int_0^1 \frac{\partial A(\lambda)}{\partial \lambda} d\lambda = \int_0^1 \langle \frac{\partial U(\lambda)}{\partial \lambda} \rangle_\lambda d\lambda\,. \tag{37}$$

In practice we proceed analogously as in TP, performing simulations for several intermediate states or windows with values of λ_i between 0 and 1. These simulations yield samples of configurations over which we can estimate the derivative of $A(\lambda)$ according to Eq. (35). The free energy difference ΔA is then evaluated by numerical integration

$$\Delta A = A_1 - A_0 \approx \sum_i \langle \frac{\partial U(\lambda)}{\partial \lambda} \rangle_{\lambda_i} \Delta\lambda_i\,. \tag{38}$$

An advantageous property of the TI method is that the formula Eq. (37) is linear in the potential energy. This allows free energy changes to be decomposed into contributions from the different terms in the potential energy function (bond stretching, angle and dihedral deformation, van der Waals and electrostatic) as well as different parts of the system (solvent, protein, donor and acceptor chains, individual residues).[87] This decomposition allows us to obtain insights into the microscopic details of the studied phenomena. A drawback of the decomposition scheme is that it is not strictly physical — the components usually do not correspond to observable quantities. Also, in simulations geared at evaluating ΔA, the accuracy of components is usually lower than that of the total free energy change.

Typical applications of free energy simulations to macromolecules involve studies of influence of point mutations on thermodynamic stability of proteins, or analysis of the differences of binding free energies in a series of chemically modified ligands at a given receptor site.[8,10,11] Free energy simulations are able to provide important insights and improve understanding of complex macromolecular processes, such as ligand binding, folding, aggregation, solvation and mutation by decomposing the total free energy change into contributions from solvent, protein, and even individual residues, chemical groups or atoms.

4.5. *Simulations of Hemoglobin Sickling*

The following discussion is based on the results of Ref. 93.

4.5.1. *Hemoglobin and Sickling*

The sickle-cell mutant (HbS) differs from the wild type hemoglobin (HbA) by a point mutation: at position 6 in each β chain Glu is replaced by Val (β6 Glu→Val). Upon deoxygenation, the HbS molecules tend to polymerize into fibers, causing stiffening and deformation of the red blood cells, and ultimately the impaired circulation characteristic of sickle-cell anemia.[12,88,89] The structure of the fibers is not known at atomic resolution. Experimental data indicate that the deoxy-HbS fiber is composed of seven double strands.[12,89–91] The fiber double strands correspond to double strands found in the deoxy-HbS crystal[92] differing only by a helical twist.[12,89–91] The specific interaction responsible for fiber formation involves a lateral contact in which the β6 Val sidechain of one HbS molecule fits into a hydrophobic pocket formed by the E, F helices and EF loop of a β chain from a neighboring HbS molecule. A parallel double strand of HbS molecules is formed, with each HbS participating in two lateral contacts, once as a donor and once as an acceptor of the Val sidechain. The stereochemistry of sickling is presented in Fig. 10.

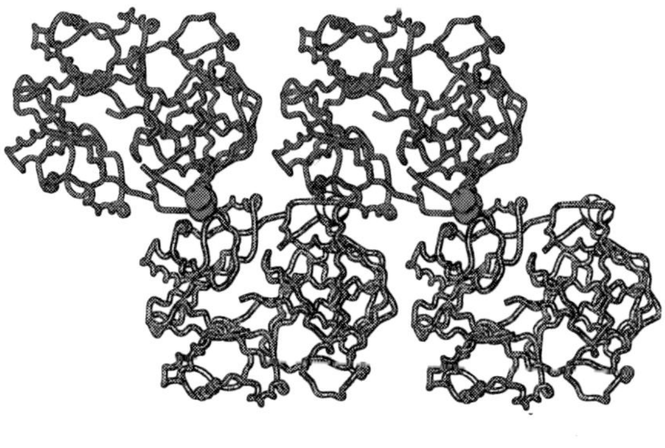

Fig. 10. Stereochemistry of the sickling aggregation. The four hemoglobin tetramers shown form two parallel single strands running horizontally. The two strands are colored white and grey, respectively. Lateral contacts at the strand interface are emphasized by displaying the β_2:6 Val sidechain atoms of each tetramer as spheres of exaggerated size (3 Å radius). Structures generated using the crystal structure of human deoxy-HbS[92] (PDB file 1hbs[114,115]) by translation along crystal a axis by unit cell length (64.33 Å). Drawing created using program MOLSCRIPT.[116]

4.5.2. *Thermodynamics of HbS Aggregation*

The basic property describing the sickling aggregation of hemoglobin is the association free energy ΔG_{assoc}, the free energy difference between two Hb molecules forming a lateral contact (hereafter called a Hb dimer) and two separate Hb molecules (hereafter called Hb monomers). Taking into account the stereochemistry of sickling and assuming that the aggregation only influences local Hb structure, we can simplify the picture by considering only the donor and acceptor regions:

$$\begin{aligned}\Delta G_{\text{assoc}} &= G(\text{Hb:Hb}) - 2G(\text{Hb}) \\ &\approx G(\text{donor:acceptor}) - G(\text{donor}) - G(\text{acceptor}).\end{aligned}$$

That is, we express ΔG_{assoc} as the difference between the free energies of the acceptor and donor regions in the dimer and separated donor and acceptor sites. All species considered are in aqueous solution.

Since HbS (β6 Val, or V in one-letter amino acid code) and HbA (β6 Glu, or E) differ only in the donor site, a further simplification occurs: assuming that the free energy of the solvated acceptor sites is the same in HbS (V) and HbA (E), then the difference between association free energies can be written as:

$$\Delta\Delta G = \Delta G^V_{\text{assoc}} - \Delta G^E_{\text{assoc}} = \Delta G^V_D - \Delta G^E_D\,,$$

where $\Delta G^X_D = G^X(\text{donor:acceptor}) - G^X(\text{donor})$ is the free energy of transferring sidechain $X(X = V$ or $X = E)$ from solution (monomer) to the

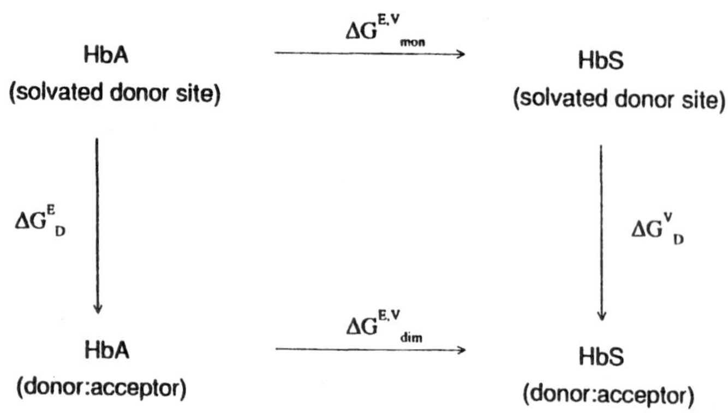

Fig. 11. Thermodynamic cycle for Hb sickling.[93] Quantities explained in text.

lateral contact (dimer) environment. This analysis leads to to the construction of the thermodynamic cycle in Fig. 11 for the evaluation of $\Delta\Delta G$. In Fig. 11 the vertical arrows describe normal chemical processes of transferring a sidechain between two environments, indicating that $\Delta\Delta G$ is an observable quantity. The horizontal arrows describe "alchemical" processes[87] corresponding to the introduction of a point mutation at position $\beta 6$ in the solvated monomer ($\Delta G_{\text{mon}}^{E,V}$ and lateral contact environment ($\Delta G_{\text{dim}}^{E,V}$). The free energy differences $\Delta G_{\text{mon}}^{E,V}$ and $\Delta G_{\text{dim}}^{E,V}$ cannot be measured directly but can be calculated from appropriate free energy simulations.

4.5.3. *Simulation Details*

Two simulation systems were created, starting from the HbS crystal structure.[92] The dimer system included all residues of the lateral contact area within a 14 Å sphere centered around the donor $\beta 6$ Val C_β, with empty spaces in the sphere filled with water. To generate the monomer system, all the acceptor chain residues were deleted and replaced with water. The dimer system consisted of 719 protein atoms and 109 water molecules, the monomer — of 271 protein atoms and 290 water molecules. Simulations were performed using the Stochastic Boundary Molecular Dynamics (SBMD) protocol,[94] with the central 11 Å sphere undergoing standard Newtonian dynamics, while in the 3 Å external spherical shell (buffer region) heavy atoms evolved according to Langevin dynamics, with protein atoms harmonically constrained to their initial positions and water molecules restrained to remain inside the 14 Å sphere by a mean-field solvent boundary potential. In the free energy simulations the mutation was performed in the $E \to V$ direction, i.e. the initial state corresponded to HbA and the final state to HbS, with the hybrid potential $U(\lambda) = (1 - \lambda)U_E + \lambda U_V$. Trajectories were generated for 9 values of the coupling parameter, $\lambda =$ 0.1, 0.2, 0.3, 0.4, 0.5, 0.6, 0.7, 0.8 and 0.9, with 5 ps equilibration and 10 ps production runs for each window (except $\lambda = 0.1$, where 10 ps equilibration was employed). Free energy differences $\Delta G_{\text{mon}}^{E,V}$ and $\Delta G_{\text{dim}}^{E,V}$ were calculated using both TP and TI protocols. The symbol ΔG is used for the calculated free energy differences. The SBMD method is expected to give free energies intermediate between Gibbs and Helmholtz values; the differences between ΔA and ΔG for the processes considered are expected to be small. The ionized (COO^-) form of the Glu sidechain was used in the simulations.

4.5.4. *Results*

The results for the Glu → Val mutation are summarized in Table 1 (TI method, the TP results are very similar.[93]) The free energy changes were $\Delta G^{E,V}_{\rm mon} = 76$ kcal/mol and $\Delta G^{E,V}_{\rm dim} = 62$ kcal/mol, i.e. the mutation $\beta 6$ Glu → Val significantly destabilizes both the dimer and the monomer, but the former less than the latter. The resulting difference between the free energies of association between HbS and HbA is $\Delta\Delta G = -14$ kcal/mol. This is in qualitative agreement with the experimental observation that sickling aggregation occurs in deoxy-HbS, but not in deoxy-HbA under normal conditions.

Table 1. Simulations of the sickle-cell mutation in hemoglobin, $\beta 6$ Glu → Val. Decomposition of the calculated free energy differences into protein and solvent contributions. This is a shortened version of Table 4 from Ref. 93. Units: kcal/mol.

Contribution	$\Delta G^{E,V}_{\rm dim}$	$\Delta G^{E,V}_{\rm mon}$	$\Delta\Delta G$
Total[a]	62.1	76.4	− 14.3
Solvent	63.1	70.6	− 7.4
Protein	− 0.1	6.4	− 6.5
Interchain	− 3.6	–	− 3.6
Interchain	3.5	6.4	− 2.9

[a]The total free energy changes include contributions from both internal strain terms in the potential (bond stretching, angle bending, ...) and from nonbonded interactions. Following the convention employed in Ref. 93 only the nonbonded terms are given for the componenets, leading to a slight discrepancy between the total free energy changes and the sum of the protein and solvent contributions.

To obtain insight into the nature of the sickling aggregation, the TI free energy changes were decomposed according to the scheme described in Ref. 87. Among the contributions from the different terms in the potential energy function the electrostatic energy was dominant, accounting for more than 95% of $\Delta G^{E,V}_{\rm mon}$, $\Delta G^{E,V}_{\rm dim}$ and $\Delta\Delta G$.[93] In the contributions from the different parts of the system, the interactions with solvent were the largest term in both dimer (63.1 kcal/mol) and monomer (70.6 kcal/mol). Although the $\beta 6$ glutamate is relatively buried in the dimer, it is apparently able to form quite favorable long-range interactions with water. The contribution from the interaction of the mutation site with the remaining protein atoms is approximately zero in the dimer, while in the monomer it is 6.4 kcal/mol, or about 10% of the solvation term. In $\Delta\Delta G$ the contributions from solvation (7.4 kcal/mol) and interactions with protein (6.5 kcal/mol) are of similar magnitude and both tend to stabilize the Val dimer. There are also self terms in $\Delta G^{E,V}_{\rm mon}$ and $\Delta G^{E,V}_{\rm dim}$ which favor Glu over Val, but these cancel out in $\Delta\Delta G$ and are not discussed further.

A further dissection of the protein interactions into contributions from the donor and acceptor chains was especially interesting (Table 1). Dimer interactions with the acceptor chain favor valine over to glutamate at $\beta 6$ ($\Delta G^{E,V}_{\rm dim}$ interchain term is -3.6 kcal/mol), while interactions with the donor chain favor glutamate over valine ($\Delta G^{E,V}_{\rm dim}$ intrachain term is 3.5 kcal/mol). In the monomer, the intrachain contribution is even more unfavorable for valine ($\Delta G^{E,V}_{\rm mon}$ intrachain term is 6.4 kcal/mol). This is not surprising, since $\beta 6$ Glu corresponds to the wild type protein. More detailed decomposition showed that the overall free energy changes were a sum of many individual terms of similar magnitude, with no single residue playing a dominant part.[93]

4.5.5. *Conclusion*

The picture of the sickling aggregation that emerges from the free energy simulation study is complex and interesting. If only the interactions of the mutation site with the acceptor chain were considered, the dimer with valine would be favored over the one with glutamate, in accord with the standard view of sickling as binding of $\beta 6$ Val to a hydrophobic acceptor pocket. However, in the overall free energy difference $\Delta\Delta G$, differential solvation of the dimer and the monomer, as well as interactions with the donor polypeptide chain are important. Two somewhat unexpected effects were discovered in this study. The first was the relatively strong interactions with solvent of $\beta 6$ Glu in the dimer, due to the long-range nature of the Coulombic force. It is only because the Glu interacts even more strongly with water in the monomer environment that the Val dimer is stabilized through solvation. The second unexpected effect was the presence of significant interactions between the mutation site and the donor polypeptide chain, which had not previously been considered. These effects were detected by virtue of the component analysis approach. It is this analytical capability that makes free energy simulations such a useful tool in the investigations of complex macromolecular systems. The results of this work clearly point out the necessity of considering the properties of the monomers as well as the aggregates, the solvent and the different parts of the solutes, in studying macromolecular aggregation.

The reliability of the simulation results is supported by good qualitative agreement with available experimental data. The sign of $\Delta\Delta G$ is predicted correctly, the magnitude is below an upper bound of -1.8 kcal/mol established from HbS/HbA cocrystallization experiments.[89,93] The value of $\Delta G^{E,V}_{\mathrm{mon}}$ is in the -70 to -80 kcal/mol range expected for the free energy of hydration of a charged carboxylate group.[95] The list of residues giving largest contributions to the free energy changes correlates well with the list of intermolecular contacts in the crystal[92] and lists of secondary mutation sites influencing HbS gelation.[12,88,89,91]

4.6. *Simulations of Hemoglobin Cooperativity*

The following discussion is based on the results of Ref. 87.

4.6.1. *Hemoglobin Cooperativity*

Hemoglobin is a tetrameric protein, consisting of two α and two β subunits. Each of the four subunits can bind oxygen (O_2), or other small ligands, like CO or NO to an iron atom of a heme group. In the unliganded, or deoxy form, the subunits exhibit unliganded tertiary structure, similar to unliganded myoglobin, and the tetramer exists in a well defined quaternary state called T. In the fully liganded form Hb·4O_2, each of the subunits is in the liganded tertiary structure, and the tetramer is in the R quaternary state.

From X-ray studies it has been determined that the $T \rightarrow R$ quaternary structure change involves mainly a relative rigid-body motion of the two $\alpha\beta$ dimers constituting the Hb tetramer. The two halves of Hb rotate relative to each other by about 15° and their centers of mass shift by about 1 Å[12] (see Fig. 12). This motion changes the interface between the two $\alpha\beta$ dimers, involving breaking and formation of contacts between chemical groups and changes in the total exposed surface of Hb and its solvation shell.[12,96,97] If the two dimers constituting the Hb tetramer are denoted as $\alpha_1\beta_1$ and $\alpha_2\beta_2$, the changes in interchain contacts occur at the $\alpha_1 : \beta_2$ and $\alpha_2 : \beta_1$ interfaces. Oxy- and deoxyhemoglobin structures are compared in Fig. 12.

The quaternary structural change causes the R state to have significantly higher affinity for oxygen than T.[12,88] Since ligand binding changes

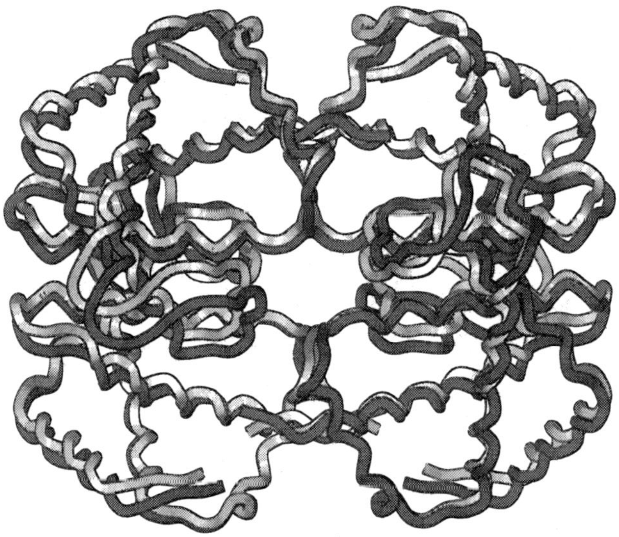

Fig. 12. Quaternary structural change in human hemoglobin. Comparison of structures of human deoxy-HbA in grey,[15] PDB file 3hhb[114,115] and oxy-HbA in white,[16] (PDB file 1hho[114,115]). View is as in Fig. 2; the vertical (y) axis is a twofold symmetry axis for each of the tetramers. It is possible to see that a small rotation around the horizontal (x) axis will overlay the two structures.

the affinity of the protein for the ligand, the four subunits do not bind oxygen independently. This effect is known as cooperativity; it plays an important role in hemoglobin biological function, enabling it to efficiently bind oxygen at the lung side of circulation at high O_2 pressure, and release oxygen to tissues at low O_2 pressure.[12,88] As the quaternary structural transition involves mainly relative reorganization of the two $\alpha\beta$ dimers of Hb, it is usually assumed that the dimer is noncooperative, and the free energy of cooperativity is defined as[87,98]:

$$\Delta G_{\mathrm{coop}} = \Delta G_4^{D\to O} - \Delta G_2^{D\to O}\,, \tag{39}$$

where $\Delta G_4^{D\to O}$ is the oxygen affinity of the Hb tetramer and $\Delta G_2^{D\to O}$ the oxygen affinity of two separated $\alpha\beta$ dimers (see Fig. 13).

For the $\alpha_1\beta_2$ interface residue β99 Asp, a series of mutants have been found with reduced cooperativity and increased oxygen affinity relative to normal hemoglobin.[12,88,89] Based on X-ray structures of oxy and deoxy HbA, it has been proposed that the essential role of β99 Asp is to stabilize the deoxy (low affinity) tetramer by making hydrogen bonds to α42 Tyr and α97 Asn.[96] Those hydrogen bonds are absent in the oxy tetramer. To gain improved understanding of the nature of hemoglobin cooperativity at the atomic level in general, and of the role played by β99 Asp in particular, free energy simulations of the β99 Asp $\rightarrow$ Ala mutation were performed for the oxy and deoxy forms of Hb.[87]

4.6.2. *Thermodynamics of Cooperativity*

The experimental quantity of interest is the difference between the free energies of cooperativity between the wild type (Asp) and mutant (Ala):

$$\begin{aligned}\Delta\Delta G &= \Delta G_{\text{coop}}(\text{Ala}) - \Delta G_{\text{coop}}(\text{Asp})\,,\\ &= \Delta G_4^{D\rightarrow O}(\text{Ala}) - \Delta G_4^{D\rightarrow O}(\text{Asp})\\ &\quad -[\Delta G_2^{D\rightarrow O}(\text{Ala}) - \Delta G_2^{D\rightarrow O}(\text{Asp})]\,. \end{aligned} \tag{40}$$

Because the β99 residue lies on the $\alpha\beta$ dimer surface and is not involved directly in oxygen binding, we can assume that the oxygen affinities of the wild

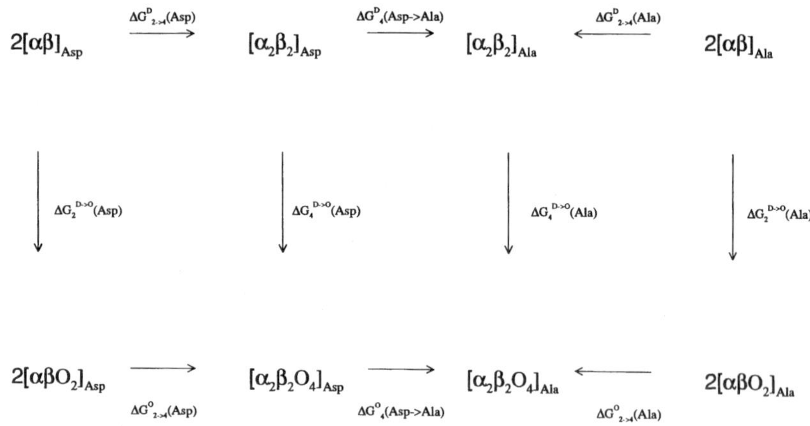

Fig. 13. Thermodynamic cycle for Hb cooperativity. $2[\alpha\beta]$ and $[\alpha_2\beta_2]$ denote two separate deoxy $\alpha\beta$ dimers and the whole deoxy-hemoglobin tetramer, respectively; $2[\alpha\beta O_2]$ and $[\alpha_2\beta_2O_4]$ denote two separate oxygenated $\alpha\beta$ dimers and the whole oxy-hemoglobin tetramer, respectively. The superscripts D, O and $D \rightarrow O$ denote properties of the oxy form, deoxy form and the oxy $\rightarrow$ deoxy transition, respectively. Subscripts 2, 4 and 2$\rightarrow$4 denote properties of the separated dimers, tetramers and the process of forming a tetramer from two dimers, repectively. Asp and Ala denote the two residues at position β99 considered in the study. (Analogous to Fig. 2 from Ref. 87.)

type and mutant $\alpha\beta$ dimers are the same: $\Delta G_2^{D\to O}(\text{Ala}) = \Delta G_2^{D\to O}(\text{Asp})$. This observation leads to the thermodynamic cycle in Fig. 13, which can be used to connect observable and calculated quantities:

$$\begin{aligned}\Delta\Delta G &= \Delta G_4^{D\to O}(\text{Ala}) - \Delta G_4^{D\to O}(\text{Asp}) \\ &= \Delta G_4^{O}(\text{Asp}\to\text{Ala}) - \Delta G_4^{D}(\text{Asp}\to\text{Ala})\,. \qquad (41)\end{aligned}$$

where $\Delta G_4^{O}(\text{Asp}\to\text{Ala})$ and $\Delta G_4^{D}(\text{Asp}\to\text{Ala})$ are the free energies of mutating β99 Asp to Ala in the oxy (O) and deoxy (D) tetramers. These quantities correspond to the "alchemical" processes described by horizontal arrows in Fig. 13, and can be evaluated from appropriate free energy simulations. The differences in oxygen affinities between the oxy and deoxy tetramers of the wild type and mutant, $\Delta G_4^{D\to O}(\text{Ala})$ and $\Delta G_4^{D\to O}(\text{Asp})$ correspond to normal chemical changes which can be measured experimentally. These processes are described by the vertical arrows in Fig. 13. In effect, Eq. (41) provides a recipe for the direct calculation of the observable quantity $\Delta\Delta G$, by performing free energy simulations of the β99 Asp $\to$ Ala mutation in the oxy and deoxy hemoglobin tetramers.

4.6.3. *Methods*

Two simulation systems were created, starting from the crystal structures of oxy and deoxy HbA,[15,16] including all residues within a 15 Å sphere around the center-of-mass of the $\beta_2$99 Val sidechain, with empty spaces in the sphere filled with water. Simulations were performed using the Stochastic Boundary Molecular Dynamics (SBMD) protocol,[94] with the central 10 Å sphere undergoing standard Newtonian dynamics, while in the 5 Å external spherical shell (buffer region) heavy atoms evolved according to Langevin dynamics, with protein atoms harmonically constrained to their initial positions and water molecules restrained to remain inside the sphere by a mean-field solvent boundary potential. In the free energy simulations the mutation was performed in the Asp $\to$ Ala direction, i.e. the initial state corresponded to wild type HbA and the final state to the β99 Ala mutant, with the hybrid potential $U(\lambda) = (1-\lambda)U_{\text{Asp}} + \lambda U_{\text{Ala}}$. Trajectories were generated for five values of the coupling parameter, $\lambda = 0.1$, 0.3, 0.5, 0.7 and 0.9, with 5 ps equilibration and 10 ps production runs for each window (except $\lambda = 0.1$ and 0.9, where 10 ps equilibration was employed). Free energy differences $\Delta G(\text{oxy})$ and $\Delta G(\text{deoxy})$ were calculated using both TP and TI protocols. The free energy changes from the SBMD method are expected to be intermediate between Gibbs and Helmholtz free energy values. Since the difference between A and G is expected to be small for the studied processes, the symbol ΔG was used to designate the calculated free energy changes.

4.6.4. *Results*

The simulation results are presented in Table 2. Both the deoxy and oxy tetramers are destabilized by the β99 Asp $\rightarrow$ Ala mutation, by 66.0 and 60.5 kcal/mol respectively per $\alpha_1\beta_2$ interface. The larger destabilization of the deoxy tetramer leads to the increased oxygen affinity and decreased cooperativity of the Ala mutant, in agreement with experimental observations. The calculated change in the free energy of cooperativity $\Delta\Delta G$ is -5.5 kcal/mol, which has the same sign and is of similar magnitude as the experimental value of -3.4 kcal/mol.[98] This agreement suggests that the simulation results are accurate enough to yield qualitative insights into the detailed interactions responsible for the free energy differences.

Table 2. Simulations of the change in hemoglobin cooperativity upon the β99 Asp $\rightarrow$ Ala mutation. Decompositions of the calculated free energy differences into protein and solvent terms. This is a shortened version of Table 1 from Ref. 87. Units: kcal/mol.

Contribution	ΔG(deoxy)	ΔG(oxy)	ΔΔG
Total	66.0	60.5	− 5.5
Solvent	46.0	68.5	22.5
Protein	20.0	− 8.0	− 28.0
β_2 Asp 99[a]	− 8.8	− 11.0	− 2.2
Interchain	2.8	− 24.4	− 27.2
Interchain	26.1	27.4	1.3

[a]Total self-energy contribution, including both nonbonded and internal strain terms.

The calculated free energy changes have been separated into parts due to the solvent and protein parts of the system, the latter divided into terms from the same subunit (β_2) and the neighboring subunit (α_1), and finally decomposed into individual residue contributions. The solvent contributions are 46.0 kcal/mol in ΔG(deoxy) and 68.5 in ΔG(oxy). The large difference is due to the fact that the Asp sidechain is more exposed to the solvent in the oxy tetramer and is partly buried in the $\alpha_1\beta_2$ interface in the deoxy form. This is consistent with recent experimental results on large changes in Hb tetramer solvation associated with the $R \to T$ transition.[97] The net contribution of the solvent interactions is to increase the cooperativity of the Ala mutant by 22.5 kcal/mol. Interactions with the protein part of the system stabilize the oxy tetramer (ΔG(oxy) term is -8 kcal/mol) and destabilize the deoxy tetramer (ΔG(deoxy) term = 20 kcal/mol) of the wild type hemoglobin relative to the β99 Ala mutant. The net effect is to decrease the cooperativity of the Ala mutant by 28.0 kcal/mol, counterbalancing the solvent effect.

The interchain contribution favors Asp over Ala at β99 by 2.8 kcal/mol in the deoxy form, and is unfavorable for Asp by 24.4 kcal/mol in the oxy tetramer. The interchain interactions of -27 kcal/mol are thus the major protein contribution to $\Delta\Delta G$. The intrachain interactions favor Asp over Ala in both oxy and deoxy tetramers by a significant amount (27 and 26 kcal/mol respectively). These terms mainly cancel out in $\Delta\Delta G$, with the net intrachain contribution of about 1 kcal/mol.

Especially interesting are the contributions of individual residues. The intersubunit interaction between β99 Asp and α42 Tyr indeed stabilizes the deoxy form, in accord with the analysis of Perutz.[12,96] For α97 Asn the mutation is unfavorable in both oxy and deoxy forms (by 10.0 and 9.7 kcal/mol, respectively). The simulations thus predict that β99 Asp – α97 Asn interactions are stronger in the oxy form, in disagreement with the structural analyses,[12,96] which focused on the Asp...Asn hydrogen bond, which is present only in the deoxy form. Energy calculations for the X-ray structures showed attractive β99 Asp – α97 Asn interactions in both quaternary states, primarily due to Asp-COO^- and Asn-NH_2 groups. The single largest contributor is α94 Asp, which has unfavorable interactions with the β99 Asp in both quaternary states, with significantly larger repulsion in the oxy (-44 kcal/mol) than in the deoxy (-22 kcal) forms. Analysis of X-ray structures shows that the Asp sidechains are closer and in an unfavorable orientation in the oxy relative to the deoxy forms. The simulations show that the unfavorable interactions are balanced by the presence of bridging water molecules; there were three water bridges in the oxy form and two in the deoxy form.

4.6.5. *Conclusion*

Free energy simulations of the β99 Asp $\to$ Ala mutation yielded an overall change in the free energy of cooperativity $\Delta\Delta G$ of -5.5 kcal/mol, close to the experimental value of -3.4 kcal/mol. Since some of the calculated contributions were as large as 60 kcal/mol, much greater than $\Delta\Delta G$, essential elements of the studied phenomenon are hidden in the measured results. Decomposition of the total free energy changes exposed contributions from the solvent and protein, inter- and intrachain interactions, as well as from individual residues. This has provided new insights into the complex thermodynamic changes due to mutations in proteins. Several effects beyond those usually considered in structural studies were detected. These include differential solvation of different molecular structures, role of removal of unfavorable interactions as well as creation of favorable ones and the importance of long-range electrostatics and microscopic solvation.

4.7. *Effectiveness of Macromolecular Free Energy Simulations*

Two applications of free energy simulation methods have been presented, concerning studies of two phenomena of chemical and biological significance — hemoglobin sickling[93] and cooperativity.[87] In both cases, the main quantities of interest — $\Delta G_{\mathrm{assoc}}$ for HbS and ΔG_{coop} for HbA — could not be calculated directly. Rather, differences between wild-type and mutant proteins were determined, and connections between calculated and observed quantities were established through the use of thermodynamic cycles. Simulations allowed the detailed investigation of the role of the solvent, and different parts of the systems, down to individual residues and chemical groups. Analysis of the simulation results revealed complex patterns of changing interactions involved in protein aggregation (sickling) and tertiary structure change (cooperativity), which were not previously evident from purely structural analysis. Maybe the most significant result was that many large calculated free energy contributions were found to add up to relatively small observable values of $\Delta\Delta G$. Thus, essential elements of the thermodynamics are hidden in the measured results, and may be exposed through simulations.[87] Overall, significant insight was gained into the studied phenomena, leading to an improved understanding of both basic physico-chemical properties and the biological function of hemoglobin.

The calculations described here involve a number of assumptions, including treating only the close surroundings (within 14–15 Å) of the mutation sites, the Stochastic Boundary Molecular Dynamics method of truncating the systems, the form of the molecular potential, calculating averages over limited configuration samples, and the numerical inaccuracies of using the TP or TI formulae. It is expected that the systematic errors of the free energies will be greater than the statistical errors reported in Tables 1 and 2, with the components being less accurate than the total free energy changes. Some of the systematic errors should approximately cancel out in $\Delta\Delta G$. The free energy simulations gave overall results in good qualitative agreement with available experimental data, and also some specific new predictions. Even if the numerical values of the free energy components exhibit significant systematic errors, the qualitative picture of aggregation and quaternary structural transition as complex phenomena with many contributing factors is unlikely to change.

Because they yield results directly comparable to experimental data, free energy simulations are becoming a powerful tool in chemistry, molecular physics and biology. For the tool to gain widespread acceptance, the accuracy of free energy simulations has to be evaluated. Besides improving the accuracy of potential energy evaluation and choosing the pathway between the initial and final state, the main problem to be solved is the insufficient conformational sampling in flexible systems. In recent years, several studies have attempted to deal with the sampling problem by performing simulations of extended length and/or comparing results from independent trajectories for the same systems[99–101] or constructing potentials of mean force for the flexible coordinates.[102] Failure to correctly describe system flexibility can lead to significant disagreement between simulation results and experimental data.[103] The overall conclusions appear to be that carefully planned simulations which take into account the nature of the studied system and its structure and dynamics yield quite reliable results.

5. Conclusion and Future Directions

Simulation methods deal with approximate computer models of real systems studied over limited time intervals. However, these methods provide highly detailed information, describing motions of individual atoms on a subpicosecond time scale. At the current stage of development, computer simulation results are complementary to experimental studies, giving a microscopic interpretation of observed phenomena and insight into their mechanisms. The continuous developments of computer hardware, software and algorithms have made possible computer simulations of biological macromolecules. Simulations have provided valuable information on basic physical properties of these systems, including structure, dynamics and thermodynamics. Correlations have also emerged between basic properties and biological function.

Emerging future trends in computer simulations of biomolecules involve more accurate description of larger and more complex systems over longer time intervals.

System size and complexity. Current simulations involve systems of 10^3–10^4 atoms: Within the next 5–10 years sizes of 10^4–10^5 atoms will become the standard. The systems themselves, from the current solvated peptides, proteins or nucleic acids, will become more complex, including buffer ions and interfacial environments. This will enable studies of many new interesting biochemical systems, including large, multisubunit proteins, proteins in membranes and protein-nucleic acid interactions.

Simulation length. The standard of simulation length has changed from 100 ps several years ago to about 1 ns currently. This trend will certainly continue and longer simulations can be expected in the future. With longer simulations, structural and energetic dynamics will be probed on longer time scales. This will also provide larger samples of configurations and yield more accurate averages. Parallel computing can be used to extend both system size and simulation length. Special schemes for molecular dynamics with very long time steps are being developed to extend simulation times.[104]

Potentials. More accurate potential energy functions are being developed. Improvements include both the description of internal structural deformations and nonbonded interactions. In the internal deformation force field off-diagonal, anharmonic and conformation-dependent terms are being added (see Ref. 105 and following articles). As for the nonbonded interactions, most of the attention is focussed on accurate description of the long-range electrostatic term. More accurate representations of the static charge distribution are being sought,[106] polarization effects are being included,[107–109] schemes for avoiding truncation effects based on multipole expansion[110] and Ewald summation[111] are being developed. Mixed quantum mechanics/molecular mechanics (QM/MM) approaches are being introduced, which allow the quantum mechanical treatment of selected parts of the system.[112,113]

Sampling. Molecular dynamics trajectories are often used to generate samples of representative configurations rather than dynamical trajectories. It such cases it is possible to significantly enhance the sampling rate by using the Locally Enhanced Sampling (LES) approach.[52,71] Usage of LES-type methods will certainly increase with the growing applications of free energy simulations.

References

1. J. A. McCammon, B. R. Gelin and M. Karplus, *Nature* (London) **267**, 585 (1977).
2. A. E. Torda and W. F van Gunsteren, K. B. Lipkowitz and D. B. Boyd, *Reviews in Comp. Chem.*, Vol. 3 (VCH Publishers, Inc., New York, 1992).
3. J. W. Petrich, J.-C. Lambry, K. Kuczera, M. Karplus, C. Poyart and J.-L. Martin, *Biochemistry* **30**, 3975 (1991).
4. R. F. Tilton Jr., U. C. Singh, I. D. Kuntz Jr. and P. A. Kollman, *J. Mol. Biol.* **199**, 195 (1988).
5. H. Li, R. Elber and J. E. Straub, *J. Biol. Chem.*, **268**, 17908 (1993).
6. O. Schaad, H.-X. Zhou, A. Szabo, W. A. Eaton and E. R. Henry, *Proc. Natl. Acad. Sci. USA*, **90**, 9547 (1993).
7. W. F. van Gunsteren and H. J. C. Berendsen, *Angew. Chem. Int. Ed. Engl.* **29**, 992 (1990).
8. C. L. Brooks III, M. Karplus and B. M. Petitt, *Proteins: A Theoretical Perspective of Dynamics, Structure, and Thermodynamics* (John Wiley and Sons, New York, 1988).
9. M. Mezei and D. L. Beveridge, *Ann. N. Y. Acad. Sci.* **482**, 1 (1986).
10. D. L. Beveridge and F. M. DiCapua, *Ann. Rev. Biophys. Biophys. Chem.* **18**, 431 (1989).
11. T. P. Straatsma and J. A. McCammon, *Ann. Rev. Phys. Chem.*, **43**, 407 (1992).
12. R. E. Dickerson and I. Geis, *Hemoglobin: Structure, Function, Evolution, and Pathology* (Benjamin/Cummings, Menlo Park, CA, 1983).
13. T. Takano, *J. Mol. Biol.* **110**, 569 (1977).
14. S. E. V. Phillips, *J. Mol. Biol.* **142**, 531 (1980).
15. G. Fermi *et al.* *J. Mol. Biol.* **175**, 159 (1984).
16. B. Shaanan, *J. Mol. Biol.* **171**, 31 (1983).
17. M. F. Perutz, G. Fermi, B. Luisi, B. Shaanan and R. C. Liddington, *Acc. Chem. Res.* **20**, 309 (1987).
18. J. Kuriyan, M. Wilz, S. Karplus and G. A. Petsko, *J. Mol. Biol.* **192**, 133 (1986).
19. B. R. Brooks, R. Bruccoleri, B. Olafson, D. States, S. Swaminathan and M. Karplus, *J. Comp. Chem.* **4**, 187 (1983).

20. L. Verlet, *Phys. Rev.* **159**, 98 (1967).
21. J. M. Sanz-Serna, *Acta Numerica* **1**, 243 (1991).
22. T. E. Creighton, *Proteins. Structures and Molecular Properties* (W. H. Freeman, New York, 1993).
23. D. A. McQuarrie, *Statistical Mechanics* (Harper and Row, New York, 1976).
24. J. P. Valleau and G. M. Torrie, *Statistical Mechanics. Part A. Equilibrium Techniques*, ed. B. J. Berne (Plenum, New York, 1977).
25. M. Toda, R. Kubo and N. Saito, *Statistical Physics I. Equilibrium Statistical Mechanics* (Springer-Verlag, Berlin, 1992).
26. R. Chang, *Physical Chemistry with Applications to Biological Systems* (Macmillan, New York, 1981).
27. E. R. Henry, W. A. Eaton and R. M. Hochstrasser, *Proc. Natl. Acad. Sci. USA* **83**, 8982 (1986).
28. E. R. Henry, M. Levitt and W. A. Eaton, *Proc. Natl. Acad. Sci. USA* **82**, 2034 (1985).
29. K. Kuczera, J. Kuriyan and M. Karplus, *J. Mol. Biol.* **213**, 351 (1990).
30. R. L. Levy, R. P. Sheridan, J. W. Keepers, G. S. Dubey, S. Swaminathan and M. Karplus, *Biophys. J.* **48**, 509 (1985).
31. P. J. Steinbach, R. J. Loncharich and B. R. Brooks, *Chem. Phys.* **158**, 383 (1991).
32. L. A. Findsen, S. Subramanian, V. Lounnas and B. M. Pettitt, in *Principles of Molecular Recognition*, ed. A. D. Buckingham (Chapman Hall, London, 1991).
33. C. L. Brooks III, *J. Mol. Biol.* **227**, 375 (1992).
34. R. Elber and M. Karplus, *Science* **235**, 318 (1987).
35. M. Frey, *Water and Biological Molecules*, ed. E. Westhof (CRC Press, Boca Raton, 1993).
36. Loncharich and Brooks, *Proteins* **6**, 32 (1989).
37. D. Ringe and G. A. Petsko, *Ann. Rev. Biophys. Bioeng.* **13**, 331 (1984).
38. U. H. Zucker and H. Schulz, *Acta Cryst.* **A38**, 563 (1982).
39. V. Schomaker and K. N. Trueblood, *Acta Cryst.* **B24**, 63 (1968).
40. R. E. Bruccoleri and M. Karplus, *Biopolymers* **29**, 1847 (1990).
41. J. Smith, B. Tidor, W. Doster, S. Cusack and M. Karplus, *Physica B* **156–157**, 437 (1989).
42. F. Parak, H. Hartmann, K. D. Aumann, H. Reuscher, G. Rennekamp, H. Bartunik and W. Steigemann, *Eur. Biophys. J.* **15**, 237 (1987).
43. F. Parak, E. N. Frolov, R. L. Mossbauer and V. I. Goldanskii, *J. Mol. Biol.* **145**, 825 (1981).
44. F. Parak, E. W. Knapp and D. Kucheida, *J. Mol. Biol.*, **161**, 177 (1982).
45. W. Doster, S. Cusack and W. Petry, *Nature*, **337**, 754 (1989).
46. W. Doster, S. Cusack and W. Petry, *Phys. Rev. Lett.*, **65**, 1080 (1990).
47. D. L. Stein, *Comments Mol. Cell. Biophys.* **3**, 373 (1986).

48. J. Smith, K. Kuczera and M. Karplus, *Proc. Natl. Acad. Sci. USA* **87**, 1601 (1990).
49. R. J. Loncharich and B. R. Brooks, *J. Mol. Biol.* **215**, 439 (1990).
50. A. Ansari, J. Berendzen, S. F. Browne, H. Frauenfelder, I. E. T. Iben, T. B. Sauke, E. Shyamsunder, and R. D. Young, *Proc. Natl. Acad. Sci. USA* **82**, 5000 (1985).
51. R. F. Tilton Jr., I. D. Kuntz and G. A. Petsko, *Biochemistry* **23**, 2849 (1984).
52. R. Elber and M. Karplus, *J. Am. Chem. Soc.* **112**, 9161 (1990).
53. R. H. Austin, K. W. Beeson, L. Eisenstein, H. Frauenfelder and C. Gunsalus, *Biochemistry* **14**, 5355 (1975).
54. E. R. Henry, J. H. Sommer, J. Hofrichter and W. A. Eaton, *J. Mol. Biol.* **166**, 443 (1983).
55. A. Ansari, J. Berendzen, D. Braunstein, B. R. Cowen, H. Frauenfelder, M. K. Hong, I. E. T. Iben, T. B. Johnson, P. Ormos, T. B. Sauke, R. Scholl, A. Schulte, P. J. Steinbach, J. Vittitow and R. D. Young, *Biophys. Chem.* **26**, 337 (1987).
56. K. Kuczera, J.-C. Lambry, J.-L. M. Martin and M. Karplus, *Proc. Natl. Acad. Sci.* **90**, 5805 (1993).
57. P. A. Cornelius, R. M. Hochstrasser and W. A. Steele, *J. Mol. Biol.* **163**, 119 (1983).
58. J. W. Petrich, C. Poyart and J. L. Martin, *Biochemistry* **27**, 4049 (1988).
59. A. Szabo, *Proc. Natl. Acad. Sci. USA* **75**, 2108 (1978).
60. D. A. Case and M. Karplus, *J. Mol. Biol.* **132**, 343 (1979).
61. H. Frauenfelder and P. G. Wolynes, *Science* **229**, 337 (1985).
62. Y. S. Bai and M. D. Fayer, *Phys. Rev.* **B39**, 11066 (1989).
63. J. Klafter and M. F Shlesinger, *Proc. Natl. Acad. Sci. USA* **83**, 848 (1986).
64. P. J. Steinbach, A. Ansari, J. Berendzen, D. Braunstein, K. Chu, B. R. Cowen, D. Ehrenstein, H. Frauenfelder, J. B. Johnson, D, C. Lamb, S. Luck, J. R. Mourant, G. U. Nienhaus, P. Ormos, R. Philipp, A. Xie and R. D. Young, *Biochemistry* **30**, 3988 (1991).
65. V. Srajer, K. T. Schomacker and P. M. Champion, *Phys. Rev. Lett.* **57**, 1267 (1986).
66. M. Lim, T. A. Jackson and P. A. Anfinrud, *Proc. Natl. Acad. Sci. USA* **90**, 5801 (1993).
67. G. Verkhivker, R. Elber and Q. H. Gibson, *J. Am. Chem. Soc.* **114**, 7866 (1992).
68. D. A. Case and J. A. McCammon, *Ann. N. Y. Acad. Sci.* **482**, 222 (1986).
69. J. Kottalam and D. A. Case, *J. Am. Chem. Soc.* **110**, 7690 (1988).
70. R. F. Tilton Jr., U. C. Singh, S. J. Weiner, M. L. Connoly, I. D. Kuntz Jr., P. A. Kollman, N. Max and D. A. Case, *J. Mol. Biol.* **192**, 443 (1986).
71. G. Verkhivker, R. Elber and W. Nowak, *J. Phys. Chem.* **97**, 7838 (1992).
72. A. Roitberg and R. Elber, *J. Chem. Phys.* **95**, 9277 (1992).
73. A. Ulitsky and R. Elber, *J. Phys. Chem.* submitted, 1994.

74. A. Ulitsky and R. Elber, *J. Chem. Phys.* **98**, 3380 (1993).
75. R. Czerminski and R. Elber, *Proteins* **10**, 70 (1991).
76. W. Nowak, R. Czerminski and R. Elber, *J. Am. Chem. Soc.* **113**, 5627 (1991).
77. G. Verkhivker, R. Elber and Q. H. Gibson, *J. Am. Chem. Soc.* **114**, 7866 (1992).
78. Q. H. Gibson, R. Regan, R. Elber, J. S. Olson and T. E. Carver, *J. Biol. Chem.* **267**, 22022 (1992).
79. A. H. Reynolds and P. M. Rentzepis, *Biophys. J.* **38**, 15 (1982).
80. J. L. Martin, A. Migus, C. Poyart, Y. Lecarpentier, R. Astier and A. Antonetti, *Proc. Natl. Acad. Sci. USA* **80**, 173 (1983).
81. E. W. Findsen, T. W. Scott, M. R. Chance, J. M Friedman and M. R. Ondrias, *J. Am. Chem. Soc.* **107**, 3355 (1985).
82. V. Srajer, L. Reinisch and P. M. Champion, *J. Am. Chem. Soc.* **110**, 6656 (1988).
83. L. P. Murray, J. Hofrichter, E. R. Henry and W. A. Eaton, *Biophys. Chem.* **29**, 63 (1988).
84. M. Ikeda-Saito, Y. Dou, T. Yonetani, J. S. Olson, T. Li, R. Regan and Q. H. Gibson, *J. Biol. Chem.* **268**, 6855 (1993).
85. J. P. Valleau and G. M. Torrie, *Statistical Mechanics. Part A. Equilibrium Techniques*, ed. B. J. Berne (Plenum, New York, 1977).
86. W. L. Jorgensen and C. Ravimohan, *J. Am. Chem. Soc.* **83**, 3050 (1985).
87. J. Gao, K. Kuczera, B. Tidor and M. Karplus, *Science* **244**, 1069 (1989).
88. H. F. Bunn and B. G. Forget, *Hemoglobin: Molecular, Genetic and Clinical Aspects* (W. B. Saunders Co., Philadelphia, 1986).
89. W. A. Eaton and J. Hofrichter, *Adv. Prot. Chem.* **40**, 63 (1990).
90. S. J. Edelstein, *J. Mol. Biol.* **150**, 557 (1981).
91. S. J. Watowich, L. J. Gross and R. Josephs, *J. Mol. Biol.* **209**, 821 (1989).
92. E. A. Padlan and W. E. Love, *J. Biol. Chem.* **260**, 8272; 8280 (1985).
93. K. Kuczera, J. Gao, B. Tidor and M. Karplus, *Proc. Natl. Acad. Sci. USA* **87**, 8481 (1990).
94. C. L. Brooks III and M. Karplus, *J. Mol. Biol.* **208**, 159 (1989).
95. P. Kebarle, *Jerusalem Symp. Quantum Chem. Biochem.* **8**, 81 (1975).
96. M. F. Perutz, *Nature* **232**, 408 (1971).
97. M. F. Colombo, D. C. Rau and V. A. Parsegian, *Science* **256**, 655 (1992).
98. D. W. Pettigrew, P. H. Rameo, A. Tsapis, J. Thillet, M. L. Smith, B. W. Turner and G. K. Ackers, *Proc. Natl. Acad. Sci. USA* **79**, 1849 (1982).
99. A. E. Mark, W. F. van Gunsteren and H. J. C. Berendsen, *J. Chem. Phys.* **94**, 3808 (1991).
100. M. J. Mitchell and J. A. McCammon, *J. Comp. Chem.* **12**, 271 (1991).
101. A. Elofsson and L. Nilsson, *Molecular Simulations* **10**, 255 1993.
102. K. Kuczera, O. Schaad, S. Edesltein and M. Karplus, Free energy simulation of sickle cell mutants of hemoglobin, in preparation.

103. Yunyu Shi, Alan E. Mark, Cunxin Wang, Fuhua Huang, Herman J. C. Berendsen and Wilfred F. van Gunsteren, *Protein Eng.* **6**, 289 (1993).
104. T. Schlick, S. Figueroa and M. Mezei, *J. Chem. Phys.* **94**, 2118 (1991).
105. N. A. Alliger, Y. H. Yuh and J.-H. Lii, *J. Am. Chem. Soc.* **111**, 8551 (1989).
106. Arieh Warshel and Johan Aaqvist, *Annu. Rev. Biophys. Biophys. Chem.* **20**, 267 (1991).
107. S. T. Russel and A. Warshel, *J. Mol. Biol.* **185**, 389 (1985).
108. T. P. Lybrand and P. A. Kollman, *J. Chem. Phys.* **83**, 2923 (1985).
109. D. Van Belle, I. Couplet, M. Prevost and S. J. Wodak, *J. Mol. Biol.* **198**, 721 (1987).
110. R. H. Stote, D. J. States and M. Karplus, *J. Chim. Phys.* **88**, 2419 (1991).
111. N. Karasawa and W. A. Goddard III, *J. Phys. Chem.* **93**, 7320 (1989).
112. M. J. Field, P. A. Bash and M. Karplus, *J. Comp. Chem.* **11**, 700 (1990).
113. J. Gao and X. Xia, *Science* **258**, 631 (1992).
114. F. C. Bernstein, T. F. Koetzle, G. J. B. Williams, E. F. Meyer Jr., M. D. Brice, J. R. Rodgers, O. Kennard, T. Shimanouchi and M. Tasumi, *J. Mol. Biol.* **112**, 535 (1977).
115. E. E. Abola, F. C. Bernstein, S. H. Bryant, T. F. Koetzle and J. Weng, *Crystallographic Databases — Information Content, Software Systems, Scientific Applications*, eds. F. H. Allen, G. Bergerhoff and R. Sievers, Data Commission of the International Union of Crystallography, Bonn/Cambridge/Chester (1987) pp. 107–132.
116. P. J. Kraulis, *J. Appl. Cryst.* **24**, 946 (1991).

CHAPTER 2

REACTION PATH STUDIES OF BIOLOGICAL MOLECULES

RON ELBER

Department of Physical Chemistry, The Fritz Haber Research Center, The Institute of Life Sciences and The Israel Center for Structural Biology, The Hebrew University, Jerusalem, 91904 Israel

Contents

1. Introduction

Numerous computational studies of biomolecules have been published in the last few years.[1,2] These computations probe the structures, dynamics and thermodynamics of large molecular systems and provide new insights into many biochemical processes. One of the most popular (and computer intensive) approaches is the Molecular Dynamics (MD) simulation technique. What is meant by "MD" is a bit vague and different studies limit or expand the definition of the Molecular Dynamics method.

In the present review I adopt a "broad range" definition of MD. MD is defined here as a collection of methods that solve initial value differential equations in *atomic detail* as a function of time. The differential equations can be Newton's equations, the Langevin equation, or variants designed to provide sampling appropriate for different thermodynamic ensembles.[3,4]

An example of initial conditions are velocities and coordinates of all the atoms in a protein molecule at time zero. From these initial conditions it is possible to calculate the coordinates and the velocities of the system at other times using Newton's equations. The result is a sequence of structures as a function of time, obtained from an iterative numerical procedure and called "a trajectory". The trajectory is used to study the dynamics and the thermodynamics of the system.

Here we include all the iterative algorithms under a single title — Molecular Dynamics. Formally, they can be written as a sequential algorithm that depends on some past information. For example, we write below the widely used velocity form of the Verlet algorithm.[3,5] The Verlet algorithm depends only on the present values of the coordinates and the velocities:

$$X(t+\delta t) = X(t) + V(t)\delta t - \delta t^2/2M \; dU/dX(t)\,, \tag{1a}$$

$$V(t+\delta t) = V(t) - \delta t/2M(dU/dX(t) + dU/dX(t+\delta t)) \tag{1b}$$

where δt is the time step, X and V are the coordinate and the velocity vectors respectively, M is the mass and U is the potential energy. The functional forms of the potentials commonly employed in macromolecules are described elsewhere.[1,2]

The popularity of the MD approach is a clear indication of its ease of use and its relative success. On the other hand, examination of the many published MD applications reveals the main limitation: The common MD simulations are far too brief to study many interesting biochemical processes. That is, all processes occurring on a time scale longer than a few nanoseconds are not accessible by this approach. For example, protein folding occurs on a time scale of (at least) milliseconds while MD can be used to investigate only processes that are millions of times shorter (nanoseconds).

Even with the rapid growth in computer technology, a speed-up of six orders of magnitude is not likely to happen soon. A "brute force" straightforward MD approach is therefore hopeless for such lengthy processes.

We note that some approximations made in the past has allowed us to stay in the framework of MD. For example, simulation of a system at high temperatures accelerates the time scales and enables the study of "long time" phenomena.[6-8] However, the range of validity of this approximation and the relevance of high temperature trajectories to room temperature behavior is not clear. It is therefore desirable to have alternative simulation methodologies that will make it possible to study much longer time scales. As a minimal requirement, the alternative approaches should employ different (and hopefully better) approximations from high temperature MD.

The search for alternative approaches to simulate proteins expanded considerably during the last few years. The expansion was motivated in large part by the progress in theoretical studies of protein folding (see chapter 4). The technological and intellectual challenges of the last problem and the difficulties in applying straightforward MD led to the development of many approximations aiming at modeling protein folding.

The approximations are based on a simplified representation of the protein molecule,[9] or on a special design of the energy surface to speed-up folding in the model system.[10] The design accelerates the search for the correct three-dimensional structure of the protein, compared to the time scale required by nature.

The "accelerating potentials" aim at rapidly folding proteins and they are designed with this purpose in mind. Thus, the goal is to find the correct 3-dimensional structure, and the pathway that leads to it is of lesser concern. Therefore mechanisms and time scales for biochemical and biophysical processes are difficult to study with potentials modified in the way described above. In this chapter we focus on an approach that maintains the atomic detail description. This approach also employs the usual potentials which are consistent with chemical intuition, and at the same time is capable of investigating long times. This is (of course) using a different set of approximations which should be examined separately.

After justifying the need for an alternative to MD, we proceed to describe the principles and the applications of the reaction path approach.

A combination of reaction coordinate computations[11] and rate theory[12] in small molecules has proven useful many times in the past. Together, these techniques do much better than straightforward MD in describing passages over high energy barriers — processes that can take very long times. It is therefore suggested that extension of some of the "path" ideas that are so widespread in gas phase calculations will also be possible in condensed matter. I want to emphasize that our interest here is **not** in "gas like" reactions in which the process is localized in space and only a few atoms actively participate, and thus the rest of the (large) system can be regarded as a perturbation.

Unfortunately the usefulness of "path like" approaches in condensed phases and in large molecules is far from being fully exploited. This is from the formal as well as the computational points of view. The adaptation of this approach is far from trivial and its applicability has its own limitations. One of the goals of this review (in addition to describing the reaction path approach) is therefore to highlight the limitations and also the possibilities. Suggestions for possible future directions will be outlined.

In the introduction I point out the main limitation of MD — the short time scales and explain the need for reaction path algorithms that do not suffer from the above restriction. In the next section I shall review approaches to reaction paths, designed originally for small molecules. Small molecules are defined as systems with 100 or less atoms (Sec. 2). The widely used definition of the reaction coordinate — the [steepest descent path] (SDP) — is discussed next. The numerical algorithms used to compute the SDP are described and the connection to rate theory is explored. I further discuss why most of these algorithms and theoretical frameworks are **not** suitable for condensed phase problems nor for analysis of conformational transitions in proteins.

Modifications in theory and in numerical algorithms are required when the focus changes from small molecules in the gas phase to proteins. This requirement delayed significantly the application of the reaction coordinate approach to macromolecules. The modifications (and the delay) are to be contrasted with the remarkable similarity of MD algorithms in both limits, studying small molecules or large macromolecules.

In Sec. 3, I follow the "small molecules" outline and discuss some practical aspects of computations in larger molecular systems. Significant emphasis is made on the computations of the potential of mean force and of time scales. I emphasize that the use of a rate theory along a given reaction coordinate (with the underlying assumption that this is the only relevant path) is a reasonably established formalism once the reaction coordinate is determined. The computational cost of this formalism is very high but at least it is quite clear what to do.

Things behave differently in calculations that determine the reaction coordinate. Proven methods of search for reaction paths in large systems are not widespread. Furthermore, justification of the use of a single reaction coordinate is far from trivial. (It is far from trivial also in some relatively small systems but more so in many multi-atom systems such as proteins.) In this paper, existing approaches are reviewed critically, and the pitfalls of these are outlined. Extension of the rate theory formula to the larger systems follows the derivation of the previous section (for small molecules) and is discussed only briefly.

In Sec. 4, I describe a number of algorithms that were designed specifically for large molecular systems. The theory and the computational implementation are very different from those of small molecules, making a smooth interpolation between "small" and "large" more difficult. Nevertheless, the above approaches are much better when processes in high dimensions are considered.

In Sec. 5, I describe a number of applications. Applications are taken from the work of my group.[13–15] They look into the feasibility of the approach and the typical questions it can address. Finally (Sec. 6) I discuss a stochastic, path integral formulation of the reaction coordinate and rates, which is a promising approach to condensed phase problems. I conclude with a detailed description of the implementation of the reaction path code on distributed and parallel computers. The ease of parallelization of the stochastic path algorithm makes it even more attractive as a modern computational tool.

A last comment concerns the focus of this chapter. This paper is not intended as a review and it is not aimed at covering all the work or related work in the field. Instead it provides a personal and biased picture of how the field looks.

2. Reaction Coordinates and Rates in Small Molecular Systems

2.1. *Reaction Coordinates*

It is intuitively clear to most chemists what is meant by the term "Reaction

Coordinate" (RC). The reaction coordinate is usually a continuous curve that smoothly interpolates between a starting point on the potential energy surface that we call "reactants" and another point on the potential energy surface that we call "products" (Fig. 1). The curve is chosen so that the energy barrier separating the two ends is as low as possible. Of course we assume that the points associated with the reactants and with the products are reasonably stable, i.e. they are found at the bottom of local energy minima.

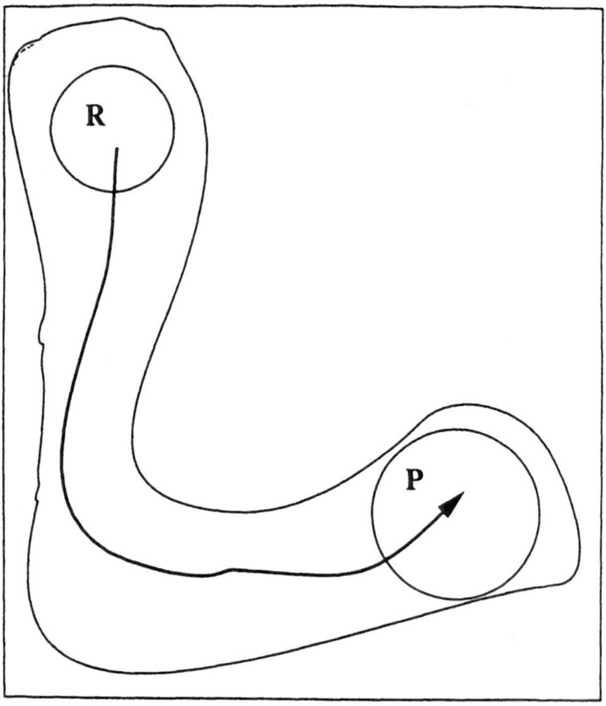

Fig. 1. A schematic diagram of a reaction coordinate on a two-dimensional energy surface going from reactants (the energy minimum on the upper left corner of the energy surface) to products (the lower right corner of the energy surface).

Ideally one would like to think of the path as a representative trajectory of sufficiently low energy, such that some key qualitative features of all the trajectories are included. This is not the case for the motion of the system in the neighborhood of the minima, but is more likely to be so at sufficiently low energies and near the energy barrier. Lowest energy reactive trajectories are most likely to pass in the neighborhood of the lowest possible barrier, even if the trajectories are different from the reaction coordinate elsewhere. This is one of the reasons for the relative success of the transition state theory that focuses on the properties of the trajectories near the barrier.

Assuming for the moment that we know how to calculate this coordinate, it is not at all obvious how to connect the RC to experimental observables. Elaborate calculations are required to obtain quantitative comparison to experimental data. For example, considerable effort is required to obtain reliable free energy profiles to compute rate constants according to transition state theory. After the rate constant is estimated, further calculations are frequently pursued to establish the validity of transition state theory and to compute corrections.

However, the RC alone is already useful in the qualitative understanding of the reaction mechanisms. It is not necessary to complete the whole lengthy, expensive and painful process of "rate computing" to obtain publishable results. For example, the study of the molecular structure at the energy barrier (the highest energy point along the path) may provide hints on how to speed up the reaction. This is by a chemical (or genetically engineered) modification that will ideally change the height of the barrier but will not affect the stable states (reactants or products). Reduction of barrier heights is one of the useful things that enzymes do to biological reactions. Connection to experiment can be therefore made by examining reaction paths in a series of similar reactions and explaining trends in the relative rates for different reactants and products. This approach usually works much better than attempts to compute absolute rates.

Let us return to the calculation of the RC. We consider a mathematical definition and computations of the RC found to be most useful in the gas phase.

The RC is defined with reference to a saddle point. A saddle point ($\mathbf{R}_{sad}$) satisfies two conditions: (i) It is a stationary point ($\nabla U(\mathbf{R}_{sad}) = 0$, where ∇ denotes the gradient operator and U the potential) and (ii) The second derivative matrix of the potential energy calculated at $\mathbf{R}_{sad}$ must have one and only one negative eigenvalue.

To ascertain that $\mathbf{R}_{sad}$ is a saddle point, we compute the second derivative matrix of the potential and diagonalize it. The (i, j)th element of this matrix is given by $\partial^2 U/\partial R_i \partial R_j$, calculated at $\mathbf{R}_{sad}$. The matrix is symmetric and therefore diagonalizable with real eigenvalues.[16] For $\mathbf{R}_{sad}$ to be a saddle point, all the eigenvalues except one must be non-negative. (If the matrix is calculated in Cartesian coordinates six of the eigenvalues that correspond to rigid body translations and rotations will be zero.)

At a minimum point on the energy surface all the eigenvalues are non-negative, making this location stable to small displacements. That is, any small displacement from the minimum position that is not a translation or a rotation results in a restoring force in the opposite direction to the displacement. At a saddle point, restoring forces are exerted along the directions that follow eigenvectors with positive eigenvalues. However, the direction with the negative eigenvalue is unstable and a small displacement along that mode yields a force directly proportional to the displacement, pushing the system away from the initial position. An example for a saddle point in two dimensions is shown in Fig. 2. Starting at the saddle point we can obtain lower energy values if we use the eigenvector with the negative eigenvalue to direct the move. The move is either forward or backward along that vector direction of the mode. Let the eigenvector with the negative eigenvalue be $\mathbf{e}_n$, and let ε be a small positive number. The solutions of the following differential equation:

$$d\mathbf{R}/d\tau = -\nabla U(\mathbf{R}) \tag{2}$$

with the two initial conditions $\mathbf{R} = \mathbf{R}_{sad} \pm \varepsilon \mathbf{e}_n$ provide the RC in the limit of $\tau \to \infty$. τ is a dummy parameter.

Pictorially we start at the saddle point and we slightly push the system either in the positive or in the negative direction. Each of the solutions that is obtained glides down the potential surface according to Eq. (2) to a corresponding minimum. The complete RC is obtained by combining the positive and negative segments at $\mathbf{R}_{sad}$. We denote the above path as the steepest descent path (SDP), to distinguish it from different definitions of the reaction coordinate.

A few comments are needed regarding the above discussion: First, we assume, of course, that the potential is bounded from below; otherwise, Eq. (2) will not converge. Second, according to Eq. (2), U is a monotonic, non-increasing function of τ. This is easy to show by multiplying both sides of Eq. (2) by $\nabla U(\mathbf{R})$ and noting that $\nabla U(\mathbf{R}) d\mathbf{R}/d\tau$ is $dU(\mathbf{R})/d\tau$. Thus we obtain the following equation which is equivalent to Eq. (2) (but harder to solve directly)

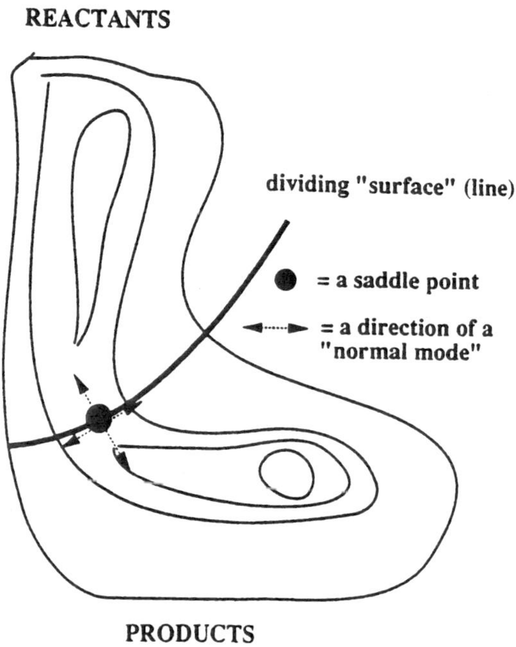

Fig. 2. A two-dimensional potential energy surface with "reactants", "products", a dividing surface and a saddle point. Also shown are the eigenvectors of the second derivative matrix at the saddle point.

$$dU(\mathbf{R})/d\tau = -|\nabla U(\mathbf{R})|^2 \,. \tag{3}$$

According to Eq. (3), the derivative of the potential with respect to τ is always non-positive: that is, U is a monotonic non-increasing function of τ. The steepest descent path is therefore the curve along which the energy decreases in a monotonic way starting from the saddle point reaching the relevant minimum. This coordinate is continuous and has the lowest possible energy barrier connecting the two minima, subject to the following condition: if only one saddle point connects the two stable states, or if this saddle point is the lowest in energy of all other saddle points that connect the two states.

Another comment concerns the possible physical interpretation of τ and of the path. One realization of Eq. (2) is of an overdamped trajectory. To show this, we start with the Langevin equation[17]:

$$Md^2\mathbf{R}/dt^2 = -\nabla U(\mathbf{R}) - \gamma M d\mathbf{R}/dt + F \,, \tag{4}$$

where M is the mass matrix, γ the friction constant (or tensor) and F the random force. The friction and the random force are interdependent via the fluctuation-dissipation theorem, to maintain a constant temperature in the system. To "derive" Eq. (2), we consider two specific limits of Eq. (4). In the first limit the temperature is set to zero. The temperature is defined by the balance between the random force (that pushes the temperature up) and the friction (that pushes the temperature down).[17] At temperature of zero degrees (Kelvin) the random force is zero too, and we therefore have

$$Md^2\mathbf{R}/dt^2 + \gamma M d\mathbf{R}/dt = -\nabla U(\mathbf{R})\,. \tag{5}$$

Taking the Fourier transform[20] on both sides we have

$$-M\omega^2 \boldsymbol{R} - i\gamma\omega\boldsymbol{R} = -\nabla \mathrm{U}(\boldsymbol{R})\,, \tag{6}$$

where we have used an *italic font* to denote a Fourier transform of a function from the time to the frequency domain

$$\mathrm{X}(\omega) = \int e^{i\omega t} X(t)dt\,. \tag{7}$$

Consider the limit of long times (or very small frequencies). In this limit one may ignore the highest power in the frequency and keep only the lower order terms. That is, we neglect the term $-M\omega^2\boldsymbol{R}$ in Eq. (6) which is quadratic in ω and keep the term $-i\gamma\omega\boldsymbol{R}$ and the potential derivatives. Finally, we transform the reduced form of Eq. (6) back to the time domain

$$\gamma M d\mathbf{R}/dt = -\nabla U(\mathbf{R})\,. \tag{8}$$

The last equation identifies τ of Eq. (2) as $t/(\gamma M)$. We found that the steepest descent path corresponds to a rather unusual trajectory, in which the classical system is moving with minimum effort to achieve its goal (the products of the energy surface).

In spite of the peculiar properties of the SDP "trajectory", it is still very useful for qualitative analysis of the reaction progress. The SDP is the lowest energy path. For sufficiently large barriers, it is expected that "real" trajectories will pass near the barrier and therefore the properties of this single path may be relevant to a significant set of trajectories in the neighborhood of the barrier.

However, I do not consider the SDP as "the intrinsic" reaction coordinate (i.e., the only legitimate definition). There is no real justification for the SDP definition except at very low temperatures. Alternative definitions to a RC can and will be suggested in this manuscript, which in some cases have better physical realization than the SDP. We devote significant portion of the discussion to the SDP since it is still the politically "winning" definition of the RC in reaction path calculations in small molecules.

It should be emphasized that while we interpret the SDP in terms of "time", this does **not** imply the time scale limitations found in MD. The "time scale" which we used (τ) is arbitrary, since γ — the friction constant — is arbitrary too. The solution of the differential equations does not show any oscillatory behavior which makes the integration much simpler than MD that includes high frequency motions.

From the above discussion it is clear that once a saddle point separating two minima is found, it is straightforward to calculate the rest of the path by solving the differential Eq. (2) in the two directions. Some techniques therefore focus only on finding the saddle points. We restrict the discussion to a number of techniques that we consider representative and also relevant to condensed phase calculations. These methods usually provide more than just a saddle point.

The most popular class of techniques to compute reaction coordinates is the collection of methods that employ what we call a "reference coordinate" (REF). This class of methods is also called adiabatic mapping.[11] The REF approach is the most straightforward, easiest to use, and most computationally efficient technique known to me. It also provides the user with the satisfaction of a real adventure, since the probability that something will go wrong in the calculations is very high. There is no guarantee that REF will provide the steepest descent path, a saddle point or even a continuous path.[11]

To somewhat reduce the effect of the discouraging comments, I must add that I personally used the REF approach too,[14] and that the method **may** yield useful approximations to the SDP. This is if the results are analyzed and tested carefully to verify that they do make sense. The problems associated with the application of the REF approach have been known for more than two decades,[11,18] so I cannot get the credit of being the first to notice the difficulties. However, the warnings seem to have been forgotten by some researchers and we therefore re-examine them at length. We start with the description of the methodology. Approaches that overcome some of the difficulties presented by the REF and at the same time take advantage of the speed and the computational efficiency of this protocol will be described in the study of conformational transitions in a tetrapeptide (Sec. 5).

In the REF approach we first select a reference coordinate, denoted by Q. We assume that Q is a good zero order representation of the reaction coordinate. Examples for Q can be the bond length for a dissociation process, e.g., $H_2O \rightarrow OH + H$, where the distance between the oxygen and the dissociated hydrogen is the reference coordinate. Another example for Q can be a torsion angle for an isomeric change ($CH(CH_3)=CH(CH_3)$ (cis) $\rightarrow C(CH_3)H=CH(CH_3)$ (trans)). Q is a scalar function of $\mathbf{R}$, but it is not necessarily a single valued function of $\mathbf{R}$. The closer Q is to the steepest descent path, the better are our chances of obtaining sensible results. This will become clearer as we proceed.

Consider for example the reaction coordinate for the conformational transition in a dipeptide (from C_7 axial to C_7 equatorial — Fig. 3). A possible reference coordinate is a straight line interpolation between the two minima. The straight line interpolation is a useful starting point for cases in which our knowledge of the energy surface of the system is minimal. This is **not** the case for the dipeptide whose conformational transitions are determined by only two soft degrees of freedom: ϕ and ψ. "Soft" means that motions along these degrees of freedom use relatively less energy. However, the straight line is a useful choice to demonstrate and to test the point which we make below. The straight line is (of course) continuous and it connects the reactants and the products. Continuous connection of reactants and products is not as obvious as it may seem and is more than what some other guesses provide (see for instance Ref. 11).

To compute an approximation to the SDP starting from Q we pursue a series of energy minimizations with different (but fixed) values of Q. Let $\mathbf{R}'$ be the subspace of $\mathbf{R}$ that is orthogonal to Q. Starting at an arbitrary position along Q, say Q_i (usually we start with the reactants or the products), we compute the minimum energy position subject to the constraint that Q is "nailed" at Q_i:

$$U(\mathbf{R}', Q) = \min\,(Q = Q_i)\,. \tag{9}$$

Computation of energy minima is by far a simpler calculation than the computation of reaction paths. I have already mentioned one possibility — we can minimize any function U using the differential Eq. (2). In the

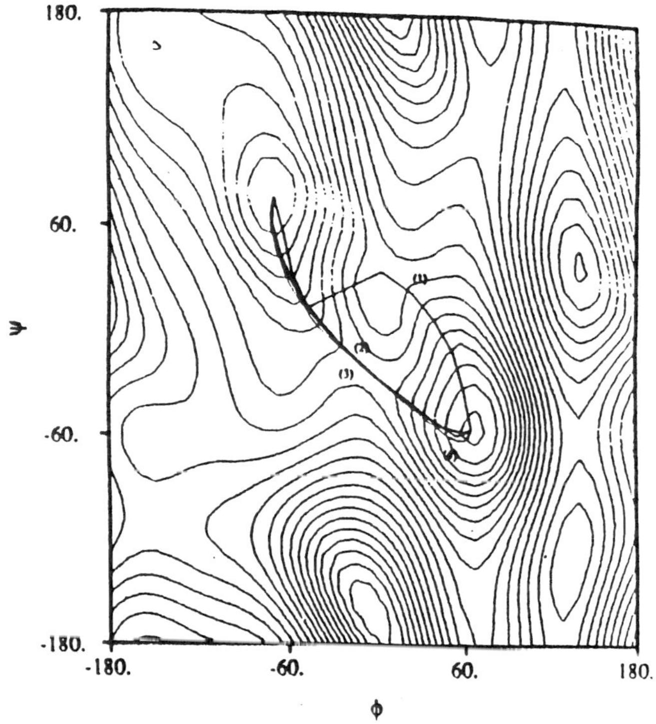

Fig. 3. A discontinuous jump in a path computation that uses the REF approach. The path computed is between the C_7 axial and C_7 equatorial in alanine dipeptide. The reference coordinate used in this calculation was a straight line connecting the "reactant" and the "product". In addition to the REF results, two paths that approximate (better) the SDP are also shown. One path is using the LUP approach and the second is a solution of Eq. (2) (Ref. 21).

limit of large τ we approach a local energy minimum regardless of what the starting point was. More efficient algorithms such as the conjugate gradient or Newton-Raphson are also in wide use.[19] However, we do not discuss all of them here. Minimization algorithms are beyond the scope of the present chapter and are described only when they are pertinent to reaction path computations.

The additional complication in Eq. (9) as compared to regular minimization is the presence of a constraint (a fixed value of Q). Equation (9) can be solved by a variety of approaches that differ in their ease of implementation and in numerical efficiency. The most popular and easiest approach is that of a penalty function. We add to the regular potential a penalty function (PF) that forces the system to remain in the neighborhood of a specific $Q - Q_i$. We then minimize the sum U + PF. For example, we look for a **minimum** of the system using

$$U(\mathbf{R}) + K(Q(\mathbf{R}) - Q_i)^2 = \text{minimum}\,, \tag{10}$$

where we chose a specific form (a harmonic potential) for the PF. Ideally the constant K should be chosen to be as high as possible to force the value of Q to be near Q_i at the minimum of the sum of the potential and the penalty function. In practice, very large values of K make the minimization space non-homogeneous and more difficult to optimize. Some compromise is usually required. A sophisticated compromise is a series of minimizations in which the value of K is increased gradually.

Once the minimum of the function that is defined by Eq. (10) is found, a new Q_i value is chosen and the minimization is repeated. Each of the minimizations for different fixed values of Q_i provides an optimized coordinate set $\mathbf{R}(Q_i)$ and an energy value — $U(\mathbf{R}(Q_i))$. The sequence of energy optimized structures with fixed (but different) Q_i-s is an approximation — $Q(\mathbf{R})$ — to the exact SDP coordinate $Q_{\mathbf{sdp}}$. The energy values provide an approximation to the continuous energy profile.

The resulting path — $Q(\mathbf{R})$ — is **never** the $Q_{\mathbf{sdp}}$, unless the initial $Q(\mathbf{R})$ is already the steepest descent path, or in other very special cases. One way of proving that $Q(\mathbf{R})$ is never equal to $Q_{\mathbf{sdp}}$ is by using an alternative definition to the steepest descent path given below. $Q_{\mathbf{sdp}}$ can be defined as a continuous path connecting the reactants and the products in which any point along it is at a constrained minimum. That is, they are minimized in all directions **excluding** the direction of the $Q_{\mathbf{sdp}}$. The path resulting from the constrained minimization described above is minimized in all directions **except** the direction of Q. Unless the direction of Q is the same as that of $Q_{\mathbf{sdp}}$ for all the path points (a very special case), the two paths cannot be the same.

Significant difficulties may occur when $\mathbf{R}(Q_i)$ is not a single valued function. This is always true when the space perpendicular to Q_i includes more than one energy minimum. In that case the optimization will depend on the initial conditions. Even the simple example of the reaction coordinate for a conformational transition in a dipeptide (simple compared to a protein) demonstrates this ambiguity. In Fig. 3 we show a discontinuous jump between two alternative paths that occurs using the reference coordinate approach. We stress that jumps become even more likely when the density of minima is large as is the case in macromolecules and condensed phase systems.

The only way to guarantee a continuous path is to make sure that the subspace in which the minimization is performed includes only one minimum and that the position of this minimum varies smoothly as a function of Q (see Ref. 65). In other cases it is necessary to check that the complete sequence of Q_i is continuous in that the displacements between sequential minimized structures can be made as small as we wish.

Nevertheless, it should be emphasized that even piecewise continuous paths can be useful in the analysis of the potential energy surface. As we shall describe in the study of conformational transitions in tetrapeptides the results of adiabatic mapping can be used as a starting point for further search of transitions states.

Further discussion of the reference coordinate approach can be found in Ref. 14. In summary, the great advantage of this technique is that it is very fast, easy to implement and use. It is therefore very effective when it works. However, it does not work at all times and it is necessary to carefully evaluate the results to ensure their validity and their usefulness as a starting point for more accurate search of transition states.

We conclude the section on adiabatic mapping by describing an alternative numerical algorithm to the penalty function technique. This alternative approach is considerably faster and more accurate. We consider constrained optimization in the Cartesian space. It is simple to manipulate coordinates in that space and it is therefore the representation of choice in most macromolecules and condensed phase simulations. However, the value of the Cartesian coordinates is affected by rigid body motions (in contrast to internal coordinates).

It is therefore necessary to fix the coordinate system during the minimization and to search for the RC without allowing for translations and rotations of the system as a whole. This is relevant if Q is represented by a Cartesian vector $\mathbf{R}(Q)$ in a fixed coordinate frame. The elimination of rigid body motions is pursued via a projection procedure. The constrained minimization is pursued using N-7 dimensions, where N is the total number of degrees of freedom. Six of the dimensions include the rotations and the translations of the rigid body and the last constraint (to make it a total of seven) is the motion along the reaction coordinate. We denote the constraints by σ_i $(i = 1,..,7)$. The individual constraints satisfy the following algebraic equations

$$\begin{aligned} \sigma_{1-3} &= \Sigma m_j(\mathbf{r}_j - \mathbf{r}_{jo}) = 0\,, \\ \sigma_{5-6} &= \Sigma m_j \mathbf{r}_{jo} \times (\mathbf{r}_j - \mathbf{r}_{jo}) = 0\,, \\ \sigma_7 &= \nabla_{\mathbf{R}}\mathbf{R}(Q_i) \cdot (\mathbf{R} - \mathbf{R}(Q_i)) = 0\,. \end{aligned} \tag{11}$$

The first three constraints are on translations. The next three are on rotations. The last constraint aims at fixing the position along the reference coordinate Q. The Cartesian representation of Q_i is $\mathbf{R}(Q_i)$. We assume that we start the minimization at the correct Q_i as the computations and the discussion below are somewhat more convenient if this is indeed the case. A simple inspection of the last constraint (Fig. 2) reveals that it allows motion only in the direction perpendicular to the reference coordinate Q_i. Hence steps that satisfy the seventh constraint cannot change the value of Q and therefore correspond to a fixed value of the reaction coordinate.

Note that all the constraints are linear ($\nabla_{\mathbf{R}}\mathbf{R}(Q_i)$ is a constant vector, it does not depend on the coordinates). Optimization with linear constraints, if correctly implemented, approaches the efficiency of optimization without constraints.

We have previously shown,[14] that the use of linear constraints instead of penalty functions speeds up the calculations by (at least) an order of magnitude. Furthermore, the computed coordinates satisfy the constraints very accurately. The accuracy was limited only by the computer round-off error, a precision which is not approachable in the penalty function approach.

The main idea of the algorithm can be illustrated as follows. We consider a deterministic optimization algorithm that is based on the solution of the differential Eq. (2). As argued above, the solution of Eq. (2) approaches a minimum in the limit of large τ (extensions to other optimization algorithms are possible but will not be discussed here). To introduce the constraints to Eq. (2), we use Lagrange multipliers. That is, the differential equations are modified to:

$$d\mathbf{R}/d\tau = -\nabla U(\mathbf{R}) + \Sigma\lambda_i\nabla\sigma_i\,. \tag{12}$$

The λ_i -s are the Lagrange multipliers. They are determined using the constraints already defined in Eq. (11). Let $d\mathbf{R}$ be the optimization step at time τ, estimated by first order integration: $d\mathbf{R} = (-\nabla U(\mathbf{R}) + \Sigma\lambda_i\nabla\sigma_i)d\tau$.

We require that the new coordinates $(\mathbf{R} + d\mathbf{R})$ will also satisfy the constraints (assuming for the moment that $\mathbf{R}$ already satisfies the constraints — $\sigma_i(\mathbf{R}) = 0$).

$$\sigma_i(\mathbf{R} + d\mathbf{R}) - \sigma_i(\mathbf{R}) = 0 = \nabla\sigma_i d\mathbf{R} = \nabla\sigma_i(-\nabla U(\mathbf{R}) + \Sigma\lambda_j\nabla\sigma_j)d\tau\,. \quad (13)$$

For finite $\mathbf{dR}$, the two last equalities are exact only for linear constraints, which is our case. The set of Eqs. (13) (for all the σ_i) is used to determine the Lagrange multipliers λ_i. For linear constraints the vectors $\nabla\sigma_i$ are constants: they do not depend on $\mathbf{R}$. Therefore the seven vectors can be orthogonalized once (using the Gram–Schmidt procedure), at the beginning of the optimization, to obtain a set of seven orthogonal vectors (and constraints) that are used throughout the calculations. We shall denote these constraints by $\eta_i(\mathbf{R})$ and their gradients by $\nabla\eta_i$; Note that $\nabla\eta_i \cdot \nabla\eta_j = \delta_{ij}$. Using the orthogonality of the gradients of the new constraints, Eqs. (13) can be reduced to

$$\lambda_i = \nabla\eta_i \cdot \nabla U(\mathbf{R})\,. \quad (14)$$

We thus project out of the force a component that is in the subspace of the constraints. Since the solution for the Lagrange multipliers is analytical it can be substituted back into Eq. (12) yielding a minimization equation in which the constraints are included explicitly

$$d\mathbf{R}/d\tau = -\nabla U(\mathbf{R}) + \Sigma[\nabla\eta_i\nabla U(\mathbf{R})]\nabla\eta_i\,. \quad (15)$$

Note that an explicit solution could not be obtained for non-linear constraints, since the set of orthonormal constraints would have to be recalculated at each step.

The above discussion of linearly constrained optimization will also be useful in the calculation of free energy profiles and in estimating the friction kernel.[13] Considering that the numerical efficiency of this approach is about an order of magnitude higher than the penalty function method, it is a bit surprising that the penalty function approach is still popular.

It is interesting to introduce at this point an approach similar in spirit to Eq. (15) which yields the steepest descent path rather than an approximation to it.[20,21] More precisely, this approach is limited only by numerical, finite difference approximations to the path that can be refined to a desired accuracy. It is based on an alternative definition of the SDP that we already discussed: The SDP is a continuous path connecting the reactants and products, such that any point along the path is fully minimized in all directions excluding the direction of the steepest descent path. In the previously described reference coordinate approach, the direction of the SDP is approximated by a fixed coordinate. In the Locally Updated Planes (LUP) approach[20,21] the "reference" coordinate is modified using iterations during the minimization. The result is a discrete representation of the path that is consistent with the above definition of the SDP.

This is done as follows: Consider a continuous reference coordinate, $Q(\mathbf{R})$, that is approximated by a set of N points $\{\mathbf{R}_i(Q)\}, i = 1, \ldots, N$. The points are not necessarily equidistant, however we do expect the distance between sequential points to be sufficiently small so that the finite difference approximation can be used to compute the slope of the path. The curvature of the path, $\mathbf{e}_i$, at $\mathbf{R}_i(Q)$ is defined by

$$\hat{e} = d\ell/d\mathbf{R} \tag{16}$$

and is estimated numerically as:

$$\mathbf{e}_i = (\mathbf{R}_{i+1}(Q) - \mathbf{R}_{i-1}(Q))/|\mathbf{R}_{i+1}(Q) - \mathbf{R}_{i-1}(Q)| \,. \tag{17}$$

Following the definition of the SDP, at any $\mathbf{R}_i(Q_{\mathrm{sdp}})$ we have

$$(\nabla U(\mathbf{R}_i) \cdot \mathbf{e}_i)\mathbf{e}_i - \nabla U(\mathbf{R}_i) = 0 \,. \tag{18}$$

Consider the following set of coupled differential equations:

$$d\mathbf{R}_i/d\tau = (\nabla U(\mathbf{R}_i) \cdot \mathbf{e}_i)\mathbf{e}_i - \nabla U(\mathbf{R}_i) \,. \tag{19}$$

The coupling is via the $\mathbf{e}_i$. Each of the $\mathbf{e}_i$ depends on $\mathbf{R}_{i-1}$ and $\mathbf{R}_{i+1}$. These equations are the essence of the LUP (Locally Updated Planes) method.[20,21]

It is clear from Eq. (18) that the steepest descent path is an asymptotic solution of Eq. (19). That is, the LUP method is formally exact in contrast to the reference coordinate approach. The direction of the slope of the "reference" path is changed at each step until the path (hopefully) converges to the steepest descent path. One difficulty in the application of Eq. (19) is that nowhere in the LUP protocol do we force the points i and $i+1$ to be close to each other. It is possible to find disjoint path pieces that satisfy Eq. (19). Hence, the SDP is not necessarily the only solution. When the number of alternative SDP paths in the neighborhood of the SDP is large, the probability of path fragmentation is high. The LUP approach is useful when the initial guess is within the radius of convergence of a single SDP.

Compared with other approaches in calculating the steepest descent paths, such as the mode following,[22–25] the LUP method has the advantage that no second derivatives are required. Second derivatives are sometimes difficult to compute (i.e. in high level *ab initio* calculations on small molecules), manipulate, invert and store (i.e. in calculations of reaction paths in condensed phases or macromolecules). We shall return to comparison between LUP and other methods that yield the SDP after introducing the approach of mode following.

The mode following approach that was introduced by Cerjan and Miller[22] and further refined by others[23] is in principle an exact technique. It is used extensively in small molecular systems. I have argued in the past that mode following is not likely to be useful in studies of large molecular systems. This is due to the large number of low energy modes and the necessity of manipulating matrices of second derivatives. However, if one works in the space of dihedral angles (thus significantly reducing the number of degrees of freedom), it is possible to investigate some reaction paths in proteins that include local structural changes.[65]

The mode following approach has been used to compute the reaction path only for the dihedral angles in Ref. 65. The [reference coordinate method (adiabatic mapping)] is used to optimize the positions of the rest of the degrees of freedom. The adiabatic mapping is adequate for the applications considered, since the potential energy terms of these coordinates (bonds and angles[65]) have only one minimum. This reduction in the number of degrees of freedom extends the range of applicability of the mode following technique beyond my previous expectation.

The mode following approach is based on ideas related to the Newton–Raphson minimization procedure. (In the same spirit the LUP approach is a follow-up on the steepest descent minimization algorithm.) It is therefore useful to briefly describe the Newton–Raphson (NR) minimization procedure. We present a pedagogical outline, while maximum numerical efficiency is of lesser concern. In the NR algorithm, we consider a quadratic expansion of the energy surface at $\mathbf{R}$. We compute a local approximation to the true potential, discarding all the terms above the quadratic ones. The goal is to calculate the coordinate vector $\mathbf{R}'$, which is the minimum of approximate potential. For a quadratic potential, the analytical solution is derived from the following equation:

$$U(\mathbf{R}') = 1/2(\mathbf{R} - \mathbf{R}')\mathbf{K}(\mathbf{R} - \mathbf{R}') + \mathbf{K}'(\mathbf{R} - \mathbf{R}') + C\,, \qquad (20a)$$

where $\mathbf{K}$ is the second derivatives matrix ($K_{ij} = \partial^2 U/\partial R_i \partial R_j|_R$) and $\mathbf{K}'$ is the vector of the first derivatives at $\mathbf{R}$. At the minimum of the approximate potential Eq. (20), the force must be zero. That is

$$\nabla_{\mathbf{R}'} U(\mathbf{R}') = \mathbf{K}(\mathbf{R} - \mathbf{R}') + \mathbf{K}' = 0 \qquad (20b)$$

which has the simple solution to $\mathbf{R}'$ — the new guess for a minimum — $\mathbf{R}' = \mathbf{R} + \mathbf{K}^{-1}\mathbf{K}'$. This is an oversimplified version of the "real life" algorithm. It is assumed that $\mathbf{K}$ is a positive definite matrix. If it is not, the straightforward solution of Eq. (20a) is singular. A very safe algorithm (with some cost in performance) is listed below. It is convenient to introduce this algorithm here since it is closely related to the mode following technique which is the next method on the agenda.

Let $|k_i\rangle$ be the ith eigenvector of $\mathbf{K}$ with an eigenvalue k_i. A possible representation of the $\mathbf{K}$ matrix is $\mathbf{K} = \Sigma |k_i\rangle\, k_i \,\langle k_i|$ and for $\mathbf{K}^{-1}$ it is $\mathbf{K}^{-1} = \Sigma |k_i\rangle\, 1/k_i \,\langle k_i|$. We use the Dirac notation for column and row vectors. The approximate expansion of the potential [Eq. (20)] means that modes with negative eigenvalues have a potential which locally is an inverted parabola. This is an unstable potential if we attempt to use this expansion globally. Calculation of a stationary point for an inverted parabola of (say) mode k_j yields (of course) a maximum for this mode, which is not where we want to be.

One way of getting around this problem is to separate the problem into two parts, one part that includes all the modes with positive k_i and another part with the rest of the modes. Let the total number of modes be N and the number of non-positive eigenvalues n. We then pursue a Newton–Raphson step only for the $N - n$ modes with positive eigenvalues $\mathbf{R}_{N-n} = \mathbf{K}_{N-n}^{-1}\mathbf{K}'$ where $\mathbf{K}_{N-n}$ is given by

$$\mathbf{K}_{N-n} = \sum_{i=1}^{N-n} |k_i\rangle\, k_i \,\langle k_i|\,. \qquad (21)$$

The rest of the modes must be treated by a procedure different from the regular Newton–Raphson scheme. One option that is currently implemented in our molecular dynamics code MOIL[51] is to optimize separately each of the modes with negative eigenvalues. This is done as follows: Each of the eigenvectors $|k_j\rangle$ $(j > N - n)$ defines a one-dimensional direction. We search for a minimum by moving **only** along $|k_j\rangle$ and repeating this process for each of the eigenvectors. This is not the cheapest way of doing things, but it is very safe and stable. Usually there are not very many negative k_j modes and they disappear as the minimization proceeds.

One more comment is in place. Since we are working in the Cartesian space it is also necessary to take care of the rigid body motions for which $k_j = 0$. The eigenvectors of rigid body translations and rotations are the gradients of the first six constraints in Eq. (11). The **K** matrix that includes these modes does not have an inverse. The simplest solution is to "shift" these eigenvalues upwards. For example, let $|k_{tx}\rangle$ be the eigenvector of the translation along X. The "shifted" matrix with respect to $|k_{tx}\rangle$, $\mathbf{K}_s$ is $\mathbf{K}_s = \mathbf{K} + |k_{tx}\rangle\, s\,\langle k_{tx}|$, where s is a very large positive number. Assigning the rigid body motions as modes with very high "eigenvalues" ensures that the rigid body displacements are negligible. From this small trick it is also clear that movements along specific modes can be manipulated as we please by shifting the eigenvalues up or down. This is the basic idea of the mode following reaction path algorithm of Cerjan and Miller[22] that was further developed by Barker.[23]

The mode following approach "walks" on the energy surface and locates stationary points. This is done as follows: Let the starting coordinate be a local energy minimum, $\mathbf{R}_m$. At the minimum the matrix $\mathbf{K}_s$ is definitely positive and no further progress in **minimization** is expected. One of the eigenvectors is then picked to define a search direction for a reaction coordinate — $|kp\rangle$. The simplest choice is the mode with the lowest eigenvalue that is not a translation or a rotation. However "simple" does not necessarily imply correct and the choice is far from trivial if we care where we go. In large systems the number of modes with small eigenvalues can be quite high. It is thus necessary to have an "educated guess" for the initial search direction. In small molecular systems (where this approach is popular) an educated guess is usually possible.

After the choice was made, we "shift" the eigenvalue of $|kp\rangle$ to a negative number $\mathbf{K}_p = \mathbf{K}_s - |k_p\rangle\,\lambda\,\langle k_p|$ where λ is a positive number larger than k_p — the eigenvalue of $|k_p\rangle$. If we perform a Newton–Raphson step now using $\mathbf{K}_p$ for a "second derivative matrix", the motion along $|k_p\rangle$ is uphill, getting the system away from the minimum with a step proportional to

$1/(p - \lambda)$. It is therefore clear that the choice of λ is another issue that requires consideration. Small $p - \lambda$ corresponds to a large change in the coordinates. Large steps can be problematic because of two reasons: First, the local quadratic expansion may be a poor approximation and other modes will not be at a minimum. Second, the relevant mode $|k_p\rangle$ may change its direction and eigenvalue significantly if the step leads to a position that is too far from the minimum. Thus, smaller steps that follow the curvature of the path may be necessary. On the other hand very small steps are computationally inefficient and one should use a step as large as possible, satisfying the above conditions.

From the definition of the Newton–Raphson step it is clear that the walks defined by Eq. (2b) have stationary points. These are the points at which the forces (and the length of the step) are zero. A minimum is a stationary point, suggesting that our previous description of the algorithm above was oversimplified. If we start exactly at the minimum we shall move nowhere.

To initiate the iterations, a small displacement along the shifted mode is made first, and only then a Newton–Raphson step is computed. The energy is minimized in all the other directions, excluding the shifted mode. We therefore expect that at most one mode can have a negative curvature — i.e., the shifted mode. We hope that the movement along $|k_p\rangle$ includes at least one maximum. A search direction with two minima and a single maximum has (of course) three stationary points **assuming** that the rest of the degrees of freedom can be minimized to a unique local energy minimum as we move along $|k_p\rangle$. However, this is not always the case.

Figure 4 shows an example in which the calculation of a reaction path for a dipeptide is considered. One of the other modes with a positive eigenvalue becomes negative along the path. This results in splitting of the path into two branches. In systems where none of the modes changes its sign valve, we find stationary points that are either minima or saddle points. This is what the "mode following" approach is all about.

There is a number of important advantages of the mode following approach compared with other techniques. Firstly, it is exact. (Locally, all directions excluding the relevant one are fully and correctly minimized.) Secondly, the minimization is performed locally, therefore the requirements of memory can be small if the molecular system considered is not large, compared to methods that provide the whole path at once. Third, this

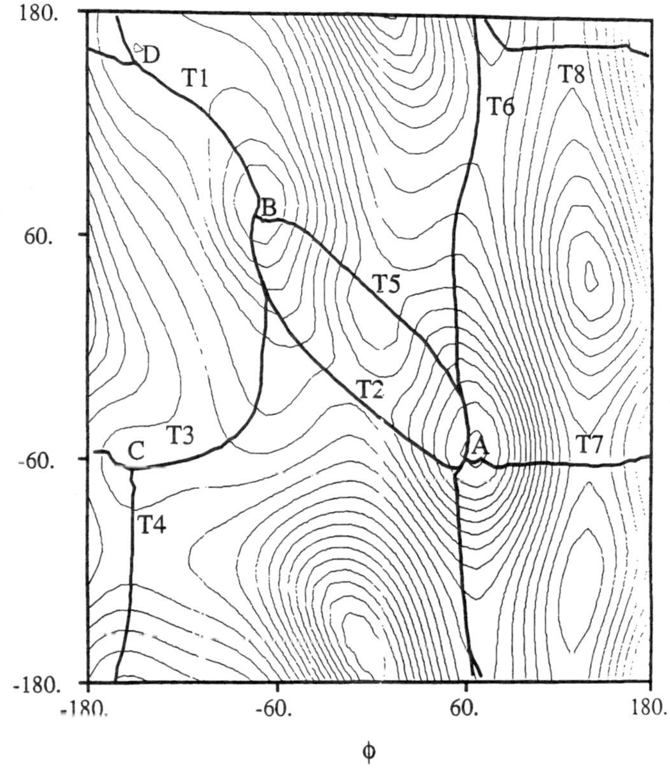

Fig. 4. A conformational change in a dipeptide studied by a reaction path approach. Eight different paths are shown between four minima. Note that paths T2 and T3 have a common ancestor at B which is split due to the change of sign of one of the modes (see Ref. 14).

approach does not require any prior guess on the shape of the minimum energy path. The technique locates stationary points along $|k_p\rangle$, very effectively, provided that stationary points exist.

Some of the advantages which we have just mentioned can also be disadvantages or limitations. For example, the fact that the path is calculated locally makes it difficult to direct the path towards the desired products. We do not know in advance where the path goes, leaving us at the mercy of the changes in direction of the $|k_p\rangle$ vector. If we are really unlucky, $|k_p\rangle$ may yield a circular path (see for example the steepest descent path T6 of Fig. 4 that connects the minimum A to itself).

The choice of $|k_p\rangle$ can be exceedingly difficult in large molecular systems in which there is a large number of low energy modes. The simple choice of walking along the mode with the lowest eigenvalue is not always fruitful. For example, a rotation around a H_3C-C bond of a CH_3 group can be very "soft" but still of little relevance to the reaction in which we are interested. A large number of modes of similar spatial extent and eigenvalues (typical of low frequency modes of proteins[24–27]) may lead to additional difficulties. Which of the modes should be picked? It is worse when some of the modes with nearby eigenvalues mix as we move along $|k_p\rangle$. That is, an "intuitive" normal mode may change to another normal mode as we proceed. The direction in which we are interested (and in which we started to move) is not necessarily the direction in which we will find ourselves moving after some propagation.

An additional complication is the necessity to frequently deal with second derivative matrices. This is to be contrasted with the LUP approach that does not require Hessian or related matrices. The use of these matrices is not a real problem for systems with up to 100 atoms. However, for larger molecules, the storage of the Hessians becomes increasingly more expensive and the manipulation of these objects becomes more difficult. Furthermore, the inversion required by the algorithm of a matrix in which many eigenvalues are close to zero (see Refs. 24–27 for an example of the density of modes in a protein) makes the inversion procedure unstable in the general case. Special tricks that are computationally demanding are required for the few cases in which this procedure still works. Therefore, unless the choice of the reaction coordinate is obvious or the number of the relevant degrees of freedom is very small (e.g. a rotation around a single bond), the method is not recommended for studying reaction paths in systems with more than 300 coupled degrees of freedom.

We proceed next to describe some results of rate theory in the gaseous phase. In contrast to the detailed description of the reaction path algorithms we shall not attain the same level of detail in the description of the rate theory. The reason is the availability of excellent reviews on the topic (see for instance Ref. 28).

2.2. *Computations of Rates based on the Reaction Path Approach*

It goes without saying that results of computer simulations must be tested against experimental data. The reaction path approach suggests several different levels of tests. A qualitative test may consist of a comparison of a series of reactions, all of a similar character with one (or a few) parameters varied systematically. For example, we may compare the reaction paths calculated for the series $CH_4+X \rightarrow CH_3X + H$, where X=F,Cl,Br. Another biological example is the effect of mutations in proteins on reaction rates.[29]

One issue of interest is comparison of the relative rates of the reactions. That is, what is the ordering of the rates. A suggested approach is to consider the energy barriers (the highest energy points along the paths for the different reactions) directly available from the reaction coordinates. Reactions with higher energy barriers are expected to be slower. This expectation is of course approximate since we correlate energy barriers with rates. Nevertheless, it is expected to be a useful comparison tool if the energy barriers are much larger than the thermal energy.

The above approach is useful and quite successful since it does not put high demands on the theory while providing useful predictions. This does not mean that more quantitative calculations should not be attempted. It is also of significant interest to be able to carry out more quantitative tests, i.e., the comparison of absolute rates for molecular processes between theory and experiment. Below we give a summary of how to calculate rates following the reaction path approach. The discussion is based on the Transition State Theory (TST[28]). We also discuss reduction of dimensionality and motions along the "effectively fewer" dimensions.

We consider a molecular system of K degrees of freedom that is moving on a single electronic energy surface, from reactants to products. It is evident that quantum effects (either for the electrons or for the nuclei) are important in a number of biologically related processes.[30] However, quantum effects are beyond the scope of the present discussion (which focuses mainly on the calculations of reaction paths). We therefore describe here rate calculations employing classical mechanics. There are many biological processes for which classical mechanics appears reasonable, including protein folding and other conformational transitions.

In the TST we assume the existence of a "perfect" dividing surface (of K-1 dimensions) between the reactants and the products (Fig. 1). A dividing surface is "perfect" if all reactive trajectories pass this dividing surface once and only once. Identification of this surface is a central problem. We are making the connection between the reaction path approach and the transition state theory by setting the highest energy point along the path as a point on the dividing surface. We further assume that the dividing surface is taken to be the orthogonal Cartesian subspace to the reaction coordinate at the relevant saddle point.

If the TST assumptions are accepted, the calculation of the rate is reduced to the count of the number of passes through the dividing surface from the reactants to the products. That is, the count is over trajectories that pass the dividing surface with velocities pointing towards the products side. The TST rate is the number of trajectories that pass the dividing surface in the "right" direction per unit time. The next assumption which we make is that the probability of finding the system on the dividing surface can be calculated using statistical arguments. This assumption is equivalent to time scale separation between the motion in the direction perpendicular to the dividing surface and the motion in the surface. We assume that the degrees of freedom of the dividing surface approach equilibrium much faster than the motion along the reaction coordinate.

A correction to the transition state theory that makes it "exact" is the application of the transmission coefficient.[31] The transmission coefficient is defined as the ratio of the number of trajectories that finally end on the product side to the number of trajectories that are assumed to be reactive according to the TST "rules",[28] that is, the trajectories that reach the dividing surface while their velocity vector points to the product side. A straightforward use of the transmission coefficient to correct the transition state theory is somewhat circular and the real question is whether it is possible to calculate the transmission coefficient accurately enough to significantly improve the numerical value of the TST rate, without solving the entire problem.

In the most straightforward approach to estimate the transmission coefficient, trajectories are initiated at the dividing surface with velocity that points to the product side and the number of trajectories that settle finally on the product side and the reactant side is counted. However, while this approach is straightforward, some difficulties remain.

First, the trajectories are initiated from a thermal equilibrium distribution on the dividing surface. The TST and the transmission coefficient are useful when both reactants and products are in true thermal equilibrium and exchange between the states is (of course) possible but extremely rare. However, one may ask if the system can be in real thermal equilibrium on relevant time scales if the motion along the reaction coordinate is that slow. If the number of re-crossings is very large (which totally invalidates the TST in its basic form), then the equilibrium assumption is more likely to be valid and the procedure employing the transmission coefficient becomes more accurate. However in this case the basic TST approach is incorrect and we may better avoid the detour of making adjustments to something that is very far from the truth.

Second, deciding when the trajectories are settled down in the product or the reactant part of the energy surface can be quite tricky and requires very clear separation of time scales. If crossing occurs on many time scales that cover a very broad range then it is not possible to clearly identify reactant and product states. The calculation of the transmission coefficient becomes similar (in numerical effort and in theory) to straightforward simulations employing individual trajectories. The computation of the transmission coefficient is justified when the probability of being on the dividing surface is small and trajectories are unlikely to cross it again on a time scale that is long compared with the simulation time but still short compared with the reaction time.

Another approach to compute the transmission coefficient is based on a model equation. Rather than solving the complete N dimensional problem, one replaces the equation by a different model that is simpler to analyze — the Generalized Langevin Equation (GLE).[32,33] Results for the transmission coefficient are readily available from an elegant theory.[32,33] However one should keep in mind that the model (a memory kernel that depends only on the time difference) is a highly simplified version of reality. As was shown by Zwanzig[34] and later by Pollak,[33] this model is equivalent to an arbitrary reaction coordinate that is coupled to a harmonic bath.

It is very well known[35,36] that a harmonic model for biological molecules (such as proteins) is a poor approximation. I therefore do not expect GLE to be a quantitative model and be able to provide better numerical estimates for the rate in large macromolecules than straightforward TST. In studies of proteins, agreement between experiment and theory employing the GLE model with one "reaction coordinate" should be considered coincidental, since the error bars associated with the basic assumptions of the TST are not expected to be necessarily larger than the errors coming from the application of the GLE model itself.

The straightforward correction to the TST rate brings us closer to straightforward MD simulation. In straightforward MD, many individual trajectories are used to compute an average quantity. This approach was found useful in the modeling of a few very fast reactions in biological molecules.[37–39] However, as I have argued above, when the typical time scale of the reaction exceeds a few hundred picoseconds, a study of the reaction in a systematic way using MD is difficult.

We now give a heuristic derivation for the rate formula. Let dN be the fraction of molecules that passes to the product side. A reaction is counted when the molecular system passes successfully an interval dQ located on the dividing surface which we denote by $\mathbf{Q}^{\#}$ (see Fig. 2). The probability of finding the system between $\mathbf{Q}^{\#}$ and $\mathbf{Q}^{\#}+dQ$ is directly proportional to the rate. We assume that this probability can be calculated, using statistical arguments, with no reference to the time history of the trajectory before it reaches the barrier. The probability is proportional to the weighted area of the phase space near $\mathbf{Q}^{\#}$, divided by the total weighted area available to the reactants.

$$dN = N\left[\int d\mathbf{R}d\mathbf{V}\delta(\mathbf{R}-\mathbf{Q}^{\#})\, W(\mathbf{R},\mathbf{V})\, dQ\right] / \left[\int d\mathbf{R}d\mathbf{V}W(\mathbf{R},\mathbf{V})\right] . \tag{22}$$

The integrals are over the phase space — the coordinates ($\mathbf{R}$) and the velocities ($\mathbf{V}$). One should emphasize that $\mathbf{Q}^{\#}$ denotes a plane and not a point. The Dirac delta function ensures that only configurations in the plane (the dividing surface) are selected. $W(\mathbf{R},\mathbf{V})$ is a weight assigned to each point. For example, if the temperature is constant, $W(\mathbf{R},\mathbf{V})$ is the Boltzmann factor — $W(\mathbf{R},\mathbf{V}) = \exp[-(U(\mathbf{R}) + E_k(\mathbf{V}))/k_BT]$, where $E_k(\mathbf{V})$ stands for the kinetic energy, k_B is the Boltzmann factor and T is the absolute temperature. To normalize the weight in order to obtain probability we divide the integral in the nominator by a similar integral which is extended over the configuration space of the reactants.

Let the velocity along the reaction coordinate (that is, in the direction perpendicular to the dividing surface) be V_q. dQ is therefore given by $dQ = Vdt$, where dt is an infinitesimal time interval that is required to pass the dividing surface. The number of molecules that react per unit time (the rate) is therefore

$$dN/dt = N\left[\int d\mathbf{R}d\mathbf{V}\delta(\mathbf{R}-\mathbf{Q}^{\#})\, W(\mathbf{R},\mathbf{V})V{q^{+}}\right] / \left[\int d\mathbf{R}d\mathbf{V}W(\mathbf{R},\mathbf{V})\right] . \tag{23}$$

Vq^+ points to the product side. In the special case of the canonical ensemble (that was mentioned above), the weight can be factored to the coordinate and the velocity parts, and the integral over the velocities is analytical. We have

$$\begin{aligned} dN/dt &= N\langle Vq^+\rangle^{Ek}\left[\int d\mathbf{R}\exp(-U(\mathbf{R})/k_BT)\ \delta(\mathbf{R}-\mathbf{Q}^\#)\right] \\ &\quad /\left[\int d\mathbf{R}\exp(-U(\mathbf{R})/k_BT)\right] \\ &= N\langle Vq\rangle^{Ek}\ \langle\delta(\mathbf{R}-\mathbf{Q}^\#)\rangle^U , \end{aligned} \tag{24}$$

where $\langle\ldots\rangle^{Ek}$ denotes an average with respect to the kinetic energy part of the weight function, and $\langle\ldots\rangle^U$ denotes an average with respect to the potential energy part.

The ratio of the two integrals on the right-hand side is a function of the position of the dividing surface. It is therefore a function of a single coordinate only, since the rest of the degrees of freedom are integrated out. An important feature of this ratio is that it can also be written as an average, i.e., an average of the Dirac delta function over the weighted configuration space. Averages are quantities that can be computed effectively with Monte Carlo or Molecular Dynamics procedures. Calculation of the average value of the delta function is of major importance in computational studies of rates. It is directly related to the free energy surface (or the potential of mean force). Up to an arbitrary constant, the potential of mean force — U_{pmf} is defined as:

$$U_{\text{pmf}}(Q) = -k_BT\ \log\left[\langle\delta(\mathbf{R}-\mathbf{R}(Q))\rangle^U\right] . \tag{25}$$

We shall discuss the calculation of the potential of mean force at length later in this chapter.

As we argued above the reaction path approach suggests $\mathbf{Q}^\#$ as the hyper plane perpendicular to the vector parallel to the reaction coordinate at the highest energy point. (The vector is the slope of the curvilinear line which is the reaction coordinate. If the curvilinear coordinate that is used to denote the path is called l, then the slope of the path is proportional to $\nabla_{\mathbf{R}} l(\mathbf{R})$.)

There is a number of assumptions made in this suggestion. First, it is assumed that the surface perpendicular to the slope of the path at the highest energy point is indeed the best choice for a surface of "no return". This is an intuitively pleasing assumption, but intuition is not a proof. Second, it is assumed that the dividing surface is everywhere perpendicular to the direction of the path slope at $\mathbf{Q}^{\#}$. The orthogonality relation holds only in the neighborhood of the reaction coordinate, and the surface can be curved in unexpected ways when the system is not in close proximity to the reaction coordinate.[40]

There is no general computational method for finding the exact shape of the dividing surface (assuming it exists in a useful way). The intuitive suggestions of the reaction path approach make it possible to compute sensible values for the rate. However it is clearly not a rigorous computational approach.

The assumption on the shape of the dividing surface becomes exact only in the limit of a "narrow" surface, that is, in the limit in which the different reactive trajectories remain very close to the computed minimum energy coordinate. This can happen at low temperatures or for stiff potentials that hold the bundle of reactive trajectories in a "narrow pipe". In this case, it may be possible to justify a different approximation — the quadratic expansion of the potential energy in the $\mathbf{Q}^{\#}$ plane. We hope that portions of the surface that are not described well by the quadratic expansion are of high energy and therefore not sampled. If our wishes are fulfilled it is not necessary to describe the high energy portion of the energy surface accurately. In particular the integral

$$\mathbf{I}_1 = \int d\mathbf{R} \exp(-U(\mathbf{R})/k_B T)\delta(\mathbf{R} - \mathbf{Q}^{\#})$$

becomes much simpler after we write

$$\begin{aligned} U(\mathbf{R}) \approx U(\mathbf{R}(Q^{\#})) & \\ & + 1/2\Sigma\ \partial^2 U(\mathbf{R})/\partial R_i \partial R_j|_{\mathbf{R}(Q^{\#})} \\ & \times (R_i - R_i(Q^{\#}))(R_j - R_j(Q^{\#}))\,. \end{aligned} \tag{26}$$

The potential is at a minimum in the directions perpendicular to the reaction path, therefore, first derivatives of the potential do not appear in the expansion. R_i and R_j are the individual components of the $\mathbf{R}$ vector and $\mathbf{R}(Q^{\#})$ is the coordinate vector at the position of the minimum energy path that crosses $\mathbf{Q}^{\#}$. Following the definition in Eq. (20) we use $\mathbf{K}$ to denote the second derivative matrix. Here we use $\mathbf{K}(Q)$ to denote the second derivative matrix excluding the direction perpendicular to the reaction coordinates. The approximate form of $U(\mathbf{R})$ is substituted in the integral of Eq. (25) and the integration is performed analytically

$$\mathbf{I}_1 = \exp(-U(\mathbf{R}(Q^{\#})/k_B T) \, [\pi/\det(\mathbf{K}(Q^{\#})/k_B T)]^{1/2} . \qquad (27)$$

The square root on the right-hand side is an entropic term that may depend on Q. The broader is the "pipe" in the energy surface at $Q^{\#}$ that allows reactive trajectories to pass through, the smaller is $\mathbf{K}(Q^{\#})$, and the larger are $\mathbf{I}_1$ and the rate. This is an example of why the highest energy point is not always the best position for the dividing surface, since the entropic factor (or the width of the reactive pipe) is not taken into account in this estimate and it may be important.

The contribution of entropy (as found in experiments) to free energy barriers rarely exceeds a few kcal/mol, provided that the path is well-defined. By a "well-defined path" we mean that the "pipe" of the reactive trajectories is not exceptionally broad and that the number of alternative pathways is small. So if the energy barrier is significantly larger than a few kcal/mol, the profile of the potential of mean force is likely to be dominated by the energy only.

We shall further denote the integral $\int d\mathbf{R} \exp(-U(\mathbf{R})/k_B T)$ as $\mathbf{I}_2$, which in a similar spirit to Eq. (27) is written as

$$\mathbf{I}_2 = \int dQ \exp(-U(\mathbf{R}(Q)/k_B T) \, [\pi/\det(\mathbf{K}(Q)/k_B T)]^{1/2} . \qquad (28)$$

$\mathbf{I}_2$ is a one-dimensional integral over the Q part of the reactants, computed using the same approximations as the ones used to estimate the value of $\mathbf{I}_1$ above. It is the partition function of the reactants, employing the harmonic approximation for coordinates other than the reaction coordinate.[41]

We deliberately do not proceed to derive a simpler closed form expression for the rate. A local quadratic expansion near a single minimum is not very useful in liquid and liquid-like phases. In these phases the anharmonic parts of the energy surface are frequently sampled. This is to be contrasted with small molecules and gas phase kinetics in which the quadratic expansion was found to be useful in numerous cases. From now onwards we focus on direct calculation of Eq. (24), including anharmonicity.

To summarize, I list below the common assumptions used to derive and finally to compute the rates according to Eq. (24):

(1) There is a dividing surface that is crossed only once during a reactive trajectory.
(2) The motion in the direction perpendicular to the reaction coordinate is in a local statistical equilibrium as the reaction proceeds.
(3) The reaction path is assumed to cross the dividing surface at the highest energy point. (This restriction can be slightly relaxed as we shall see below.)
(4) The dividing surface is further assumed to be always orthogonal to the reaction coordinate at the point shared by it and by $Q^{\#}$.

I would like to conclude the discussion on calculations of rates by considering a different approach to the rate theory that is also based on the identification of the reaction coordinate. In this alternative approach we assume that the motion along the reaction coordinate is slower than other time scales in the system and that the system is very close to equilibrium. In contrast to the TST that focuses only on the dividing surface, here we are interested in computing the dynamics along the continuous "slow" coordinate.

It may seem that the underlining assumptions in the slow coordinate approach are stronger than those of the transition state theory, since at any point along the reaction coordinate the system is assumed to be in a local thermal equilibrium. This is in contrast to the TST that assumes equilibrium only at the free energy barrier. However the last point (the barrier) is usually the position at which equilibrium is most difficult to obtain. Furthermore, recrossing in the "slow" coordinate approach presents no problem. Therefore, the approximations in the slow coordinate approach are somewhat weaker than the approximations made in the TST.

It is possible to formally write the equations of motion for the slow coordinate Q within the linear response framework[17]:

$$\begin{aligned} d/dt[m(Q)dQ/dt] = & -dU(Q)/dQ \\ & - \int \gamma[Q(t), Q(t'), dQ(t)/dt, dQ(t')/dt', t, t'] \\ & \times dQ(t')/dt' \, dt' + F \, , \end{aligned} \tag{29a}$$

where $m(Q)$ is the conjugate mass for moving along the Q coordinate. The mass in principle can be Q dependent. $-dU(Q)/dQ$ is the systematic force. In some formulations the systematic force is buried as a part of the memory function, and here we prefer to identify it phenomenologically as the derivative of the potential of mean force[31]

$$U(Q) = U_{\mathrm{pmf}}(Q)\,, \tag{29b}$$

where $U(Q)$ is determined up to an irrelevant additive constant.

The most complex part of Eq. (28) is the memory function γ that in principle depends on all the previous history of the trajectory — time, coordinates and velocities. F is a random force that is related to the memory function via the fluctuation-dissipation theorem. It is far beyond the focus of this chapter to go into the details of how Eq. (28) was constructed and we refer the interested reader to the literature.[17] However, it is important to note the difference between Eq. (28) and the so-called Generalized [Langevin Equation] (GLE) in which the memory function is assumed to be a function of the time difference only $\gamma \equiv \gamma(t - t')$. Neglecting all other dependencies is a serious approximation. Therefore the GLE cannot be considered a systematic theory that is based on a microscopic derivation, but rather a phenomenological model. The evaluation of the approximations made is difficult. There are only a few calculations of other dependencies of the memory function[42,43] and even fewer exact calculations of γ.[17] Without the simplifications the theory is complicated. The calculation of the memory tensor is difficult even in the absence of explicit spatial dependence. The use of the tensor in practical studies of the dynamics is also far from trivial. I am aware of a single computation employing the tensor in biological molecules.[44] Nevertheless, the GLE is still a popular approach, probably because of its attractive analytical form that enables derivation of closed expressions for some special cases.[32–34]

Computationally, a starting equation for many phenomenological theories is not the (still complex) GLE but rather a simpler variant in which $\gamma = \gamma_o \delta(t - t')$ (short range memory). With this choice of the memory function (which is phenomenological) and after setting the mass to a constant, Eq. (29) is reduced to the Langevin equation

$$m d^2Q/dt^2 = -dU(Q)/dQ - \gamma_o dQ/dt + F\,. \tag{30}$$

Computations of observables of a system that is described by Eq. (30) can be pursued using a variety of techniques. The most straightforward one is calculation of stochastic trajectories and averaging of quantities of interest over an ensemble of trajectories obtained in the simulation. The reduced dimensionality and the "slow" character of Q that was assumed to begin with make the direct integration of the equations of motion much easier than straightforward molecular dynamics. Another equivalent approach is based on a solution of the equation for the density — the Fokker–Planck equation.

The difficulties in determining the phenomenological parameters of the alternative approach are very significant and the connection to microscopic theory is still not known in a satisfactory way. Therefore the phenomenological parameters must be fitted independently against experimental data. In the discussion of rates throughout the remainder of this chapter I shall consider examples employing only the TST — an approach on which much work has been done in connecting the experimental observables to microscopic parameters.

3. More on Reaction Coordinates and Computations of Mean Force Potentials

As noted above, the question of locating relevant reaction coordinates in systems with many atoms is far from trivial. Most of the calculations in many atom systems are based on the reference coordinate approach which was described in the previous section. In some studies a reaction coordinate was assumed with no further refinement. The choices were based on intuition and some prior information on the process. Choices of reaction coordinates that are based on intuition include (for example) the distances between pairs of atoms,[45] a rotation around a single bond or a few bonds,[46] and the location of an atom along a preferred axis.[47] We shall denote these coordinates by IC (Intuitive Coordinates). The most common models for reaction coordinates in condensed phases are therefore simple compared to a number of the elaborate computational methods used at present in studies of (the simpler) small molecules (see previous section).

It is unlikely that the additional complexity of larger systems as compared with small molecules will improve the accuracy of ad-hoc approaches to select reaction coordinates. Popular justifications for the IC approach are: (a) we do not have an alternative and (b) it does not really matter. Here, I shall argue in favor of more elaborate computations of reaction paths in the condensed phases. I shall proceed to discuss the difficulties in the application of common approaches (in the gas phase) to problems that include many atom multiple minima systems. In a follow-up section

I shall discuss alternatives to the computations of minimum energy paths that differ significantly from the approaches employed in few atom systems.

The IC paths (e.g., a rotation around a bond without further energy and structural refinement) may yield energy barriers that are too high due to the probable presence of close van der Waals contacts. This limits the usefulness of IC paths in rough estimates of rates that are based on barrier heights. The energy profile of an IC can be and sometimes is refined using the REF coordinate approach. However, the correction is subject (of course) to all the difficulties of adiabatic mapping, such as discontinuous paths.

Thus, direct application of an IC path or the use of a REF refinement are both problematic. Discontinuous paths are considered a more serious problem when free energy profiles are of interest. Applications of IC to condensed phase problems where free energy profiles are of interest are therefore pursued without energy relaxation.

This is a good point to reconsider the application of the transmission coefficient (see previous section). From the formal point of view the use of the transmission coefficient indeed makes the Transition State Theory exact.[31] (I consider here only the explicit calculation of the transmission coefficient, starting from the top of the barrier and not application of the GLE model). It has therefore been suggested that the specific choice of the dividing surface is not very important. That is, whatever the dividing surface is, the calculation of the transmission coefficient provides the exact answer. However, an exact formalism does not necessarily explain or provide guidelines on how the theory should be applied to obtain numbers that can be compared to experimental data. It is therefore useful to assess the numerical efficiency of the corrected form of the transition state theory and the computational effort necessary to converge the results to the desired accuracy.

The correction is calculable if the trajectories employed in the computations of the transmission coefficient relax rapidly to either the reactant or the product side. This is regardless whether the asymptotic value of the coefficient is large or small. If the trajectories do not relax much faster than the typical time scale of the reaction, then the whole correction scheme is in doubt from a computational point of view. Moreover, this suggests that the assumed separation of time scales in the TST is incorrect, which puts in question the whole scheme of estimating the rate. An inappropriate choice of a dividing surface (or of a reaction coordinate) will have a significant influence on the separation of time scales of different coordinates and on practical applications of the TST.

After this side remark on the transmission coefficient we return to the potential of mean force. As discussed in the section on "small" molecules it is better to consider the profile of the potential of mean force (Eq. (25)) along an IC path, rather than the energy profile, since the profile of the mean force potential also includes entropic effects on the thermal probability of finding the system at different Q's. It must be emphasized that for different reaction coordinates, one obtains different dividing surfaces and different PMF-s.

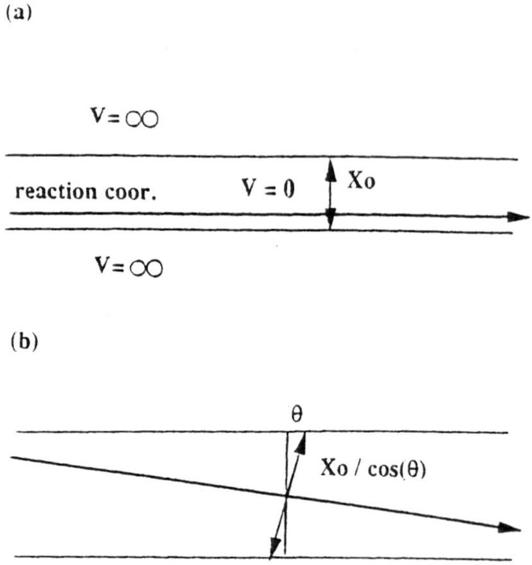

Fig. 5. An example of the dependence of the potential of mean force (pmf) on the reaction coordinate: Two-dimensional energy surfaces with hard boundaries and two possible reaction coordinates (a) and (b).

As an analytical example for the PMF dependence on the choice of the reaction coordinate, consider a model potential of a channel — zero inside the walls and infinity outside (Fig. 5). In addition to a picture of the channel the figure also includes two alternative choices for the reaction coordinate. The PMF's for each of the reaction coordinates are plotted in Fig. 6 and the analytical expressions are also provided. It is evident that

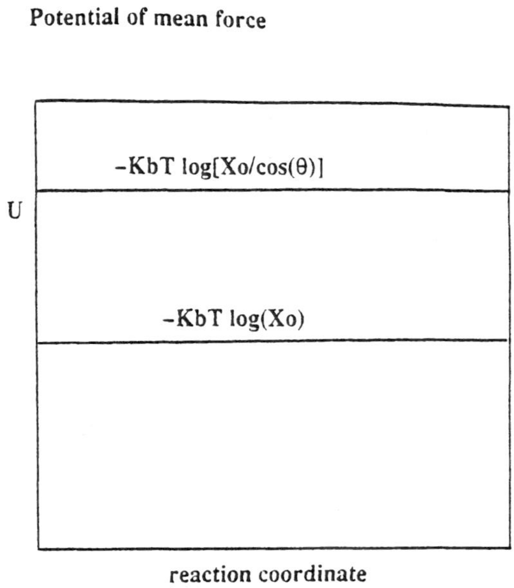

Fig. 6. An example of the dependence of the potential of mean force (pmf) on the reaction coordinate: The potential of mean force of the two cases (5a) and (5b).

the potentials are significantly different. There is no reason to assume that in the more complex condensed matter studies the dependence of the PMF on the choice of the coordinate will be weaker.

The most popular approach to locate SDP's in small molecular systems is the mode following approach. However, as I argued in the previous section the necessity of dealing with large matrices and the rather arbitrary choice of the mode to be followed makes this technique difficult to use in large molecules. Nevertheless, difficult is not impossible, and with some reduction in the number of degrees of freedom (e.g. using adiabatic mapping for the bonds and the angles only), useful results can be obtained.[65]

There is another question at a more theoretical level regarding the significance of a single SDP in large and flexible molecular systems. In highly flexible systems there is a large number of minima which (in a "worst case" scenario) grows exponentially with the number of degrees of freedom. The relation between the number of degrees of freedom and the number of paths is not expected to be more favorable. For a continuous potential, the number of saddle points is bound from below by the number of minima minus one. In practice it is expected to be much larger (Fig. 7). Even in the neighborhood of a well defined three-dimensional structure (i.e., a structure of a folded protein) the number of minima is surprisingly large. It was estimated that the number of "near X-ray" minima sampled during a 300 ps, room temperature simulation of the protein myoglobin is 2000.[36]

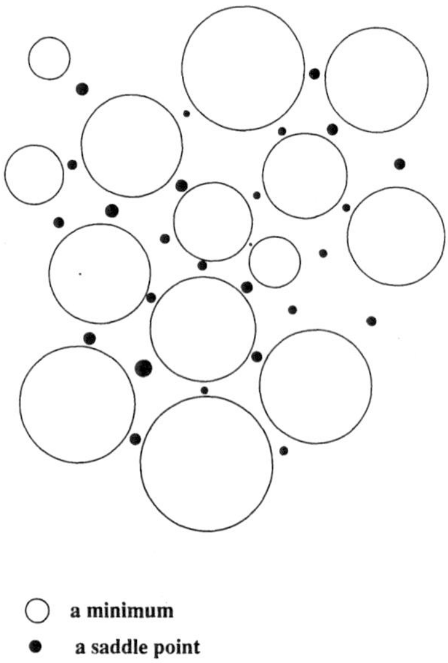

Fig. 7. A potential energy surface with a very large number of minima and an even larger number of saddle points.

The significant number of minima is expected to be connected by an even more significant number of SDP's. As argued in the previous section, these paths can be regarded as low temperature, high friction trajectories. Individual Steepest Descent Paths are therefore useful for interpretative purposes and for qualitative analysis of plausible paths in the same spirit that individual trajectories are used to investigate "how things happen". How to connect the many SDP's to a rigorous rate theory is, however, a non-trivial question which to the best of my knowledge is as yet unresolved. A clear paper on the topic is the study of Harris and Stillinger.[48]

We considered in the past[14] the validity of a single reaction coordinate by examining saddle points between a collection of minima of a tetrapeptide (we found about 500 saddle points). The significant number of minima and saddle points clearly indicates that the assumption of a single reaction path is questionable in flexible systems. This is not very surprising but it opens the ground to further development of theories of rate,[77] based on approximate (and simpler) computations of the distribution of barrier heights.[49] Consider for example a conformational change in a polymer which can be described as a sequence of rotations around individual bonds. If the rotations are not strongly coupled, then it is usually possible to change the order of the rotations and still get reasonable paths. Picking a single sequence as a single reaction coordinate drastically reduces the rate by approximately the number of different combinations of bond rotations which connect the reactant and the product.

The situation is better when there is only one significant step and the motion is spatially localized (for example [Cl + $CH_3Cl \rightarrow ClCH_3$ + Cl] in water). It also helps if a significant energy (or free energy) barrier exists. If the energy barrier is larger than k_BT and larger than other barriers in the system then the separation to a single coordinate which is not at equilibrium (the reaction coordinate) and the rest of the system which is at equilibrium is reasonable. In such systems with separation of barrier heights the method of "activated dynamics" is a useful approach.

We consider some of the ideas behind "activated dynamics", since this is an essential step in computing rates once the reaction coordinate is at hand. We do not go into all possible details since numerous papers on the topic have been published elsewhere.[31] In Eq. (24) we are requested to provide the numerical value of the following average $\langle \delta(\mathbf{R} - \mathbf{Q}^{\#}) \rangle^{\cup}$ where $\mathbf{Q}^{\#}$ represents the dividing surface, i.e. the set of points in the plane $Q^{\#}$. The delta function is "on" if $\mathbf{R}$ is in the plane and is "off" otherwise.

In condensed phase systems or macromolecules the calculations of this average are far from trivial. Previously we estimated the value of the integral using a quadratic expansion of the potential. This expansion is valid for solids and for some small rigid molecules. It is not that great for floppy systems and it is bad for liquids in which a tremendous number of alternative minima exists. In Eq. (27), a single local minimum is assumed.

It is impossible to pursue analytical integration in Eq. (24) for an arbitrary potential. We therefore use numerical integration. A common approach to pursue the integration is via Monte Carlo (MC) or Molecular Dynamics (MD) procedures. Both approaches are based on generation of a sequence of structures. A sequence of configurations is generated according to a "rule". The rule can be the "real" equations of motion (MD) or random steps that satisfy detailed balance criteria (MC). The set of configurations (if large enough) should provide an equilibrium distribution.[3,4]

The argument in favor of MD is that it produces configurations from a microcanonical ensemble and the rule (Newtons equations) has physical meaning. Monte Carlo procedure is more flexible in the choice of the rule. However, not all structures in MC are accepted. A new structure in MC is accepted if its energy is not much larger that the energy of the previous structure (measured in $k_B T$ units). Both MC and MD are useful approaches to sample probable configurations in equilibrium. However probable configurations are not that useful in computing the average in the following equation

$$\langle \delta(\mathbf{R} - \mathbf{Q}^{\#}) \rangle^{\cup} = \int \delta(\mathbf{R} - \mathbf{Q}^{\#}) \exp[-U(\mathbf{R})/k_B T] \, d\mathbf{R} / \int \exp[-U(\mathbf{R})/k_B T]/k_B T] \, d\mathbf{R} \tag{31}$$

which, using either of the sampling procedures described above, is estimated by

$$\langle \delta(\mathbf{R} - \mathbf{Q}^{\#}) \rangle^{\cup} = \Sigma \Delta(\mathbf{R} - \mathbf{Q}^{\#})/L \,, \tag{32}$$

where Δ is a discrete approximation to the Dirac delta function in (31) and L is the number of sampling points. If $\mathbf{R}$ is sufficiently close to $\mathbf{Q}^{\#}$ (within a preset cutoff), then Δ is one, otherwise, it is zero. There is a serious difficulty in computing the above average for high energy barriers: It is unlikely that a normal sequence of structures generated by MD or MC will provide sufficient statistics to obtain an accurate average (Fig. 8). It is more likely that ordinary runs will end with no sampling at all at the dividing surface. A different algorithm to overcome this problem is needed.

There are two popular approaches to solve this problem: (a) umbrella sampling, and (b) free energy perturbation. Here we describe only the second option (see Ref. 50 for a description of the umbrella sampling procedure). The umbrella sampling requires more manual work and is conceptually more complex than the free energy perturbation. In addition to the usual statistical errors one must also consider the human factor in the manual analysis of the results. Nevertheless, we found the umbrella sampling to be numerically more stable in a few examples. It therefore can be a useful alternative to the more automated free energy perturbation.

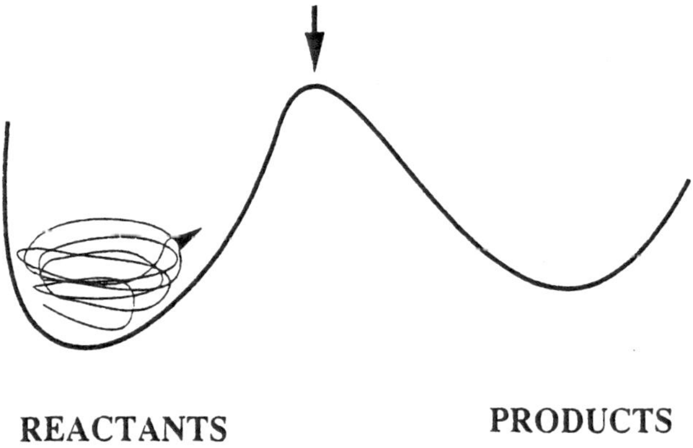

Fig. 8. MC or MD do not provide proper sampling at the transition state.

The average in Eq. (32) is the probability density at $q^{\#} - \rho(q^{\#})$. We further relate the probability density to the potential of mean force: $\rho(q^{\#}) = A\ \exp(-F(q^{\#})/k_BT)$, where A is a normalization constant and $F(q^{\#})$ is the potential of mean force (or the free energy profile at $q^{\#}$) and is defined by

$$F(q^{\#}) = k_BT \log\left[\int \delta(\mathbf{R} - \mathbf{Q}^{\#})\ \exp[-U(\mathbf{R})/k_BT]\ d\mathbf{R}\right]$$
$$F(q^{\#}) = -k_BT\ \log\left[\int \exp[-U(\mathbf{R}', q^{\#})/k_BT]\ d\mathbf{R}'\right]\ . \tag{33}$$

$\mathbf{R}'$ is the complementary space to q. The second formula in (33) is more convenient as a starting point to derive the free energy perturbation formula. Consider the difference:

$$F(q_1) - F(q_2) = -k_BT \log\left[\int \exp[-U(\mathbf{R}', q_1)/k_BT]\, d\mathbf{R}' / \int \exp[-U(\mathbf{R}', q_2)/k_BT] d\mathbf{R}'\right]. \tag{34}$$

The ratio of the two integrals can also be written as

$$\int \exp[-U(\mathbf{R}', q_1)/k_BT] d\mathbf{R}' / \int \exp[-U(\mathbf{R}', q_2)/k_BT] d\mathbf{R}'$$
$$= \int e^{-\{[U(\mathbf{R}',q_1) - U(R',q_2)] + U(\mathbf{R}',q_2)\}/k_BT} d\mathbf{R}'$$
$$/ \int e^{-U(\mathbf{R}',q_2)/k_BT} d\mathbf{R}'$$
$$= \langle e^{-[U(\mathbf{R}',q_1), -U(\mathbf{R}',q_2)]/k_BT\}} \rangle\, U(\mathbf{R}', q_2) \tag{35}$$

where the average is pursued over configurations sampled from the canonical ensemble for a potential $U(\mathbf{R}', q_2)$. The formula in (35) is exact. In principle, it is now possible to compute the free energy value and the probability density at $q\#$ (substitute $q\#$ for q_1) using a trajectory at the reactant position (say q_2).

The difficulty in Eq. (35) is that the average requested is an average over an exponent. If the difference between the two energies, $U(\mathbf{R}', q_1) - U(\mathbf{R}', q_2)$, has a broad distribution (as is likely to be the case if the difference between the two q's is large) then the average over the exponent of the difference will be extremely difficult to converge. For example, if we accidentally sample a single point such that the energy difference is -18 kcal/mol, the weight of this point at 300K is $e^{30} = 1.07 \times 10^{13}$. To reduce the weight of this point as compared to (probable) configurations of weight one, we need at least 10^{13} additional points. The computation and the storage of this number of points in ordinary computer simulations is very difficult. Thus, the above type of averaging (which is called free energy perturbation) is very sensitive to the width of the distribution of the energy differences. To make the computations practical, it is necessary to maintain a narrow distribution of the differences. This is achieved by considering only q_2's that are very close to q_1.

Thus we consider the free energy difference between q and $q + \delta q$, where δq is small.

$$\begin{aligned} F(q+\delta q) - F(q) &= -k_B T \log[\langle e^{-\{U(\mathbf{R}',q) - U(\mathbf{R}',q+\delta q)/k_B T\}}\rangle_{U(\mathbf{R}',q+\delta q)}] \\ &\approx -k_B T \log[\langle 1 - \{(dU/dq)\delta q/k_B T\}\rangle_{U(\mathbf{R}',q+\delta q)}] \\ &\approx dF/dq'|q' = q\,\delta q\,. \end{aligned} \tag{36}$$

The free energy difference between the reactant (q_1) and the dividing surface (q_2) is therefore given by a summation of the small differences defined by (36). The expansion of the exponent is not really necessary but it reduces further the fluctuations during the averaging procedure. This is using the MD or the MC protocols. Another commonly employed step is the expansion of the log to a first order

$$[\log(1 - \langle dU/dq\rangle)\delta q/k_B T)] \approx \langle dU/\delta q)\rangle dq/k_B T\,. \tag{37}$$

After substituting (37) into (36) and expressing the difference $F(q_2) - F(q_1)$

as an integral we have

$$F(q_2) - F(q_1) = \int_{q_2}^{q_1} dq dF/dq = \int dq \langle dU/dq \rangle \rangle . \tag{38}$$

The final formula is called Thermodynamic Integration (TI). It is more convenient to use than (34) since the averaging is performed in small steps and the distribution of the quantity that requires averaging is significantly narrower than in the perturbation theory approach. We comment that in spite of our heuristic derivation of Eq. (38), it is exact and there are no approximations made.

The last comment pertains to the calculation of Eq. (38) using Molecular Dynamics. In the calculations of the average $\langle dU/dq \rangle \rangle$ we need to fix the reaction coordinate q at a given value. If the reaction coordinate was computed (as opposed to being assumed), then it is unlikely to be a single internal degree of freedom such as an angle or a torsion. It is more likely that the computed reaction coordinate is a function of many internal coordinates. For example, a reaction involving rotation around a bond results in some small changes in bond lengths and bond angles, among others.

The internal coordinates are conceptually easier to employ when we investigate reaction mechanisms. However, the use of **many** internal coordinates is not advantageous to the use of the simpler Cartesian representation. The manipulations of the Cartesian coordinates on the computer are much simpler. We therefore use only Cartesian coordinates in our code.[51]

The reaction coordinate q is represented by a set of discrete (Cartesian) points along the reaction path — $\{\mathbf{R}_{qi}\}_{i=1,N}$. We now need to sample configurations at a fixed value of q using ordinary molecular dynamics trajectories. This is achieved with the use of constraints — the seven constraints that were defined in relation to Eq. (11) (also see Ref. 52). In contrast to (12) which is the steepest descent mapping under the influence of the constraints, we now solve the Lagrange equations, that is

$$\underline{\mathbf{M}} d^2\mathbf{R}/dt^2 = -\nabla U + \Sigma \lambda_\ell \nabla \sigma_\ell . \tag{39}$$

A standard algorithm to solve the equations of motion in the condensed phases is that of Verlet (Ref. 5 and Eq. (1)). We provide below an iterative algorithm that is used to solve the differential equations with the constraints. It is based on the velocity version of the Verlet integrator,[3] that is

$$\begin{aligned}\mathbf{R}_{i+1} &= \mathbf{R}_i + \mathbf{V}_i\Delta t + (1/2)\underline{\mathbf{M}}^{-1}\Delta t^2[-\nabla U(\mathbf{R}_i) + \Sigma\lambda_\ell \nabla\sigma_\ell(\mathbf{R}_i)]\\ \mathbf{V}_{i+1} &= \mathbf{V}_i + (1/2)\underline{\mathbf{M}}^{-1}\Delta t[-\nabla U(\mathbf{R}_i) + \Sigma\lambda_\ell\nabla\sigma_\ell(\mathbf{R}_i) - \nabla U(\mathbf{R}_{i+1})\\ &\quad + \Sigma\lambda_\ell\nabla\sigma_\ell(\mathbf{R}_{i+1})]\,. \end{aligned} \tag{40}$$

The orthogonalization of the constraints is carried out in a slightly different way than in Eq. (15). That is, we set $\nabla\sigma_\ell\underline{\mathbf{M}}^{-1}\nabla\sigma_m = \delta_{\ell m}$. The objective is to simplify the equations for the Lagrange multipliers λ_ℓ.

$$\begin{aligned}&\sigma|(\mathbf{R}_{i+1}) = \sigma|(\mathbf{R}_i) = 0 = \nabla\sigma|(\mathbf{R}_i \cdot (\mathbf{R}_{i+1} - \mathbf{R}_i)\\ &= \nabla\sigma|(\mathbf{R}_i)\cdot\left[\mathbf{V}_i\Delta t + (1/2)\underline{\mathbf{M}}^{-1}\Delta t^2[-\nabla U(\mathbf{R}_i) + \Sigma\lambda_m\nabla\sigma_m(\mathbf{R}_i]\right]\,. \end{aligned} \tag{41}$$

Note that the Lagrange multipliers are time dependent. (We did not denote explicitly their time dependence to save some space.) The last equation provides an explicit solution for λ_ℓ

$$\lambda_\ell = -(2/\Delta t^2)\nabla\sigma|(\mathbf{R}_i)\left[\mathbf{V}_i\Delta t - (1/2)\underline{\mathbf{M}}^{-1}\Delta t^2\nabla U(\mathbf{R}_i)\right]\,. \tag{42}$$

For the velocities we use a different set of constraints. σ_ℓ are linear in coordinates. The time derivatives of the seven coordinate constraints provide seven constraints that depend only on the velocities. These new constraints are linear in the velocities in the same way that the coordinate constraints are linear in the coordinates. The velocity constraints are solved using equations analogous to (41) and (42) (see Ref. 52). Therefore Eqs. (40)–(42) provide a method to sample configurations at any fixed point of the curvilinear reaction coordinate, q. This trick is useful for the computation of the average for arbitrary q in (38), using MD with linear constraints.

Once the free energy difference between the reactants and the transition state is known, it is possible to compute the probability of finding the system at the top of the barrier. The average over the velocities is the same as outlined for small systems. This completes the description of the computations required to obtain the numerical value for the reaction rate using the transition state theory.

4. Reaction Path Algorithms Tailored for Large Molecular Systems

As argued previously, the steepest descent path is a very special type of trajectory, one in the limit of very high friction and very low temperatures.

Clearly, this definition is not necessarily better than other possible definitions of the path. Alternative definitions for reaction paths are also valid. The key requirement for a good reaction coordinate is an adequate representation of real reactive trajectories of a system and a possibility to estimate experimental observables such as the rate. This can be on a relative scale (comparison between systems) or on an absolute scale (computation of rates).

The idea that the steepest descent path is not necessarily the best approach in condensed phases led to a number of suggestions of plausible alternative definitions for the reaction coordinate.[53–57]. Essentially, all the approaches described below are based on an integral representation of the path, and on a variational principle that is used to compute the path. The theoretical basis of the choice of the variational principle varies from qualitative arguments,[55] to approximate classical action[57] and to arguments based on the diffusion equation or on other stochastic equation.[53,54,56]

One of the first algorithms to compute reaction paths in gas phase systems[58] was based on an integral representation. Only later the local approaches to define reaction coordinates were adopted. In the representation of Empedocles,[58] one minimizes the integral of the potential energy along the path between the reactants and the products. We denote the integral (which is a functional of the path) by S. The functional depends on the complete path (denoted l) between the reactants ($\mathbf{R}_r$) and the products ($\mathbf{R}_p$)

$$S[\ell] = \int_{\mathbf{R}_r}^{\mathbf{R}_p} U(\mathbf{R})d\ell\,. \tag{43}$$

Empedocles investigated the collinear reaction H_2+H H + H_2. To find the path ℓ Empedocles assumed a functional form for the path. The path ℓ was expanded in a series and the expansion coefficients were computed from the condition that S should be minimal. The above functional minimizes the value of the potential weighted by the length of the path: a shorter path (assuming for the moment a constant potential) will have a lower value of S. As mentioned above, this is a suggestion based on qualitative considerations which does not have a simple realization of a physical process.

Independently (and much more recently), Elber and Karplus (EK)[55] proposed a very similar functional. EK were not aware at that time of Empedocles's investigation. In spite of the similarity, EK's work is different in two important aspects: First, the functional was not defined to include the length of the path (that is, no preference was given to shorter paths) — $S[\ell] = 1/L \int U(\mathbf{R})d\ell$, where L is the length of the path $-L = \int d\ell$. The functional of EK therefore corresponds to the average value of the potential energy along the path. Another important difference in the EK method is the way in which the path is computed. During the years that passed since Empedocles proposed his functional, a revolution occurred in computational sciences, namely, the introduction of much faster (and cheaper) computers. EK employed a discrete representation of the path which is more flexible than the basis set approach employed by Empedocles, and which is more suitable for computer modeling. That is, they replaced the integral by a summation over a set of points along the path:

$$S(\{\mathbf{R}_k\}_{k=1,\ldots,K}) = \Sigma U(\mathbf{R}_k) + \text{Constraints}\,. \tag{44}$$

The flexible discrete form of S makes it possible to study large molecules such as proteins[55,59] and other systems.[61] The summation on the right-hand side of the equation is the part that mimics Eq. (40), however, it does not include the weighting by the path length.

Instead of being a functional, S is now a function of all the $\mathbf{R}_i$, and we optimize S as a function of $N \times K$ variables. N is the number of degrees of freedom of a single copy of the system, and K is the number of system copies that are used to make the discrete approximation for the path. The "constraints" part of Eq. (41) includes two sub-parts. The first is a constraint that forces all the points to be equally spaced along the path. Without this constraint, an attempt to minimize S as a function of all the $\mathbf{R}_k$ would result in the collapse of all the structures to the nearest minimum and a continuous path would not be obtained. To force the structures to follow a continuous path, we use a constraint of equal distances. Let $d_{i,i+1}$ be the distance between $\mathbf{R}_k$ and $\mathbf{R}_{k+1}(d_{i,i+1} = |\mathbf{R}_{k+1} - \mathbf{R}_k|)$. For K points we have $K - 1$ distances and the average distance is defined as $\langle d \rangle = \Sigma d_{i,i+1}/(K-1)$.

The first constraint — C_1 is implemented as a penalty function:

$$C_1 = \gamma \Sigma (d_{i,i+1} - \langle d \rangle)^2\,. \tag{45}$$

The penalty function is at a minimum when all the distances are equal to the average distance. That is, they are all the same. Since we calculated the distances in Cartesian space, it is convenient to compute the difference vector $\mathbf{R}_{i+1} - \mathbf{R}_i$ in a relative orientation of i and $i+1$ such that the norm of the above difference is minimal. Hence, we need to add constraints on rigid body translations and rotations. There are six constraints for each of the path structures. The constraints are identical to the first six constraints listed in Eq. (7) — σ_{1-6}. We orthonormalize the constraints in a similar fashion to what we did before [(Eqs. (13) & (14)]. We denote the orthonormal constraints by $\sigma_i^k(\mathbf{R})$ and their gradients by $\nabla\sigma_i^k$. The normalization requires that $\sigma_i^k(\cdot\nabla\sigma_j^k = \delta_{ij}$. The index k denotes the structure number; i and j are indices of constraints. Hence Eq. (41) can be rewritten more explicitly as

$$S(\{\mathbf{R}_k\}_{k=1,\dots,K}) = \Sigma_i U(\mathbf{R}_k) + C_1 + \Sigma_{i,k}\lambda_i^k\sigma_i^k\,,$$
$$\lambda_i^k = \nabla\sigma_i^k \cdot \nabla[\Sigma U(\mathbf{R}_k) + C_1]\,. \tag{46}$$

To find an optimal path between the two structures $\mathbf{R}_1$ and $\mathbf{R}_K$ we now minimize directly the "action" S. The optimization requires an initial guess for the path. If we do not have a "chemically" sound guess we generate a straight line interpolation between the reactants and the products in which

$$\mathbf{R}_k = \mathbf{R}_1 + (k - 1/K - 1)(\mathbf{R}_K - \mathbf{R}_1)\,. \tag{47}$$

In the first application of the above definition for the path the constraints on rigid body motions were handled via a penalty function.[55] The introduction of the Lagrange multipliers method and the use of the linearity of the constraints speeds up the calculations by a factor of 10.[57]

As noted by Czerminski and Elber,[57] and later by Fischer and Karplus,[62] the path in (43) tends to aggregate in the vicinity of the minima, and the sampling of the neighborhood of the transition state may be inadequate. The reason is the lack of constraints (or penalty) on the length of the step — $|\mathbf{R}_k - \mathbf{R}_{k+1}|$. Let $\mathbf{R}_r$ be the reactant structure and $\mathbf{R}_K$ the structure of the products, there is a solution of the form $\{\mathbf{R}_r, \mathbf{R}_K, \mathbf{R}_r, \mathbf{R}_K, \mathbf{R}_r, \dots, \mathbf{R}_K\}$, jumping from reactants to products in one step. This solution is (obviously) of little interest and is removed by an additional constraint. Czerminski and Elber introduced an additional term to the effective energy — a repulsion — first between all the neighbors,[57] and later only between next nearest neighbors.[51] The constraint function is:

$$\text{Repulsion} = (\rho/\lambda)\exp(-\lambda d_{ij}^2/\langle d\rangle^2)\,. \tag{48}$$

We have shown that in the limit of small λ, fixed $\langle d\rangle$, and repulsion between next nearest neighbors only, it is possible to correlate the repulsion with the kinetic energy.[57] This can be understood as follows: In the high repulsion limit the chain of structures arranges itself so that the distances between the next nearest neighbors is maximal. The separation between the next nearest neighbors is maximal in a straight line path. A straight line path also represents a trajectory with high kinetic energy (the set of structures is linear in time). In the low repulsion limit, the stiffness of the polymer is reduced and it tends to aggregate at the low energy minima. This is similar to a low energy trajectory that stays most of the time near the potential minima.

The analogy between the new S that includes the repulsion and a classical trajectory introduces more physics to the ad-hoc form of the functional S. We call this functional Self Penalty Walk or SPW. This method, which is computationally very stable and attractive, has already been used in a number of reaction path calculations. They include conformational transitions in a tetrapeptide,[57] ligand (carbon monoxide) diffusion through a protein (leghemoglobin) matrix,[60] the B to Z transition in DNA,[15] and ion migration through a channel.[43] Others used S or variants of it to investigate a conformational transition in citrate synthase[59] and transitions between local minima of bulk water.[61] Thus, SPW was shown to work for a reasonably large number of biologically or physically relevant examples. Only the adiabatic mapping (which as I already argued is difficult to apply correctly) has been applied to a larger number of non-trivial cases.

Among other useful ideas on functionals to compute reaction paths, I note the PEM approach of Smart. Smart[62] proposed to replace the average of the potential energy by an average of a higher moment of the potential energy (i.e. $\langle U^m \rangle$ instead of $\langle U \rangle$). It is assumed that m is odd and that the potential is positive which in practice is of course possible via the addition of a large irrelevant constant. The modification of the integrand gives considerably more weight to the barrier location. Let U_b be the energy barrier and U_m the minimum. We can always redefine the potential $U \to cU$ so that $|U_k - U_{k'}| > 1$. This scaling does not change the position of the stationary points. Once the stationary points are located, it is simple to recover their original values by dividing by c. We therefore have $|U_k - U_{k'}| < |U_k - U_{k'}|^m$. Minimization of S (at large m) is therefore equivalent to the minimization of the potential energy in the neighborhood of the highest energy points. Compared with the SPW this is more in the spirit of the transition state theory that focuses on the vicinity of the potential energy barrier.

As explained above, the SPW is an approximation to a single trajectory following Newton's mechanics. By studying an SPW "trajectory" we obtain some qualitative insight into the mechanisms of the process. However, in classical mechanics it is not possible to find a "most probable" trajectory. The elements of phase space with the same energy but with different initial coordinates and momenta can follow very different paths and still have the same statistical weight. This puts into question the meaning of the few approximate trajectories calculated by the SPW method. However, the applications of the SPW usually aim at approximations to "low energy" trajectories. When low energy trajectories pass the barriers, they are similar in the neighborhood of the saddle point — the highest energy point passed by a steepest descent trajectory. Therefore a study of a single low energy trajectory can provide general information on the properties of trajectories in the neighborhood of the barrier.

How to define a reaction coordinate that also has a clear physical meaning was the subject of numerous investigations. In the gas phase everybody "knows" that the SDP is a clear winner. However, the difficulties in computing and interpreting SDP's in condensed phases made the search for alternative definitions easier (sociologically).

Berkowitz *et al.* proposed a formalism based on maximizing the reactive flux obtained in the diffusion equation.[53] Pratt proposed to use Markov chains as a computational algorithm to optimize paths.[54] Wolynes[56] argued that the Onsager and Machlup action[63] is another useful definition for a reaction path. We shall reargue in favor of the last proposal in the following section. The above suggestions are of considerable general interest. However, excluding the Onsager–Machlup action, it is not yet clear how the above approaches will perform in computations. No numerical examples were provided in Refs. 53, 54 or 56. Of all the approaches to reaction paths that are based on "functionals", only the suggestions of Empedocles,[58] Elber and Karplus,[55] Czerminski and Elber[57] and Smart[62] have actually been applied to molecular systems.

Approaches different from the functionals and closer to the widely accepted "gas phase" algorithms were also used and they include the mode following approach[64] and its variant by Lazaridis *et al.*[65] Another algorithm that focuses on the search for saddles was proposed by Fischer and Karplus.[66] Also of interest is the directed dynamics of Durup *et al.*[59] This is an original and simple idea in which a molecular dynamics trajectory is biased towards the products. Finally, also of interest is the use of the LUP method[20] by Huston and Marshall for a conformational transition between two forms of a helix.[67]

It is encouraging that useful results can be obtained for biologically relevant systems by applying currently available methodologies to reaction path calculations. In the next section we shall discuss a few examples.

5. Reaction Path Studies of Biological Molecules: Examples

In this section I describe a number of reaction path calculations in large or flexible molecules. The goal is to provide the reader with some feeling for what can be done with these algorithms and for possible connections to experiments. The examples are taken from studies in the author's group.

5.1. *Alanine Tetrapeptide*

Our first example is of a relatively simple molecule, alanine tetrapeptide, which was investigated using the REF coordinate approach by Czerminski and Elber.[14] The alanine tetrapeptide is the shortest chain that can form a complete helical turn. Therefore, the path (or paths) leading from an extended or random chain to a "folded" helical configuration is (or are) of considerable interest.

In a study of such a flexible molecule a number of questions can be addressed. For example: (a) Is the reaction path unique? (b) If the reaction coordinate is not unique can we suggest relevant experimental observables? In the description below, we provide a "historical" outline of how we approached the problem. Briefly, at the beginning we hoped to find a single dominant reaction coordinate. Once we discovered that this was not the case, we designed a method to map as many as possible of the available reaction coordinates. Finally we considered how path multiplicity influences observables such as thermal relaxation and line shape.

5.1.1. *Is the Path Unique?*

This section of the study employed the REF coordinate approach. It is therefore also an opportunity to discuss some of the difficulties of the REF method together with possible solutions. We started the calculation from minimized conformations of the extended chain and the "helix" and computed the path between the two. The calculations employed the reference coordinate approach. We tried different choices of reference coordinates. The simplest one was a straight line interpolation between the two configurations. That is, let the coordinate set of the helical turn be $\mathbf{R}_t$ and the extended chain coordinate set be $\mathbf{R}_e$. The reference coordinate $\mathbf{R}(Q) = (1 - f)\mathbf{R}_e + f\mathbf{R}_t$, where f is varied linearly from zero to one. This reference coordinate is linear, which makes the refinement of the reference coordinate exceptionally efficient (computationally), as outlined previously in the discussion on linear constraints. We also tried another reference coordinate in which the following ratio defines Q:

$$|\mathbf{R}(Q) - \mathbf{R}_e|/(|\mathbf{R}(Q) - \mathbf{R}_t| + |\mathbf{R}(Q) - \mathbf{R}_e|) \equiv Q\,.$$

The last REF coordinate is a sphere with a variable radius between the reactants and the products.[14]

The last constraint is non-linear and therefore the calculations of the REF path are more expensive than in the first algorithm. Nevertheless, when using the REF coordinate approach, it is always useful to repeat the calculation with another choice of a reference coordinate and to compare the results. When we computed the path from an extended chain to a turn we obtained discontinuous paths (using either of the above definitions for an adiabatic coordinate). This is a very common result when the adiabatic mapping is employed.

An obvious question is what to do with a discontinuous path. One approach is to give up at this point the reference coordinate method and to seek an alternative. Another scheme which is more constructive if we already computed so far is to build on the results that we have. It is of course not possible to use a discontinuous path "as is".

We have designed an automated protocol to analyze the REF coordinates obtained and to refine them as described below. We based our analysis on the observation that the paths are piecewise continuous. That is, at most a few points of discontinuity are usually observed in a path. It is therefore possible to identify local maxima and minima on the continuous branches of the paths that we obtained.

As the first step in the refinement of the path we identified local exterma, that is the maxima or minima of the energy along the continuous portions of the path. We then optimized these maxima or minima using the mode following approach. The search was set in such a way that if one of the eigenvalues of the modes is negative we propagate upwards in energy along this mode until a stationary point (a saddle point) is found. If all the eigenvalues are positive then a Newton–Raphson minimization is executed. Once the calculation is completed, we obtain a set of stationary points computed from the REF path.

In the second step of the refinement procedure, we reconsidered the saddle points. From each saddle point we computed the steepest descent path using Eq. (2), that leads from the saddle point to the corresponding minima. This calculation is quite straightforward and provides us with an identification of the minima that are connected by the present saddle point. It is not certain if the saddle point which we found connects the two minima we used to start the path computation.

When we tried to connect directly the extended chain with the helix, we found new minima along the pathway. To further explore possible connections between the extended chain and the helix we then computed all the possible paths between the new minima and between the new minima and the old ones. For example, suppose that we started from the path between minima A and D and we found two new minima — B and C. In the second step we also computed the paths between A and B,C, between D and B,C and between B and C. Thus, we have four minima and $4(4-1)/2 = 6$ paths. Moreover, each of the paths was computed four times, using the two algorithms mentioned above. Each of the algorithms was used twice, once from the reactant to the product and second from the product to the reactant. Hence, we finally obtained $2 \cdot 4(4-1) = 24$ paths or more generally $2N(N-1)$ paths, where N is the current number of minima.

If the reaction coordinate between the extended chain and the helix was unique, or close to that (as we had assumed in the beginning of the calculation), then the additional calculations between the newly discovered intermediates should fall on the already computed path. For example, if we compute the path from A to D and find that A is going first to B and then to C and only finally to D, then computing the path from B to D should have C as an intermediate if the above path is unique. This condition is of course insufficient to prove that the path is unique but it is necessary. If B is going to E and then to D then the previous path is clearly **not** unique.

The second cycle of computation of paths that include the newly discovered minima provides us with a larger number of (piecewise continuous) paths that we refined as we did for the original single path. In some of the path computations we found minima that were new. That is the path that we started with was not unique.

5.1.2. *Exploration of Systems with Multiple Paths*

The new minima were connected in a more complex network than a single reaction coordinate and therefore we continued to explore alternative pathways. We repeated the calculation, computing all the paths as before. That is, all possible path combinations using new minima as either "reactants", "products", or both were computed. The resulting REF coordinates were refined to obtain new saddle points and new minima. Convergence was assumed (and computations stopped) when the complete set of $2N(N-1)$ computations of paths and the refinement of paths did not provide new minima. At this point of the calculation all the known energy minima and saddle points form an interconnected graph in which the branches (the minima) are known[14] (also see Fig. 7).

Of course we cannot be sure that the search for minima and saddle points is complete. Nevertheless the self-consistency check which we impose — that computations of paths including the new minima will not result in additional stable intermediates (new minima) — results in a rather extensive search.

We found 138 minima and 490 saddle points. The comparable number of saddle points and the number of minima suggest a highly dilute connectivity. On the average, each minimum has only 3.5 neighbors. Moreover, the distribution of the number of nearest neighbors has a significant width. The paths include a dominant contribution from restricted rotations, i.e. rotations around only one or two bonds. The rotations around two bonds are very well-known in polymer physics to be probable structural "moves". However, in the tetrapeptide the moves are significantly more restricted than in polyhydrocarbons, since in the allowed "moves" the tetrapeptide also tries to preserve as much as possible the number of hydrogen bonds. Moves that simultaneously break two hydrogen bonds are not likely.

5.1.3. *Multiple Paths, do they have Kinetic "Fingerprints"?*

As was shown in Ref. 14, the existence of many paths with varying barriers suggests the existence of many relaxation times that can differ appreciably from each other. This is easy to understand: If we can assign independent kinetic pathways (and we can do that by diagonalizing the properly weighted kinetic matrix,[14]), each pathway will have its own rate constant. A specific rate (say from A to B) will consist of a linear combination of exponential relaxations associated with the "independent paths", each with its own rate.

To demonstrate the effect we analyzed the kinetics of conformational relaxation in a T-jump experiment, demonstrating that the relaxation process in the tetrapeptide is expected to be highly non-exponential (due to path multiplicity) and is therefore unlikely to be described properly using a theory that is based on one path (and one exponential relaxation).

The above study on alanine tetrapeptide provided a clear picture of what type of minima and paths one finds when considering conformational transitions in peptides as opposed to conformational transitions in hydrocarbon chains. It also demonstrated that the REF coordinate approach can be put to good use, provided that the results are analyzed further and not employed "as is".

It should be stated though, that the relative success of the REF coordinate approach in this case is partially due to the rather limited number of minima and saddle points. We tried to use the same approach also in the next example (of the B→Z transition in DNA) but failed miserably. During the iterations we recovered more and more local minima without seeing even a glimpse of light at the end of the tunnel. We therefore needed to resort to a new approach. In fact the next problem is the one that initiated the development of the SPW algorithm.

5.2. *The B to Z Transition in DNA*[15]

In contrast to proteins that either have a unique or a few structure or no structure at all, DNA can have numerous (widely different) stable conformations. The different conformations can be induced by changes in salt concentration as well as by temperature changes. For example, the distances between the charged phosphate groups are smaller in the Z conformation than in the B state. The Z state is more stable than B at high concentrations of salt,[68] presumably, due to the screening of the charges by the ion clouds. In this investigation we considered the transition from the right-handed DNA (B DNA) to the left-handed DNA (Z DNA).

The dynamics of the conformational transition between the two is quite intriguing since there is no obvious pathway leading from one conformer to the other. This is due to the necessity of rotating the basepairs by 180° in addition to the changes in the helical sense. There are two models of how this transition might occur. In one model, the hydrogen bonds of the Watson–Crick pair (B form) dissociate, the two basepairs glide to the solvent, rotate freely in solution and glide pack to their closely packed position in the double helix, this time in a Z orientation.

In the above mechanism the stabilizing energy of the hydrogen bonds is lost along the pathway but regained on the product side. One of the contribution to the barrier height is therefore the energy cost of breaking the hydrogen bonds. The second model for the B→Z path assumes that the cost of dissociating the hydrogen bonds is too high and seeks a coordinate that preserves hydrogen bonds as much as possible during the transition. The reaction coordinate becomes strongly dependent on the internal flexibility of the DNA molecule. In order to make it possible for the rings to rotate, a sufficiently large "cavity" must be simultaneously formed along the double helix such that the van der Waals "collision" of the rotating ring with the upper and the lower basepairs will be "softened". A scheme of how such a path could proceed was proposed by Harvey.[69] Harvey employed a wire model for two basepairs, testing for acceptable covalent structures and frame clashes. Czerminski and Elber[15] employed the SPW algorithm to compute the above path using the AMBER force field.[70]

The SPW algorithm requires an initial guess to begin the computations. The initial guess is refined by optimizing the SPW "action" to a better coordinate. Since the SPW action is a nonlinear function of the path it is possible that paths of energies lower than the one found exist. Furthermore, our search for alternative paths was limited. We did not explore all the space of plausible pathways. We started with a path similar to the coordinate suggested by Harvey. After the optimization we still had a coordinate that overall was similar to the mechanism proposed by Harvey. The energy and many structural details were different but during the minimization procedure we never really tested for the other extreme model, that is, the model of complete hydrogen bond dissociation and free ring rotation in the solvent. Our computations are therefore restricted to the neighborhood of one model. Therefore, we cannot make assessment of the relative importance of the two suggestions.

It should also be noted that the first model strongly depends on solvent and solution conditions. The basepairs are embedded in the solvent for the duration of a significant portion of the reaction path (when they execute the free ring rotation in the solvent). This is a process that we do not describe well in our vacuum calculations. Our description of the solvent is extremely crude and employs distance dependent dielectric ($\varepsilon = r_{ij}$, where r_{ij} is the distance between particles i and j). To mimic screening effects of salt we naively reduced the phosphate charges.

In contrast to the first model we have a better chance to evaluate conformational transitions described by internal DNA flexibility. These transitions depend less on the solvent and on the proper treatment of electrostatic and more on the correct description of the covalent structure and the van der Waals "collisions".

We therefore focused on tests to see if the model of Harvey is feasible. That is, when using an atomic detail model and a reasonably well established energy function (AMBER[70]), are the energy barriers that we compute absurdly high compared with the values estimated from experiments or are they within a reasonable range?

This example demonstrates a possible qualitative application of the reaction coordinate approach, that is a test of feasibility of a suggestion for a mechanism. In large molecular systems, assessment of feasibility of a pathway is sometimes all we can hope to achieve.

We considered a DNA molecule with 6 basepairs and computed the reaction coordinate from the B to the Z state. We started with an ideal model of the B and the Z structures and optimized both structures by the conjugate gradient optimization method. The next step was to build an initial guess for the reaction coordinate. Due to the complexity of the path, the initial guess was constructed in steps. To obtain an initial reaction coordinate we first constructed "intermediates" along the pathway. The intermediates are denoted by B_6, ZB_5, $Z_2B_4 \ldots Z_6$ where B denotes a basepair in the right-handed conformation and Z a basepair in the left-handed structure. To generate them we rotated the corresponding basepair by 180°. The covalent structure of the backbone and local distortion were then corrected by direct energy minimization of the newly generated "intermediate". The seven structures B_xZ_y are then connected sequentially by SPW paths from state B_xZ_y to $B_{x-1}Z_{y+1}$.

A plot of the complete energy profile along the reaction coordinate is provided in Fig. 9. It is evident that the energy barriers are high. Nevertheless, the reaction is known (experimentally) to be slow. Taking into consideration unavoidable errors in the force field and the errors in the model (e.g. the crude representation of the solvent), we cannot conclude that the above mechanism is not possible.

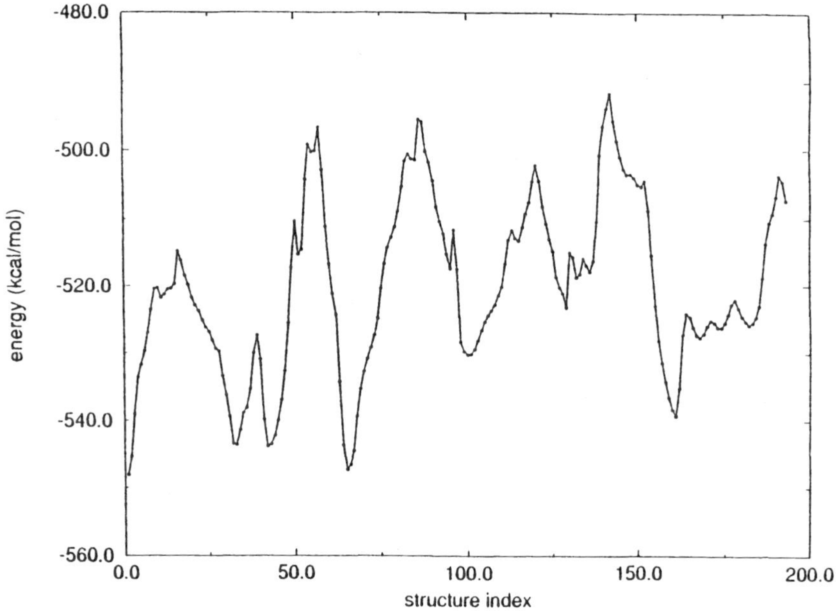

Fig. 9. The energy profile of the B→Z transition in DNA. The energy is in kcal/mol and the reaction coordinate in angstrom. This is a vacuum calculation with no screening of the phosphate charges.

Another question of interest is how the reaction coordinate changes from one basepair to another. Since we do not have an infinite double helix, we can examine the difference between the basepairs at the edge of the double helix and the basepairs in the center. Figure 10 shows the transition state structure of the first basepair rotation. The Watson–Crick hydrogen bonds are preserved throughout the transition of the first base pair. As we proceed to the center of the six basepair DNA we find that the pressure on the backbone becomes more significant and the two aromatic rings slide and adopt a stacked conformation. Hence, the Watson–Crick hydrogen bonds are not preserved even while following the "internal" B→Z reaction coordinate.

It is known that the transition is strongly influenced by the ionic strength, an effect we explored only qualitatively by changing the phosphate charge. Possible approaches to include better description of solvation is to compute solvation effects on the energy of the reaction coordinate, while assuming that the reaction coordinate is not changed. Some of the solvation effects may be computed by continuum techniques such as the nonlinear Poisson–Boltzmann approach.[73]

5.3. *Ligand Diffusion through Leghemoglobin*

The final example that is discussed is the calculation of the reaction path[60] and the rate[13] of ligand diffusion through the matrix of the protein leghemoglobin.

Similar to other proteins of the globin family, the active site of binding of a small diatomic ligand (e.g. oxygen, carbon monoxide) is buried in the protein matrix, and there are no obvious pathways for the ligand to penetrate or to exit the protein matrix. This problem has been promoted in the past as an example of the importance of protein dynamics. Without protein fluctuations the energy barriers for the exit of the ligand were too high to enable ligand escape during the life time of the protein.[71]

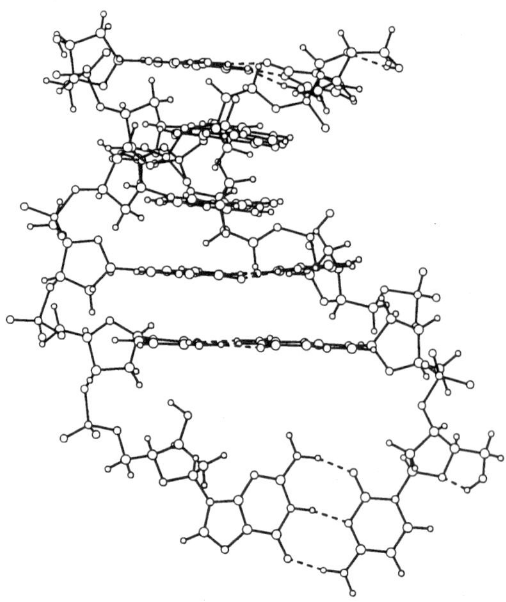

Fig. 10. (a) The transition state structure for the first basepair rotation in the B→Z transition in a six basepair DNA segment (see text for more details).

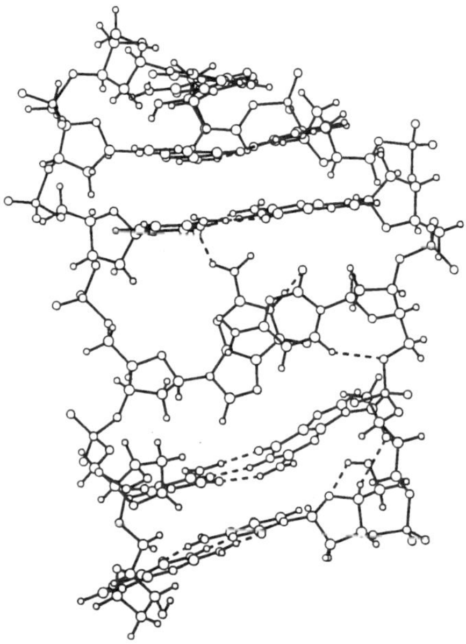

Fig. 10. (b) The transition state structure for the third basepair B→Z rotation.

Following the pioneering work of Austin *et al*,[35] the process of ligand binding to the globins is separated into at least two steps: The first step of diffusion into the protein matrix and a second step of bond formation with the heme. In each of the above steps, leghemoglobin behaves in a surprising way when compared with mammalian proteins.

Leghemoglobin (more precisely, the protein which we considered is lupine leghemoglobin) is a plant protein. It functions in an environment that is sensitive to the presence of oxygen. Free oxygen is a poison since it oxidizes the metals at the nitrogen fixation centers. As a precaution, leghemoglobin binds the transported oxygen very strongly. The binding energy is significantly stronger than the corresponding binding energies of oxygen to mammalian proteins such as myoglobin or hemoglobin. The mechanism through which leghemoglobin achieves the larger binding energy is not known. The same chemical group (the heme) is used to bind the small ligand in leghemoglobin and in mammalian proteins. Furthermore, the heme is bound to the peptide chain in the same way (it is liganded to the proximal histidine). This is one unsolved puzzle presented by the leghemoglobin family in general and by lupine leghemoglobin in particular.

The second puzzle is the diffusion to the active site. In leghemoglobins in general and in lupine leghemoglobin in particular the diffusion is exceptionally rapid and approaches the diffusion rate in pure water. The question is how this fast diffusion is possible in a protein with a "closed" equilibrium structure. We have computed the reaction coordinate, the free energy profile and the rate for the diatomic diffusion in leghemoglobin. The results compare favorably with experiment. Specifically the results of the computations were consistent with a kinetic experiment that was done after the computations.[13]

The computations of the ligand escape pathways, started from "hot" mean field trajectories.[72] (In the mean field simulations a number of ligand copies is allowed to share a single trajectory of the same protein. Each of the ligand copies feels the full force of the protein, while the protein feels the average force of all the ligands. The ligands do not "see" each other. The multiple trajectories of ligands provide significantly more statistics.) The mean field trajectories provide an initial guess to be used in the reaction path calculations.[15] Three trajectories were employed in SPW calculations. The use of three trajectories instead of one helps us to explore the variance of the optimized paths as a function of initial conditions.

The three optimized SPW paths that we obtained were not the same. However, they included many common structural and energetic features that recurred from pathway to pathway. For example, all the diffusion paths could be separated into two major steps: (a) ligand exit from the heme pocket to an alternative cavity in the protein matrix (the cavity at the contact between the CE loop and the G helix) and (b) diffusion out of the second cavity to the exterior of the protein.

Step (a) is very localized, with only one residue in addition to the ligand (phenylalanine 29) moves significantly along the diffusion pathway. (Fig. 11). Thus it is possible to disregard the rest of the protein and to concentrate on the reaction core — an appealing simplified approach. However, in step (b), we find a significantly different picture. Rather than a localized motion that opens the gate we find a global movement of the G helix that opens the last gate for ligand escape. This is a very extended motion along the reaction coordinate that includes hundreds of atoms. It cannot be accounted for by a localized reaction core.

As all of the three paths included many common features we proceeded to compute the free energy profile of only one of these paths.[13] Umbrella sampling was employed to compute the potential of mean force. The related method, free energy perturbation, was described in the previous section. The potential of mean force suggests further insights on steps (a) and (b) discussed above. Step (a) has a free energy barrier. The barrier is low (about 2.5 kcal/mol) and is associated with the tilt of phenylalanine 29 — the residue that forms the "reaction core" of this step.

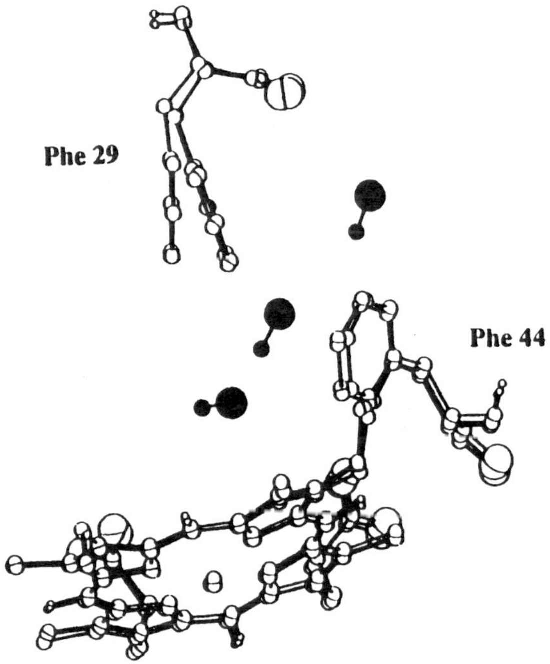

Fig. 11. Local displacement in the first step of ligand escape from the protein leghemoglobin.

The potential of mean force for step (b) has no free energy barrier at all. The motion to the entry is diffusive and barrierless. We emphasize that the motion involves a significant portion of the protein and not only the motion of the ligand. The shape of the free energy profile (Fig. 12) explains the very rapid entry of the ligand to the "closed" protein matrix: Once the thermal fluctuations of the protein are taken into account the diffusion of the ligand into and out of the matrix is on a fairly flat free energy surface with a sink at the heme end.

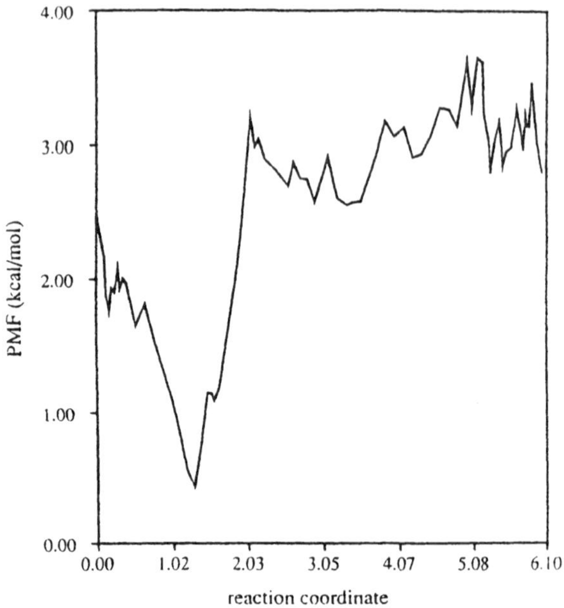

Fig. 12. The free energy profile for ligand diffusion in the protein leghemoglobin along the reaction pathway.

This completes the set of examples of applications of the reaction path approach. They were intended to give the reader a glimpse of possible applications and analyses.

6. An Advanced Topic: The Onsager–Machlup Action

The formal definition and the search for reaction coordinates in macromolecules are far from trivial as we already discussed at length in the previous sections. There is of course a philosophical question that asks if it is meaningful at all to search for a minimum energy path in the soup of minima and transition states typical of condensed phase problems.

Here we discuss an alternative computational approach for the determination of reaction coordinates which is not based on a minimum energy path. This approach is not "sensitive" to the objections that can be raised to applications of TST-type calculations in flexible, multi-minima systems. Thus, this approach is expected to work well for the latter type of systems too. The computational approach is based on the formalism of Onsager and Machlup.[63] A related suggestion to use Markov chains to compute reaction paths was made by Pratt.[54] The present approach has, however, a clearer physical realization. In the next paragraph we present some of the results of Onsager and Machlup, with an emphasis on the discrete representation of the path which is appropriate for our needs. We then describe the algorithm and the implementation on a distributed computing system.

Consider a molecular system that obeys the Langevin equation of motion, that is:

$$Md^2\mathbf{R}/dt^2 + \gamma d\mathbf{R}/dt + \nabla U + F = 0\,, \tag{49}$$

where M is the mass (represented by a matrix), $\mathbf{R}$ the coordinate vector of all the atoms in the system, γ is the friction constant, U is the microscopic potential (which provides the systematic force), and F is the random force vector. The random force is sampled from a Gaussian distribution such that $\langle F\rangle = 0$ and $\langle F(t')F(t)\rangle = 2k_BT\gamma\delta(t-t')$. The above formula provides trajectories at a given temperature T and is a useful starting point for studying the dynamics of many stochastic processes. As will become clearer later it is not necessary to use the phenomenological random force and friction. However, they are useful for the derivation, and from a conceptual point of view. It is straightforward (if desired) to take the zero limit of the friction and the random force. This can be done in the final expression, providing a solution corresponding to the usual classical mechanics.

Consider a set of initial coordinates that represent the reactants $\mathbf{R}_r$ and a set of final coordinates $\mathbf{R}_f$ that represents the products. We ask what the probability of observing a specific trajectory $\mathbf{R}(t)$ between the reactants and the products is. Recently this question was examined using the classical action.[75,76] The classical action is however a poor basis for a computational tool. Computation of paths via functionals is most easily done by minimization. The discretized classical action does not support the existence of a minimum (only extermum) for an arbitrary discretization of time.

It is useful to consider first a solution for a short time interval, dt. We employ a finite difference equation in which the time is denoted by a discrete subscript. The coordinate vector at time i is $\mathbf{R}_i$. It is also convenient to forget for the moment the random force. Without the random force we can obtain an explicit solution to the next position in the coordinate and velocity space using an algorithm similar to Verlet [Eq. (1)]. We denote the coordinates and the velocities by a single vector $\mathbf{X}_i$ and the solution for the next (small) step by $\mathbf{f}(\mathbf{X}_i) = \mathbf{X}_{i+1}$, where $\mathbf{f}(\mathbf{Y})$ is generated by a rule for the next deterministic step. Existence of a deterministic solution is equivalent to saying that the conditional probability density is a Delta function: $P(\mathbf{X}_{i+1}|\mathbf{X}_i; dt) = \delta(f(\mathbf{X}_i) - \mathbf{X}_{i+1})$.

It is also clear what the random force does to the probability density. The random force adds uncertainty to the final position of $\mathbf{X}_{i+1}$. This is because $\mathbf{f}(\mathbf{X}_i)$ is now replaced by $\mathbf{f}(\mathbf{X}_i) + \underline{F}(\underline{F} = (dt^2/2m)F)$. We obtain $P(\mathbf{X}_{i+1}|\mathbf{X}_i; dt) \propto \exp[-(\mathbf{f}(\mathbf{X}_i) - \mathbf{X}_{i+1})^2/2\langle \underline{F}^2\rangle]$, which is the solution for short times. To obtain a long time trajectory we accumulate many small steps. We are looking for all possible combinations of small steps providing all possible paths that lead from the reactants to the products

$$P(\mathbf{X}_{i+N}|\mathbf{X}_i; t) = C \int (\Pi_j d\mathbf{X}j) \exp[-1/\langle \underline{F}^2\rangle \Sigma_k (\mathbf{f}(\mathbf{X}_k) - \mathbf{X}_{k+1})^2], \quad (50)$$

where C is an unknown normalization constant. The difference $\mathbf{f}(\mathbf{X}_k) - \mathbf{X}_{k+1}$ is $\underline{F}_k$. However according to Eq. (1) it is also equal to the deterministic part of the equation (up to a factor of $dt^2/2m$). We therefore have

$$\begin{aligned} P(\mathbf{X}_{i+N}|\mathbf{X}_i; t) = C \int & (\Pi_j d\mathbf{X}_j) \\ & \times \exp\left[-(dt^2/2m\langle \underline{F}^2\rangle)\Sigma_k (m d^2\mathbf{R}_k/dt^2 \right. \\ & \left. + \gamma d\mathbf{R}_k/dt + \partial U/\partial \mathbf{R}_k)^2\right] . \end{aligned} \quad (51)$$

Equation (51) is a typical starting point to obtain a path integral. All that is required is to take an infinite number of points N and an infinitesimal dt, such that $N\,dt$ is the time length of the trajectory. After algebraic cancellations only one dt remains that is used in the resulting action integral, yielding

$$\begin{aligned} P(\mathbf{R}_f|\mathbf{R}_r; t) = C \int D\mathbf{R}(t) \exp[-1/\langle F^2\rangle \int dt (m d^2\mathbf{R}/dt^2 \\ + \gamma d\mathbf{R}/dt + \partial U/\partial \mathbf{R})^2] , \end{aligned} \quad (52)$$

where we return back to $\mathbf{R}$ from $\mathbf{X}$. The velocities do not appear in Eq. (51) and therefore are not required or of interest. D denotes a path integral. This is the result of Onsager and Machlup.[63] For computational convenience we retain the form Eq. (51). We further approximate the time derivatives by finite differences, and replace $\mathbf{X}$ by $\mathbf{R}$ as in (4):

$$P(\mathbf{R}_{i+N}|\mathbf{R}_i;t) = C\int(\Pi_j d\mathbf{R}_j)\exp[-(dt^2/2m\langle \underline{F}^2\rangle)S]\,, \qquad (53a)$$

$$S = \Sigma_k(m(\mathbf{R}_{k+1}+\mathbf{R}_{k-1}-2\mathbf{R}_k)/(dt)^2 + \gamma(\mathbf{R}_{k+1}-\mathbf{R}_{k-1})/2dt + \partial U/\partial \mathbf{R}_k)^2\,. \qquad (53b)$$

The conditional probability is given by a sum of all stochastic trajectories that start at $\mathbf{R}_i$ and end at $\mathbf{R}_{i+N}$ after time t. This formula can be used to compute the conditional probability.

A number of points are worth stressing in the context of computations:

First, the integrand is always positive, which is convenient for path integral calculations employing sampling methods such as Molecular Dynamics or Monte Carlo. This is a profound difference with respect to computations that employ the classical action.

Second, the "action" S is always non-negative. It has therefore a clear minimum as a function of a set of structures $\{\mathbf{R}_k\}$ that represent the paths. Thus, the minimum of S provides the most probable trajectory that connects reactants and products. It is therefore a suggestive definition for a reaction coordinate in the condensed phases.

Third, since computation of the most probable path is equivalent to "energy" (the action S) minimization, the calculations are very stable and the stability is independent of the size of the time step dt. This is to be contrasted with the solution of the differential equations for the trajectories, which are very sensitive to the size of the time step. Clearly, the resulting trajectories are approximate, and depend on the resolution in time that was used. However, for motions on relatively flat energy surfaces, in which the system displacements are diffusive, the above approach is quantitative. For systems with large energy barriers, the above approach can provide qualitative information. Quantitative information is more difficult to obtain since the short residence time in the neighborhood of the barriers makes it difficult to properly sample these events if a large time step is used.

Fourth, path integrals are very easy to parallelize. The parallelization can be carried out in a general way independent of the specific form of the potential. Only minimal communication between processors is required. The kth processor sends and accepts messages from processors $k \pm 1$ and possibly also $k \pm 2$. The communication is therefore restricted to a total of (at most) four neighbors to a processor, independent of the total number of nodes. We implemented a parallel program for the above action on a cluster of workstations. The calculations are pursued in a distributed computing environment with very high efficiency. (Above 90 percent of the real time is used by each node for computations. Thus, there is little time spent on communication.) The algorithm will be presented in the next section. This type of efficient parallelism is impossible to pursue employing ordinary propagation in time with small time steps. The last algorithm is inherently sequential in the time domain and it also depends on the potential employed.

Fifth, the use of the new formalism is more natural than minimum energy paths. The effect of the temperature (a useful thing to have in condensed phase calculations) is already built in. The number of paths extracted from the Onsager–Machlup action is much smaller than the number of minimum energy paths in flexible systems. As argued before, studies of flexible systems may put the whole concept of minimum energy paths in doubt.

The use of the Onsager and Machlup action also has the disadvantages:

First, it requires storage of the complete trajectory on computers. This is because the complete trajectory is optimized all at once. This can be a significant computational burden. In iterative algorithms only the current step and perhaps several previous steps are kept.

Second, computation of analytical derivatives of the action S requires calculations of second derivatives of the potential energy. These derivatives require additional storage (in the worst case scenario about $N^2/2$ elements; N is the number of the degrees of freedom). Calculation of second derivatives constitutes computational overhead in addition to space requirement.

Third, the calculation is clearly approximate and requires careful evaluation of accuracy. It is possible (with little effort) to obtain a stable solution even with time steps of hours (something that is clearly impossible using ordinary MD). In calculations employing such a time step many molecular details are of course lost. It is therefore important to have some understanding of the process under investigation. If the process is diffusive and proceeds on a relatively flat energy surface, the results of the Onsager–Machlup action with very large time steps are sound. For processes with high energy barriers, it is unlikely that the neighborhood of the barrier will be sampled properly if a large time step is used. The time that the system spends on the top of the barrier is usually very short and it is expected to be shorter than the time step which we use. It is therefore impossible to obtain a qualitative picture of what is happening on the barrier. Since large barriers determine the time scale, the rate in a system with a large and steep barrier cannot be accurate either. Furthermore, we cannot (of course) be more accurate than the time step which we use. Exploration of factors that influence the accuracy of the Onsager–Machlup action is still under investigation.

In spite of the difficulties and the uncertainties, the possibility of pursuing trajectories from reactants to products with time steps orders of magnitude larger than in MD is exciting and may open new possibilities for atomic detail simulations. A major ongoing research project in the author's group is the computation of the R→T transition in hemoglobin, using computational implementation of the Onsager–Machlup action described below.[74] In many cases we are mostly interested in qualitative ideas on the mechanism of the process and at least for that purpose the Onsager–Machlup action is likely to be adequate. We proceed with the description of the computational implementation of the algorithm on a cluster of workstations.

Let the number of structures be N and the number of processors (or workstations) P. The reactant and the product positions are fixed and do not require optimization. We therefore divide the $N-2$ "active" structures between P processors. Each processor is responsible for the optimization of $(N-2)/P$ structures. The minimization is pursued for a preset number of optimization steps L. We call the $(N-2)/P$ structures that are assigned to a given processor "the local path segment". The optimization of the structures is done in a standard way, using either the conjugated gradient algorithm or simulated annealing.

To compute the energy terms that contribute to the forces at the lth time, we need to study the sum in Eq. (50b)

$$\Sigma_k[m(\mathbf{R}_{k+1}+\mathbf{R}_{k-1}-2\mathbf{R}_k)/(dt)^2+\gamma(\mathbf{R}_{k+1}-\mathbf{R}_{k-1})2dt+\partial U/\partial\mathbf{R}_k]^2 .$$

For simplicity, we denote the sum by $\Sigma_k W_k^2$. The terms $W_{l\pm2}, W_{l\pm1}$ and W_l depend on $\mathbf{R}_l$.

A processor is assigned to carry out the optimization of a path segment that includes $(N-2)/P$ structures. To compute all the W_k terms associated with these structures we need four more coordinate sets: two coordinate sets preceding the path segment and another two sets following the end of the segment. The additional structures are not modified by the processor that requires them. They are used only for energy calculations at the edges of the local segment. We therefore communicate to each of the processors four additional structures (to the first processor and to the last processor we send only two structures). This completes the information needed to calculate the local S terms (see Fig. 9). Obviously, a coordinate set of a segment is modified by one processor or workstation only.

The calculation is pursued as follows: A UNIX shell initiates FORTRAN programs on each of the workstations. Each workstation is concerned with one path segment (fixed in size) and is provided with names of workstations responsible for a preceding and a following path segment to which it will communicate later. A workstation optimizes at least two structures and therefore there is no need for more than two neighboring workstations, one preceding or one following.

After all the workstations have initiated their path segment (either by reading a pre-prepared file or by an internal calculation of an initial guess), they continue to optimize their sets of structures for L steps. This is

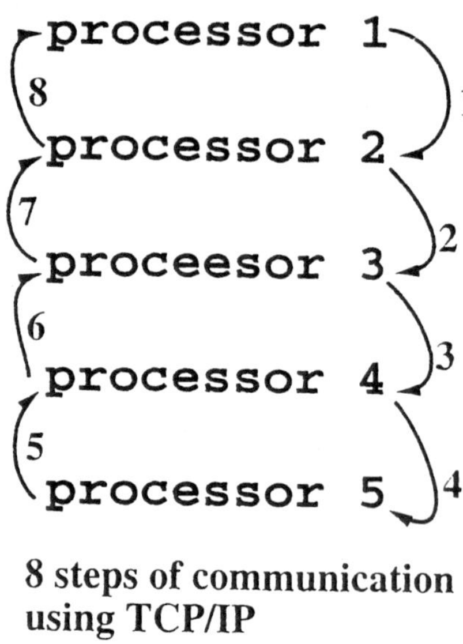

Fig. 13. Set-up of the communication protocol for the distributed computations of the Onsager action. Working versions are available for HP 7xx, SGI and IBM R/S 6000 workstations. Note that the numbers near the arrows denote sequence in time for the communication.

independent of the other workstations/processors. Once the L steps are completed the edges of the path segments are updated.

The communication is pursued in a ring structure (Fig. 13). After the first processor completes its L minimization steps, it sends its last two active structures to the second processor and waits for the reply of the second processor. The second processor first accepts the two structures and then proceeds. It sends coordinates forward in the ring to the third processor. The third computing node gets the coordinates and sends the required set to the fourth, and so on, until the last structure is reached. The last processor, after accepting its own share (two structures), sends its first two active structures to the one before the last, initiating a sequence of backward messages closing the ring in the reverse direction.

Clearly this approach is not optimal, since only two processors communicate at a time. This makes the communication time linear with the number of processors. Ideally, all pairs could communicate at the same time in parallel, providing us with a communication scheme that is independent of the number of processors. However, this is not possible using TCP/IP, which is the most widely used communication protocol between workstations. In TCP/IP, only one message can be transmitted at a time. Thus, there can be a significant penalty in communication performance in our application. However, since the communication required is very low, the ring structure based on TCP/IP is nevertheless used for my purpose. The important advantage is the generality. The code can be employed on a variety of machines with minor adjustments.

The minimization of the action and the communication are continued until convergence criteria are satisfied. The criteria include the norm of the gradient, the total number of steps and the small value of S.

Our simulation package for biomolecules — MOIL,[95] which includes the code for the Onsager–Machlup action, the SPW algorithm and general molecular dynamics applications, is available at present for Silicon Graphics, HP 7xx and IBM R/S 6000 workstations and can be obtained via anonymous ftp site 128.248.186.70 and 132.64.96.20.

Acknowledgments

I thank the following: The National Institute of Health, The Israel Science Foundation, The Alon Fellowship, and the West Scholarship of the University of Illinois for their generous support. (The Fritz Haber research center is supported by the Minnerva fund);

Prof Peter Wolynes for his numerous insightful discussions and the direction to the Onsager–Machlup action; Prof Henri Orland for his fruitful and entertaining discussions on the stochastic processes;

The researchers in my group who contributed over the years to the path algorithms: Ryszard Czerminski, Wieslaw Nowak, Gennady Verkhivker, Alex Ulitsky, Chyung Choi, Jing Zhang and Roberto Olender. The B→Z transition which was computed by Ryszard Czerminski also includes in this manuscript the energy profile which is published for the first time.

References

1. J. A. McCammon and S. C. Harvey, *Dynamics of Proteins and Nucelic Acids* (Cambridge Univ. Press, 1987).
2. C. L. Brooks III, M. Karplus and B. M. Pettitt in *Adv. Chem. Phys.* LXXI (John Wiley and Sons, 1988).
3. D. W. Heermann, *Computer Simulation Methods in Theoretical Physics* (Springer, Berlin, 1986).
4. M. P. Allen and D. J. Tildsey, *Computer Simulations of Liquids* (Clarendon Press, Oxford, 1987).
5. L. Verlet, *Phys. Rev.* **159**, 98 (1967).
6. V. Daggett and M. Levitt, *J. Mol. Biol.* **232**, 600 (1993).
7. A. E. Mark and W. F. van Gunsteren, *Biochemistry* **31**, 7745 (1992).
8. A. Caflisch and M. Karplus, *Proc. Natl. Acad. Sci. USA* **91**, 1746 (1994).
9. S. Myazawa and R. Jernigan, *Macromolecules* **18**, 534 (1985) ; J. Skolnick and A. Kolinski, *Science* **250**, 1121 (1990).
10. R. A. Goldstein, Z. A. Luthey-Schulten and P. G. Wolynes, *Proc. Natl. Acad. Sci.* **89**, 9029 (1992).
11. K. Mueller and L. D. Brown, *Theoret. Chim. Acta* **53**, 75 (1979).
12. S. Glasstone, K. J. Laidler and H. Eyring, *The Theory of Rate Processes* (McGraw-Hill, New York, 1991).
13. G. Verkhivker, R. Elber and Q. H. Gibson, *J. Am. Chem. Soc.* **114**, 7866 (1992).
14. R. Czerminski and R. Elber, *J. Chem. Phys.* **92**, 5580 (1990).
15. R. Czerminski, A. Roitberg, C. Choi, A. Ulitsky and R. Elber, *Conformational Transitions, AIP Conf. Proc.* **239**, eds. R. Lavery, J. L. Rivail and J. Smith (AIP, 1991).
16. J. Mathews and R. L. Walker, *Mathematical Methods of Physics*, (The Benjamin/Cummings Publ. Co., California, 1970) Chap. 6.
17. R. Kubo, M. Toda and N. Hashitsume, *Statistical Physics II: Nonequilibrium Statistical Mechanics* (Springer Verlag, Berlin, 1985).

18. P. Mathias and W. A. Sanders, *J. Chem. Phys.* **64**, 3898 (1976); M. J. Rothman and L. L. Lohr, *Chem. Phys. Lett.* **70**, 405 (1980).
19. T. L. Saaty and J. Bram, *Nonlinear Mathematics*, (Dover, New York, 1981) Chap. 2.2.
20. A. Ulitsky and R. Elber, *J. Chem. Phys.* **96**, 1510 (1990).
21. C. Choi and R. Elber, *J. Chem. Phys.* **94**, 751 (1991).
22. C. J. Cerjan and W. H. Miller, *J. Chem. Phys.* **75**, 2800 (1981).
23. J. Baker, *J. Comput. Chem.* **7**, 385 (1986).
24. N. Go, N. Noguti and T. Nishikawa, *Proc. Natl. Acad. Sci. USA* **80**, 6571 (1983).
25. B. R. Brooks and M. Karplus, *Proc. Natl. Acad. Sci. USA*, **80**, 6571 (1983).
26. M. Levit, C. Sanders and P. S. Stern, *J. Mol. Biol.* **181**, 423 (1985).
27. R. Elber and M. Karplus, *Phys. Rev. Lett.* **56**, 394 (1986).
28. D. G. Truhlar and B. C. Garrett, *Ann. Rev. Phys. Chem.* **35**, 159 (1984).
29. Q, H. Gibson, R. Regan, R. Elber, J. S. Olson and T. E. Carver, *J. Biol. Chem.* **267**, 22022 (1992).
30. A. Kuki and P. G. Wolynes, **236**, 1647 (1987).
31. D. Chandler, *J. Chem. Phys.* **68**, 2959 (1978).
32. R. F. Grote and J. T. Hynes, *J. Chem. Phys.* **73**, 2715 (1980).
33. E. Pollak, *J. Chem. Phys.* **95**, 533 (1991).
34. R. Zwanzig, *J. Stat. Phys.* **9**, 215 (1973).
35. R. H. Austin, K. W. Beeson, L. Eisenstin, H. Frauenfelder and I. C. Gunsalus, *Biochemistry* **14**, 5355 (1975).
36. R. Elber and M. Karplus, *Science* **235**, 318 (1987).
37. H. Li, R. Elber and J. E. Straub, *J. Biol. Chem.* **268**, 17908 (1993).
38. O. Schaad, H. X. Zhou, A. Szabo, W. E. Eaton and E. Henry, *Proc. Natl. Acad. Sci. USA* **90**, 9547 (1993).
39. R. Henri, M. Levitt and W. A. Eaton, *Proc. Natl. Acad. Sci. USA.* **82**, 2034 (1985).
40. P. Pechukas, *Dynamics of Molecular Collisions*, Part B, ed. W. H. Miller, (Plenum New York, 1976) Chap. 6.
41. McQuarrie, *Statistical Mechanics* (Univ. Science, California, 1973).
42. B. J. Berne, M. E. Tuckerman, J. E. Straub and A. L. R. Bug, *J. Chem. Phys.* **93**, 5084 (1990).
43. R. Elber, D. Chen, D. Rojewska and R. S. Eisenberg, Sodium in gramicidin: An example of a permion, submitted to *Biophys. J.*
44. K. Kostov and K. F. Freed and A. Perico, *Biopolymers* **33**, 1423 (1993).
45. D. J. Tobias, S. F. Snedden and C. L. Brooks III, *J. Mol. Biol.* **227**, 1244 (1992).
46. W. L. Jorgensen, *J. Chem. Phys.* **77**, 5757 (1982).
47. B. Roux and M. Karplus, *Biophys. J.* **59**, 961 (1991).
48. J. G. Harris and F. H. Stillinger, *Chem. Phys.* **144**, 63 (1990).
49. J. E. Straub and J. K. Choi, *J. Phys. Chem.* **98**, 10978 (1994).

50. J. P. Valleau and G. M. Torrie in *Statistical Mechanics*, Part A., ed. B. J. Borne (Plenum Press, New York, 1977), pp. 169–194.
51. R. Elber, A. Roitberg, C. Simmerling, G. Verkhivker, R. Goldstein, A. Ulitsky and H. Li, MOIL: A molecular dynamics program with emphasis on conformational searches and reaction path calculations in large biological molecules, *NATO Conf. Proc. on Statistical Mechanics and Protein Structure*, Cargese, June, 1994, ed. S. Doniach (Plenum Press), 165–191.
52. R. Elber, *J. Chem. Phys.* **93**, 4312 (1990).
53. M. Berkowitz, J. D. Morgan, J. A. McCammon and S. H. Northrup, *J. Chem. Phys.* **79**, 5563 (1983).
54. M. L. Pratt, *J. Chem. Phys.* **85**, 5045 (1986).
55. R. Elber and M. Karplus, *Chem. Phys. Lett.* **139**, 375 (1987)
56. P. G. Wolynes, Chemical reaction dynamics in complex molecular systems, in *Complex Systems, SFI Studies in the Sciences of Complexity*, ed. D. Stein, (Addison-Wesley, Massachussetts, 1989).
57. R. Czerminski and R. Elber, *Int. J. Quant. Chem.* **24**, 167 (1990) and *The Proc. of Sanibel Symposia*, 1990.
58. M. A. Ech-Cherif El Kettani and J. Durup, *Biopolymers* **32**, 561 (1992).
59. P. Empedocles, *Int. J. of Quant. Chem.* **3S**, 47 (1969).
60. W. Nowak, R. Czerminski and R. Elber, *J. Am. Chem. Soc.* **113**, 5627 (1991).
61. H. Tanaka and I. Ohmine, *J. Chem. Phys.* **32**, 561 (1992).
62. O. S. Smart, *Chem. Phys. Lett.* **222**, 503 (1994).
63. L. Onsager and S. Machlup, *Phys. Rev.* **91**, 1505 (1953); L. Onsager and S. Machlup, *Phys. Rev.* **91**, 1512 (1953).
64. D. T. Nguyen and D. A. Case, *J. Phys. Chem.* **89**, 4020 (1985).
65. T. Lazaridis, D. J. Tobias, C. L. Brooks III and M. E. Paulaitis, *J. Chem. Phys.* **95**, 7612 (1991).
66. S. Fischer and M. Karplus, *Chem. Phys. Lett.* **194**, 252 (1992).
67. S. E. Huston and G. R. Marshall, α-3/10 Helix transitions in α-methylalanine homopeptides: Conformational transition pathway and potential of mean force, *Biopolymer*, in press.
68. W. Saenger, *Principles of Nucleic Acid Structure* (Springer-Verlag, New York, 1984).
69. S. C. Harvey, *Nucleic Acid Research* **11**, 4867 (1983).
70. S. J. Weiner, P. A. Kollman, D. A. Case, U. C. SIngh, C. Ghio, G. Alagona, S. Profeta and P. Weiner, *J. Amer. Chem. Soc.* **106**, 765 (1984).
71. D. A. Case and M. Karplus, J. Mol. Biol. **132**, 343 (1979).
72. R. Czerminski and R. Elber, *Proteins* **10**, 70 (1991).
73. B. Hoing, K. Sharp and A.-S. Yang, *J. Phys. Chem.* 1101 (1993).
74. R. Elber and J. Zhang, The R to T transition in hemoglobin: Computations of atomic detail trajectories using the Onsager-Machlup action, in preparation.
75. R. E. Gillian and K. R. Wilson, *J. Chem. Phys*, **97**, 1757 (1992).
76. A. E. Cho, J. D. Doll and D. L. Freeman, *Chem. Phys. Lett.* **229**, 218 (1994).
77. J. D. Bryngelson and P. G. Wolynes, *J. Phys. Chem.* **93**, 6902 (1989).

CHAPTER 3

OPTIMIZATION TECHNIQUES WITH APPLICATIONS TO PROTEINS

JOHN E. STRAUB

Department of Chemistry, Boston University
Boston, Massachusetts 02215, USA

Contents

1. The Rugged Energy Landscape of Proteins

Proteins are complex macromolecules which exhibit a variety of interesting structural and dynamical properties. While they typically consist of thousands of conformational degrees of freedom, proteins fold into well-defined three-dimensional structures which allow them to carry out functions such as reaction catalysis with great precision. Studies on protein folding have demonstrated that the number of conformations available to a protein is far too large to allow for the folding to occur by a random search.[1] Still, proteins fold. Apparently, the sequence of a protein has been designed to fold rapidly into a stable native state reliably. Unraveling exactly how that is done has been recognized as one of the most interesting and important scientific challenges of our time.[2,3]

It is clear that proteins fold into compact low energy states with remarkably harmonious designs. This property has been expressed in the "consistency principle" of Gō,[4] the "principle of minimal frustration" of Bryngelson and Wolynes,[5] and in the studies of the importance of the "stability gap" between the lowest energy state and the manifold of higher energy native-like states[5,6] and the "energy gap" between the lowest energy and first excited conformations of model proteins.[7] If the native state of the protein is assumed to be the global free energy minimum, the problem of identifying the native state is reduced to the problem of finding the global minimum of a complicated many dimensional energy function. This view has been made popular principally through the work of Scheraga who has also contributed a fantastic number of creative approaches to the difficult problem of optimizing protein structure.[8] The difficulty of finding the global minimum is closely related to the character of the multidimensional energy function — the energy landscape — of the protein.

Experiment and computer simulations have had some success in providing the details of the energy landscape of proteins. For example, there is some fuzziness to the definition of the native state of a protein. The dominant folded states of proteins are often thermally stable, but at times appear to be metastable states.[9] Moreover, even when a protein structure is well-defined, X-ray crystal structures as well as solution NMR studies have demonstrated that there is some variety in the ensemble of structures making-up the sample of the native protein. This disorder, often important to the function of the protein, is in the form of conformational substates — conformations of the protein which are similar in structure and energy but which represent distinct conformations of the protein. The seminal experiments of Frauenfelder and coworkers on the kinetics of ligand rebinding in myoglobin following photodissociation have shown that these substates can be biologically active but may carry out their function with varying ability.[10] Conformational substates have also been identified in the computer simulation study of Elber and Karplus who found that in a 300 ps trajectory of myoglobin the protein underwent many transitions between substates (or local energy minima).[11] Similar studies by Noguti and Gō of a trypsin inhibitor have drawn the similar conclusion that during an equilibrium simulation of less than 1 ns a variety of substates are sampled.[12] They classed the transitions between substates as (1) local conformational changes involving a few residues which lead to elastic deformations of the protein and (2) cooperative rearrangements which induce plastic deformations in the whole molecule.

Simulation studies of Straub and Thirumalai involving ribonuclease A have demonstrated the existence of conformational substates which are sampled in a dynamics trajectory of less than 100 ps at room temperature.[13] It was determined that these substates are separated by barriers of a few kcal/mol. Their studies have demonstrated the existence of a broad distribution of energy barriers ranging from a tenth to tens of kcal/mol.[13] It is likely that the distribution of barrier heights is intimately connected to the broad distribution of relaxation times found in proteins. More insights into the gross features of the energy landscape have been provided by studies of protein folding based on simple "minimal" models of the protein. Camacho and Thirumalai have studied the thermodynamic and kinetic properties of model heteropolymers.[14] They found that there exists a relatively small number of compact states of low energy and that these states are separated by significant energy barriers. Detailed analysis of the density of states has been carried out by Sali, Shakhnovich and Karplus for lattice Monte Carlo simulations of a simple model protein in three dimensions.[15]

These studies demonstrate that the thermodynamics and kinetics of proteins are best described in terms of the details of the potential energy hypersurface — the energy landscape.[16] The picture of the energy landscape of a protein which is consistent with the experimental and simulation evidence is a rugged one with a large number of minima and a distribution of energy scales. The general features of the landscape are dominated by a relatively small number of low lying energy minima. The lowest energy state in many cases is believed to represent the native state of a protein. This picture is in harmony with the frustration observed in low temperature kinetics experiments as well as the important features of protein folding observed in the simulation of folding of model heteropolymers. Superimposed on this gross structure is a substructure of many local minima. This substructure explains the minor disorder observed in X-ray crystal structures and solution NMR structures of proteins as well as the frequent transitions observed between substates on a picosecond time scale in room temperature computer simulations. Landscapes of this kind have been used to explain many properties of spin glasses and a variety of theories developed for their study have been profitably extended to the study of proteins, most notably by the group of Wolynes,[5] and by Shakhnovich and coworkers.[17]

The topological details of these rugged energy landscapes are currently being defined using computer simulations of minimal models of proteins as well as all atomic models in terms of the distribution of energy minima, the connectivity of the minima, and the distribution of energy barriers separating the minima. In the study of protein folding, more than the distribution of energy minima, it is important to identify the lowest lying energy states of the compact protein which are likely to include folding intermediates and the subset of native protein configurations. Therefore, it is important to develop reliable methods for finding the compact, lowest lying energy states of the protein. From this point of view, an important part of the protein folding problem is a global energy minimization problem — a search for the lowest energy compact states of the system. This chapter will review some of the more promising algorithms for finding low lying energy minima of complex biomolecules.

2. Strategies for Global Energy Minimization

Problems centering on the global optimization of proteins share many characteristics of a large class of "hard" problems known as $\mathcal{NP}$-

complete.[18,19] These problems can be solved by a straightforward, exhaustive search of the set of possible solutions, such as the set of all possible conformations of the protein. Unfortunately, for this class of hard problems, the complexity increases faster than any polynomial in N, where N is a measure of the system size such as the number of atoms. This makes the solution of a problem, such as finding the lowest energy conformation of a large protein and folding the protein *in machina*,[20] a currently intractable problem — one which cannot be solved in a realizable time using an exhaustive search of configuration space.

This is good and bad news. The bad news is that we will have to be very clever to make progress in finding the lowest energy configuration for a protein of even 50 residues. The good news is that because this is a common problem found throughout the physical sciences, there are many techniques being developed which have provided very promising results. In this chapter I try to provide a unified way of viewing what I believe to be the most promising approaches to the global optimization problem for proteins.

All problems of global optimization may be reduced to the same problem how best to find the extremum of a "cost function" which is a measure of how well the problem has been solved. In the best known optimization problem, that of the "traveling salesman," the goal is to find the shortest path which connects a set of cities that the salesman must visit.[21,22] For a small number of cities, the number of paths is small and the solution is easily found using an exhaustive search. However, as the number of cities increases, the number of possible paths rapidly exceeds the number of paths one can consider, and an exhaustive search is rendered impotent. In the protein folding problem, one might assume that the native folded state of the protein is the thermodynamically most probable configuration. The cost function is the potential energy (or more accurately and ambitiously the free energy) of the protein. To find the folded configuration of the protein one must find the global minimum of the cost function — the potential or free energy of the protein.[23] Other examples include the problem of identifying the ground state configuration of a spin glass or very large-scale integrated circuits,[24] the optimum reaction coordinate in a many dimensional chemical reaction,[25–27] or the optimum configuration of a molecule subject to distance constraints provided by X-ray diffraction or NMR experiments.[28–31] Having abandoned the exhaustive search, we must find more intelligent strategies which can simplify the problem without loss of essential details.

2.1. *Simulated Annealing (SA)*

A breakthrough was made in 1983 by Kirkpatrick, Gellat and Vecchi[24] who proposed what is now known as the Simulated Annealing (SA) algorithm. This method is based on an analogy with statistical mechanics where the cost function may be thought of as a rugged potential energy surface. The temperature of the system is at first very high, and barriers in the system are easily crossed. The temperature is then slowly reduced according to a "cooling schedule" until the temperature is small or zero.[18] The slow cooling of the system has the effect of annealing the system, reducing strains by finding lower and lower energy basins on the vast potential energy hypersurface. While the "optimal" cooling schedule[32] cannot be realized in most problems of interest, many realistic cooling schedules have provided fabulous results on a wide variety of problems ranging from the design of circuits to the global energy minimization of molecular clusters. However, we are interested in proteins, and the results of a number of careful studies indicate that for proteins SA is an ineffective strategy for finding the lowest energy configuration.[33,34] I believe this is due to the wide array of energy scales in the biopolymers which tends to frustrate SA by making the optimal cooling schedule particularly difficult to realize. This point is discussed in detail in Sec. 4.

2.2. *Potential Smoothing and Coarse Graining*

Hoare and McInnes[35] noted that softer potentials tend to favor a regular crystal configuration (as the global minimum) while shorter range potentials tend to encourage amorphous structures. In an elegant analysis of the nonlinear optimization problem, Stillinger and Weber[36] recognized that while the number of local energy minima is a strongly (exponentially[37]) increasing function of the system size, how fast the absolute number of minima increases has a great deal to do with the length scale of the interparticle interactions.[35] Systems with short range potentials dominated by nearest neighbor interactions have large numbers of local minima. When the range of interaction is increased, the number of minima can be drastically reduced. Several detailed studies have shown how dramatic this effect can be.[38,39] The program of smoothing the potential energy hypersurface to remove high lying local minima, and to deepen and broaden the global energy minimum, has been was referred to as the "antlion strategy."[36,38] Berry has written a fine review of the related work done examining the character of the rugged energy landscapes of atomic and molecular clusters.[40]

The structure of this chapter rests on the foundation of these two important advances. I hope to convince you that the most promising methods for global optimization of proteins can be understood as variations on the simulated annealing and antlion paradigms. The next section is a selective review of some useful "tailored" optimization strategies which have been developed for specific applications. The most interesting of these algorithms are at heart potential smoothing algorithms. The rest of the chapter will concentrate on "general" algorithms for global optimization which may be

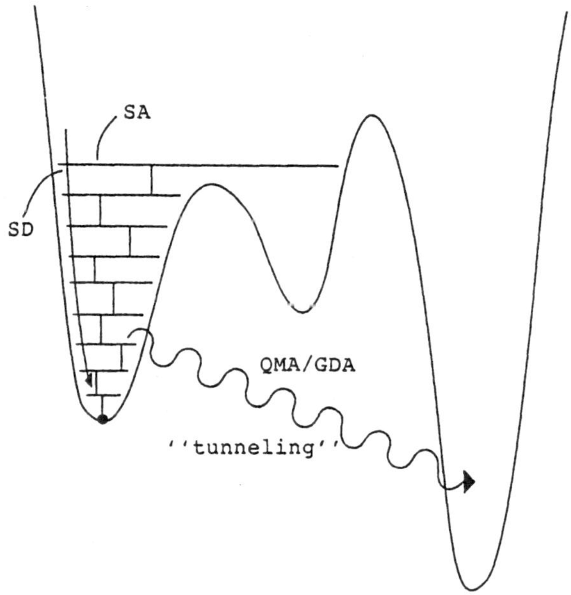

Fig. 1. A schematic of the paths taken by three optimization algorithms on a rugged potential energy surface. The local steepest descent minimizer (SD) locates the nearest local minimum. The simulated annealing (SA) algorithm searches a larger region of configuration space, but with rapid (more practical) cooling schedules, one may fail to clear important barriers and become trapped in a local minimum. The quantum mechanical annealing (QMA) method, as well as the GPP and GDA algorithms, has the advantage of working with a coarse grained potential energy surface where the wave packets can tunnel through potential barriers.

applied to any molecular structural optimization problem and to a variety of problems outside of molecular energy minimization. The third section describes simulated annealing in time using an approximate classical dynamics which results from an attempt to anneal simultaneously an ensemble of systems rather than a single point on the vast configuration space. In the third section, I discuss an unconventional annealing strategy (Gaussian Density Annealing, or GDA) where the system is followed in temperature rather than time, avoiding the problematic cooling schedule. Alongside this method, I mention two related techniques which are in the same spirit. The fourth section covers a variety of methods for performing "quantum mechanical annealing" (QMA) where the annealing process can benefit from tunneling (which dominates the system relaxation at low temperatures) as well as activated barrier crossing (effective only at high temperatures). A graphical overview of these techniques is provided in Fig. 1.

Several excellent overviews of local optimization techniques exist which are complimentary to the methods discussed in this chapter.[41–44] While many of the algorithms discussed here have been tested on proteins, some have not. The global energy minimization of clusters of Lennard-Jones atoms has emerged as a standard test case for algorithms designed for molecular optimization problems and many results presented here deal with this useful system with known results.[40,45–47]

3. Tailored Algorithms

In a broad sense, tailored algorithms refer to optimization methods which are designed *ad hoc* for a specific problem. In the cases I will discuss, what is tailored about an algorithm is that it only applies to potential functions of a certain symmetry or functional form. For example, an algorithm which exploits the functional form of a centrosymmetric potential is not easily generalized to treat covalently bonded systems, and vice versa. However, as will be clear, many of these algorithms, while tailored to specific problems, may be applied to the broad class of problems involving molecular energy minimization.[48] Moreover, they are recognizable as specific forms of the more general class of potential smoothing algorithms.

3.1. *Potential Shift and Bond Scaling*

Piela and coworkers have presented a method for global energy minimization designed for centrosymmetric potentials.[49] The method is based on a potential deformation where the range of the attractive interaction is effectively increased relative to the length scale of the repulsive (excluded volume) interaction. In an application to atomic clusters, the Lennard-Jones potential was transformed as[49]

$$V_{LJ}(r_{ij}) \rightarrow V_{LJ}(r_{ij} + ar_{\text{min}}), \tag{1}$$

where $0 \leq a \leq 1$ and r_{min} is the minimum of the initial potential energy function r_{min} (see Fig. 2). When $a = 1$ the potential is transformed such that the potential minimum is moved to the origin $r_{ij} = 0$ and the portion of the potential with a repulsive force is eliminated — the potential is purely attractive. As a decreases to zero the physical potential is restored.

In practice, the algorithm is implemented by (1) finding the minimum of the scaled potential with $a = 1$ (in the case of atomic clusters all atoms are

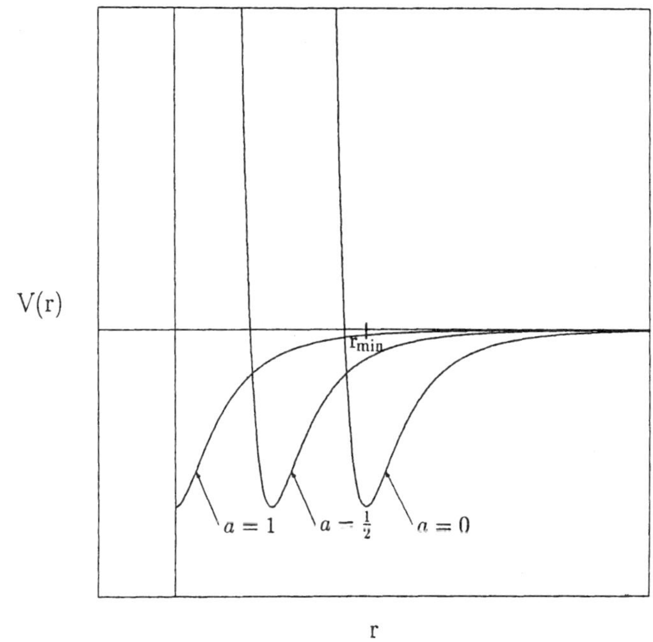

Fig. 2. A schematic of the potential energy surface for a Lennard-Jones interaction $V(r)$. The physical potential energy has a minimum at r_{min}. When the potential is transformed through the shift of $r \rightarrow r + ar_{\text{min}}$ the potential minimum moves towards the origin, increasing the relative importance of the attractive interaction.

superimposed at the origin), (2) changing $a \to a - \delta$ followed by (3) energy minimization to track the minimum on the new surface, and (4) a return to step (2) until the energy minimum is mapped back to the physical potential where $a = 0$. This method has been successful in finding the global energy minimum for a series of Lennard-Jones clusters ranging from 5 to 19 atoms but appears to be quite CPU intensive (reported to take 100 hours on an IBM 386 for the 19 atom cluster).

Loop closure is a fundamental problem in protein modeling.[50] DeLisi and coworkers[51] have introduced an effective method for loop closure which has been applied to model loop segments (of seven residues), including α-helical, β-strand and surface loops, in a variety of systems[52] as well as to bound peptides in class I major histocompatibility complex receptors.[53] The algorithms are based on a simple scaling of the bond lengths and van der Waals interactions coupled with energy minimization. Suppose the goal is to model a peptide segment into a site with a known end-to-end distance d. (1) The peptide segment is extended using random values of the backbone dihedrals leading to an initial end-to-end length of d_0. (2) The bond lengths are then uniformly scaled by adjusting the equilibrium distance as

$$r_0 \to r_0(d/d_0)^{(N-i)/N} \tag{2}$$

such that the end-to-end distance equals the known value (see Fig. 3). (3) The energy of the peptide is then minimized (if possible in the external field of the protein) where the repulsive van der Waals interactions are also scaled by a factor of $(d/d_0)^{2(N-i)/N}$. (4) Returning to step (2) the cycle is repeated N times until the peptide bonds are scaled back to their equilibrium lengths and the segment is energy minimized (hopefully) globally. This procedure should be repeated with a variety of initial conditions leading to a final ensemble of possible configurations. While this method has been employed with a local minimizer in step (3) it might be improved by substituting a more effective energy minimization method such as a form of simulated annealing discussed in the next section.

The scaling of the peptide bond is quite similar to the bond shift algorithms of Piela and coworkers.[49] Note that a primitive version of the bond shift algorithm of Eq. (1) is

$$V(r_{ij}) \to V(r_{ij} + ar_0)\,, \tag{3}$$

where $1 - d/d_0 \leq a \leq 1$ and r_0 is the minimum of the bond stretching potential energy function. When $a = 1 - d/d_0$ the potential is transformed

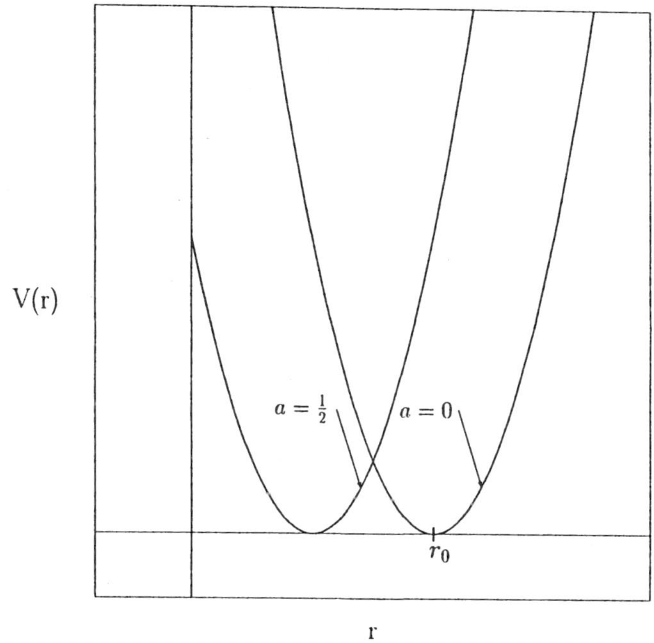

Fig. 3. A schematic of the potential energy surface for a bond stretching interaction $V(r)$. The physical potential energy has a minimum at r_0. When the potential is transformed through the shift of $r \to r + ar_0$ the potential minimum moves towards the origin, increasing the relative importance of the attractive interaction.

such that the potential minimum is moved toward the origin $r_{ij} = 0$ and the portion of the potential with a repulsive force is reduced (when $a = 1$ the potential is purely attractive). As a decreases to zero the physical potential is restored. A combination of the primitive bond shift algorithm with the nonbonded shift of Piela and coworkers discussed here is a variation of the successful loop closure algorithm of DeLisi and coworkers worth investigation.

3.2. *Sigma and Torsion Potential Fluctuation*

An inventive method for enhanced conformational space sampling has been presented by Liu and Berne.[54] This method rests on the very general idea of allowing constants which determine the length and energy scales of the interaction potential to fluctuate during a simulation according to deterministic (or stochastic) equations of motion. Their method is in the spirit of the extended dynamical methods of Andersen,[55] and Nosé,[56] and Hoover[57] for applying external constraints on temperature or pressure. While their applications involve Monte Carlo studies of a binary Lennard-Jones fluid and chain folding in normal alkanes, they provide a clear prescription for applications involving molecular dynamics which may be more suitable for biomolecular systems.[54]

For example, consider a Lennard-Jones fluid of particles with physical diameter σ. We allow the particle diameters to fluctuate between the values of $\sigma_{\min}$ and σ_0. Writing the Lennard-Jones diameters σ as variables which depend on a parameter w we have

$$\sigma(w) = \sigma_0 + (\sigma_{\min} - \sigma_0)s(w)\,, \tag{4}$$

where $s(w)$ is a sigmoidal function which varies between 0 and 1 switching σ between σ_0 and $\sigma_{\min}$. By defining a "kinetic energy" for the parameters w_k of the kth atom with fictitious mass m_k the generalized Hamiltonian energy function becomes[54]

$$\mathcal{H} = \frac{1}{2}\sum_{k=1}^{N} M_k \dot{r}_k^2 + \frac{1}{2}\sum_{k=1}^{N} m_k \dot{w}_k^2 + V(\mathbf{r}, \mathbf{w}) + F(\mathbf{w}) \tag{5}$$

with corresponding equations of motion

$$M_k \ddot{r}_k = -\nabla_{r_k} V(\mathbf{r}, \mathbf{w}) \tag{6}$$

$$m_k \ddot{w}_k = -\nabla_{w_k} V(\mathbf{r}, \mathbf{w} - \nabla_{w_k} F(\mathbf{w})\,. \tag{7}$$

$F(\mathbf{w})$ is a cost function which constrains the fluctuations of the variables $\mathbf{w}$ during an otherwise standard molecular dynamics simulation. The first set of equations is for molecular dynamics on a potential which depends on the fluctuating parameter $\mathbf{w}$. Fluctuations in $\mathbf{w}$ will deform the potential and occasionally lower barriers allowing infrequent events to occur with greater frequency. This leads to enhanced sampling of the configuration space (see Fig. 4).

Similarly, in a polymer the torsion angle potential can be written[54]

$$V(\phi, w) = V_0(\phi) + [V_m(\phi) - V_0(\phi)]s(w)\,, \tag{8}$$

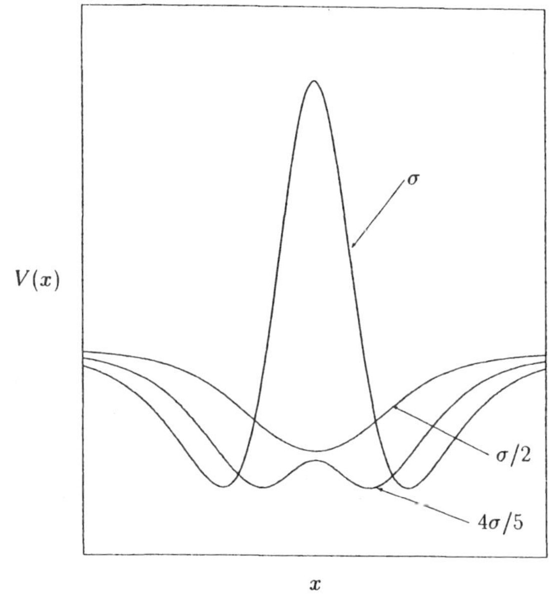

Fig. 4. The potential energy for the interaction of a single atom passing along the x-axis between two atoms fixed at ($x = 0, y = \pm\Delta$). When the diameter σ of the interior atom is allowed to fluctuate to smaller values, the potential surface for moving between the two particles changes from a separated "double well" with a high barrier, to a double well with significantly lower potential barrier, to a minimum with no barrier. This mechanism allows for enhanced barrier crossing and configurational sampling in the sigma fluctuation algorithm.

where $V_0(\phi)$ is the physical potential and $V_m(\phi)$ is a potential with lower barriers than $V_0(\phi)$. The fluctuations of the Lennard-Jones atomic diameters and torsion angle potential have been shown to lead to greatly enhanced sampling in Monte Carlo simulations of n-pentane and C_{50} H_{102}.[54] For the Lennard-Jones simulations, the effect of the sigma fluctuations is to soften the interactions allowing for closer contacts and smaller amplitude oscillations in the structure (radial distribution function) indicating a lowering in the barriers of the potential of mean force which leads to enhanced sampling. The qualitative trends are similar to those seen in the approximate dynamical annealing methods discussed in the next section.[58] The sigma fluctuation method shares many of the features of the potential smoothing algorithms featured in this chapter. Moreover, it is also possible to compute averages in the ensemble of the physical potential (fixed σ) using the enhanced sampling trajectories (fluctuating σ) via umbrella sampling techniques.[54] Similar results are expected in simulations of proteins where the use of molecular dynamics, rather than Monte Carlo, may be more attractive. This enhanced conformational space sampling should lead to more rapid folding as well as information on the distribution of low energy conformations of the molecule.

3.3. *The Antlion Method applied to Proteins*

A slightly different approach which combines potential smoothing with a novel implementation of secondary structure prediction information has been developed by Stillinger and coworkers in the form of the "antlion method." This method has been applied to a blocked alanine dipeptide and tetrapeptide[59] as well as the α-helical protein melittin.[60] In each case the potential surface was deformed by applying constraints on the (ϕ, ψ) angles which bias the configuration towards a given local secondary structure. In the case of melittin, the deformation included (1) a bias towards the *trans* peptide backbone configuration, (2) a bias towards the L conformer of each amino acid using an improper torsion potential around the C_α, (3) a penalty function to encourage a set of ideal (ϕ_0, ψ_0) angles of the form

$$V_{\phi,\psi} = k_\phi[1 - \cos(\phi - \phi_0)] + k_\psi[1 - \cos(\psi - \psi_0)], \tag{9}$$

where the values of the force constants k_ϕ and k_ψ are determined by a neural network trained on a learning set, (4) a penalty function in the form of a Coulombic potential which encourages hydrogen bonding between the backbone carbonyl O of residue i and the backbone amide H of residue $i+4$ where the charges were determined using a neural network, and (5) setting to zero the side chain atomic charges.

Following the determination of parameters for the penalty functions (to encourage secondary structure formation) the conformation of melittin was minimized from an initial extended conformation on the smoothed potential surface. (For the application to melittin, the ideal (ϕ_0, ψ_0) angles were those of an α-helix ($-57°$, $-47°$).) This was followed by minimization on the physical potential surface until convergence. This method succeeded in finding a structure within an overall 2.45 Å root-mean-square difference from the minimized crystal structure.[60]

Some protein optimization algorithms, such as that of Friesner and coworkers which was applied to myoglobin,[61] rely on an initial step of secondary structure prediction. If it is possible to identify regions of secondary structure, extended regions of protein substructure can be fixed, thereby greatly reducing the number of degrees of freedom which must be included in the search for the global energy minimum. With regions of secondary structure frozen in, folding is expected to occur as described in the diffusion-collision model of Karplus and Weaver.[62] However, secondary structure prediction methods are less than perfect (usually around 70% accuracy per residue) and any folding optimization algorithm which assumes this knowledge will have difficulties with the accumulating error. The antlion method overcomes this problem by using a penalty function which biases the residue towards a given set of idealized (ϕ, ψ) angles rather than rigidly fixing the backbone (ϕ, ψ) angles of a given residue. However, one must specify the ideal angles (ϕ_0, ψ_0) about which the penalty function will constrain the residue. In the case where the secondary structure is of one kind, such as melittin[60] or even the example of myoglobin,[61] this might succeed. However, in cases where there is a known mix of secondary structure types, the problem of best choosing (ϕ_0, ψ_0) must be addressed. Nevertheless, this tailored algorithm appears to be very promising.

The methods discussed in this section have been tailored to exploit detailed aspects of specific problems. In the remainder of this chapter, I will describe three classes of general algorithms for the optimization of molecular systems which have much in common with the successful tailored methods discussed here.

4. Classical Simulated Annealing in Time

The paradigm of global optimization of complex systems is simulated annealing.[18,24,63,64] This method makes use of a powerful analogy between the statistical mechanical process of annealing and the challenge of global optimization of a complex cost function. In global energy minimization, where our cost function is the potential energy surface, this requires no stretch of the imagination. The system is simulated at an initially high temperature using molecular dynamics or Monte Carlo. The temperature is slowly lowered according to a cooling schedule, annealing the system to a low or zero temperature. The final configuration of the system is the guess at the global energy minimum. If the cooling schedule is chosen properly, we can expect to find the global minimum with a high probability.

The difficulty comes in finding a practical cooling schedule which can guarantee a high probability of finding the global energy minimum.[65] A typical cooling schedule is defined by decreasing the temperature by a factor $\xi < 1$ from one step to the next so that[18]

$$T_{k+1} = \xi T_k \,. \tag{10}$$

While this is convenient and faster for smaller values of ξ, it is arbitrary. In many complex systems, we know that there are a variety of energy scales on the rugged energy landscape.[10,12,13] It is reasonable to expect that the optimal cooling schedule will be intimately related to the disparity between the largest and smallest energy scales in the system. A simple heuristic analysis in one dimension demonstrates this by estimating the lower bound on the simulation time $t_{\rm sim}$ necessary for finding the global minimum with a reasonable probability.

Taking the global energy minimum to be the zero of energy, $E^*_{\rm max}$ is the highest barrier on the potential surface (see Fig. 5). If we choose the initial

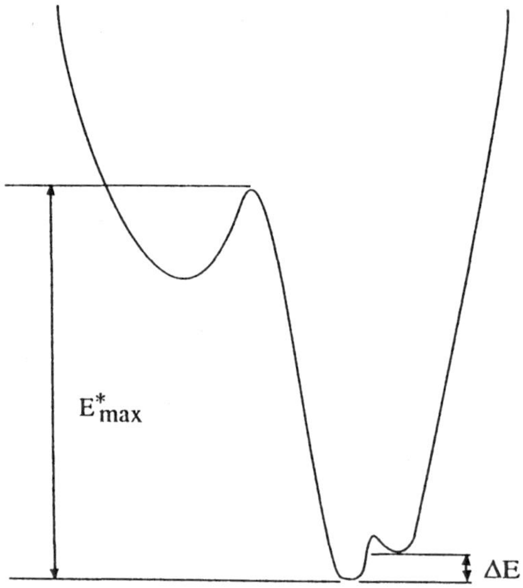

Fig. 5. A schematic of an energy landscape showing the important energy scales which determine the effectiveness of the simulated annealing algorithm. At the high end is the largest energy barrier which must be crossed $E^*_{\rm max}$ for a trajectory to explore all of the configuration space. At the low end is the energy gap ΔE separating the global and next lowest energy minimum. For many problems, the ratio of energy scales $E^*_{\rm max}/\Delta E$ defines the difficulty of the optimization problem.

maximum value of the temperature such that

$$k_B T_{\max} = 1/\beta_{\min} \simeq E^*_{\max}$$

then

$$e^{\beta_{\min} E^*_{\max}} = \mathcal{O}(1)\,. \tag{11}$$

At this high temperature, the trajectory will be free to cover the entire configuration space unhindered by the energy barriers in the system.

On the other hand, the lowest temperature we need to consider is one where $k_B T_{\min} = 1/\beta_{\max} < \Delta E$ where $\Delta E = E_1 - E_0$ is the "energy gap" between the global energy minimum and the next lowest energy minimum. This inequality should be satisfied so that the mole fraction of being found in the global energy minimum X_0 (compared to the mole fraction for all other minima $1 - X_0$) is large. For many problems, at low enough temperatures there is an effective two level system dominated by the global and next lowest energy minimum, so we simply demand that

$$e^{\beta_{\max} \Delta E} = \frac{X_0}{1 - X_0} \tag{12}$$

be reasonably large (meaning that X_0 is reasonably close to 1).

Ideally, at any point along the annealing trajectory, the system will be at thermal equilibrium. To effectively sample the system at a given β the trajectory should spend a time of at least $t = \tau_s \exp(\beta E^*_{\max})$ required to cross any barrier in the system; τ_s is a characteristic time scale for vibrational motion. For that to be true, the time spent (dt) in a given temperature interval $(\beta, \beta + d\beta)$ should be at least

$$dt = E^*_{\max} \tau_s e^{\beta E^*_{\max}}\, d\beta\,. \tag{13}$$

τ_s is a short time scale fixed by the highest frequency motion in our system which sets the upper bound on the time step we can take in the integration of our system.

The *lower bound* on the total simulation time t_{sim} can be found by integrating Eq. (13) between the initial $\beta_{\min}$ and final $\beta_{\max}$ leading to a number of simulation steps $N_s = t_{\text{sim}}/\tau_s$ which scales as

$$N_s \sim \left(\frac{X_0}{1 - X_0}\right) e^{E^*_{\max}/\Delta E}\,. \tag{14}$$

This result can be used to estimate the simulation time required in molecular dynamics to find the global energy minimum with a probability X_0. The exponential dependence of N_s on the ratio of energy scales $E_{max}/\Delta E$ shows that the most difficult optimization problems involve systems with a great disparity in energy scales. It is not possible to explain the difficulty of a given minimization problem solely in terms of a small energy gap separating the lowest and next lowest energy minima, or the presence of high barriers separating regions of configuration space. *It is the ratio in the coarsest and finest energy scales that dictates the simulation time needed.* Similar results hold for more commonly employed cooling schedules.

When the rate of lowering temperature in time is fixed, the simulation time required will be set by the limits of the initial and final temperatures. In the global optimization of biomolecular systems there is a range of energy scales, the ratio $E^*_{max}/\Delta E$ is large, and the necessary simulation time will be significantly longer than for simple systems. A rough estimate of this time, taking $X_0 = 0.01$ and $E^*_{max}/\Delta E = 10$, indicates that such a simulation is out of the question at the present time. In fact, reasonable estimates should be close to the real time required to simulate the dynamical folding of the protein *in machina* without the help of simulated annealing which now can only be performed for model systems.[66–68]

For simulated annealing to be an effective strategy for optimization of proteins, some changes are required. Think of applying the simulated annealing algorithm using an ensemble of systems in parallel. One may then explore a variety of mean field approximations for the approximate integration of an ensemble of systems. I will summarize a few of these techniques in this section. Unlike an *ad hoc* potential transform, these methods are rooted in the approximate integration of the classical Liouville equation which describes the time evolution of the classical density distribution (an ensemble of classical systems) just as Newton's equations employed in classical simulated annealing describe the time evolution of a single trajectory.

4.1. *Classical Dynamical Annealing using the Liouville Equation*

A method for global optimization which is becoming increasingly popular involves the use of straightforward ensemble integration on massively parallel computers. In the simplest application, an ensemble of initial configurations is generated (representing copies of the system) and each copy is simulated independently, for example, using simulated annealing. When it is determined that a trajectory has reached its lowest energy, or has moved outside of the region of interest in configuration space, the system is removed and replaced by a new copy of the system. In this way an ensemble of systems — a density distribution — is simulated. However, this is computationally demanding and the time over which the system dynamics must be followed is very long (according to the argument given above).

In the development of new methods, it is often desirable to describe the time evolution of an ensemble of systems (to derive the benefits of enhanced sampling) while treating the dynamics approximately (to reduce the computational cost of simulating so many systems in parallel). One place to start is the classical Liouville equation which defines the exact dynamics of the classical density distribution $\rho(r,p,t)$.

In the time evolution of an ensemble of classical systems,[69,70] where each system follows the classical mechanics defined by Newton's equations of motion,[71] the time evolution of the distribution $\rho(r,p,t)$ is described by the Liouville equation[72,73]

$$\frac{\partial}{\partial t}\rho(r,p,t) = -\mathcal{L}_0\rho(r,p,t) \tag{15}$$

where $\mathcal{L}_0$ is the Liouville operator

$$\mathcal{L}_0 = \frac{p}{m}\cdot\frac{\partial}{\partial r} + F(r)\cdot\frac{\partial}{\partial p} \tag{16}$$

and $F(r)$ is the force and m is the mass. $F(r)$, r and p are d-dimensional vectors.

An exact description of the dynamics of $\rho(r,p,t)$ is provided by the equations of motion for the average position and momentum[74]

$$\frac{d\langle r\rangle}{dt} = \frac{\langle p\rangle}{m} \qquad \frac{d\langle p\rangle}{dt} = \langle F\rangle \tag{17}$$

and for the higher-order moments of position and momentum

$$\frac{dM_{n,k}}{dt} = \frac{n}{m}M_{n-1,k+1} + kW_{n,k-1}\,. \tag{18}$$

The moments of the distribution are defined as $M_{n,k} = \langle (r-r_0)^n(p-p_0)^k\rangle$ and $W_{n,k} = \langle (r-r_0)^n(p-p_0)^k(F-F_0)\rangle$ where $r_0 = \langle r\rangle$, $p_0 = \langle p\rangle$, and $F_0 = \langle F(r)\rangle$. The brackets $\langle\ldots\rangle$ imply an average over the density distribution[69]

$$\langle A\rangle = \int drdpA(r,p)\rho(r,p,t)\,. \tag{19}$$

Integration of this hierarchy of equations provides an exact description of the dynamics of the ensemble. However, in practice this is intractable since for most anharmonic potentials the moments are coupled (up to infinite order moments). Therefore, one must either truncate the moment expansion (which can be numerically unstable) or approximate $\rho(r, p, t)$.

As a first approximation, Ma, Hsu and Straub[74,75] have taken $\rho(r, p, t)$ for a many body system to be a product of single particle distributions represented by d-dimensional spherically symmetric Gaussian phase packets (GPP).[76,77] Each GPP is completely defined by the first and second moments — the packet center (r_0, p_0) and widths $(M_{2,0}, M_{1,1}, M_{0,2})$ in phase space. It is straightforward to derive equations of motion for each Gaussian using Eq. (18).

The appeal of these simple equations of motion comes from their origin as an approximate solution of the Liouville equation for an ensemble of systems, while requiring little more computational investment than is required to integrate Newton's equations for a single representation of the system. The equations of motion are variationally optimized, meaning that the error between this approximate solution and the exact solution has been minimized in a least squares sense. While being approximate, the algorithm leads to significantly enhanced sampling in both applications to equilibrium averaging and global optimization.[74]

In general

$$m\ddot{r}_0 = -\nabla_{r_0}\langle V\rangle \tag{20}$$

which has the form of Newton's equation for the Gaussian center r_0 moving on a coarse-grained effective potential $\langle V\rangle$. When $M_{2,0} = 0$ the equations of motion reduce to the usual equations of classical molecular dynamics. The second moment equations depend on the Laplacian of the effective potential. It has been found[74] that the average value of the second moment in position $\bar{M}_{2,0}$ is related to the generalized Einstein frequency[78,79]

$$\Omega_E^2 = \frac{1}{dm}\langle\nabla^2 V\rangle \sim dk_BT/2m\bar{M}_{2,0}\,. \tag{21}$$

This provides some insights into the dynamics of the approximate Gaussian density distribution. The center of the distribution moves to minimize the force while the widths adjust to the curvature of the effective potential. Moreover, they adjust in a way that instantaneously provides the ensemble of states represented by a fluctuation about the average structure. In this sense it is similar to an effective harmonic or quasiharmonic model of the protein where the configuration space is represented by a given conformation of the protein (the center of the distribution) and the harmonic fluctuations around this configuration. In the Gaussian phase packet representation, this effective harmonic representation follows the evolution of the centers. At each instantaneous configuration the density distribution represents a distribution of protein structures and no normal mode analysis or time averaging of fluctuations (quasiharmonic analysis) is required. Furthermore, as the centers evolve, the density distribution carves out a tube in phase space representing a volume of many single protein trajectories. It is this enhanced sampling which provides the ability to find the global energy minimum most effectively during simulated annealing.

4.1.1. *Evaluating the Effective Potential for Proteins*

Any difficulty associated with the application of the GPP dynamics comes in calculating the effective potential $\langle V \rangle$ for a protein using a general empirical energy function of the ECEPP[80,81] or CHARMM[82] form. Once $\langle V \rangle$ is known, the GPP dynamical trajectory is found by solving the first order differential equations in the fashion of molecular dynamics.[71,82,83] For covalently bonded systems such as polymers and proteins this is a simple matter while, in contrast, defining effective Monte Carlo moves can be quite challenging.

How do we calculate $\langle V \rangle$ for a protein? To start we need to calculate the effective potential between a pair of interacting atoms. For a many body system it is convenient to approximate the N body distribution function as a Hartree product of the single particle distribution functions[84]

$$\rho(r^N, p^N, t) = \prod_{k=1}^{N} \rho_k(r_k, p_k, t) . \tag{22}$$

This approximation makes it particularly easy to evaluate the pair potential

$$\begin{aligned} \langle V \rangle &= \sum_{i>j=1}^{N} \int dr^N \int dp^N \; \rho(r^N, p^N) V(|r_i - r_j|) \\ &= \sum_{i>j=1}^{N} \int dr_i \int dp_i \int dr_j \int dp_j \, \rho_i(r_i, p_i) \rho_j(r_j, p_j) V(|r_i - r_j|) . \end{aligned} \tag{23}$$

The final effective potential is the interaction energy of the Gaussian densities of the two atoms. That is, the probability of the two atoms being a certain distance apart is smeared out over a range of distances. The result is an averaged potential energy with the features of a coarse grained potential — raised and fewer minima and lower barriers.

The effective pair potential $\langle V \rangle$ is a function of the separation of the centers of the two interacting density packets (as well as the widths of those packets). It turns out that we can reduce the $2d$-dimensional integral above to a one-dimensional integral over the interpacket separation.[74] The result is that the potential is averaged as

$$\langle V \rangle = \int_0^\infty dr G(r) V(r) \tag{24}$$

over a distribution function

$$G(r) = (\zeta/\pi)^{1/2} \frac{r}{r_{ij}} \left[\exp(-\zeta(r - r_{ij})^2) - \exp(-\zeta(r + r_{ij})^2) \right] \tag{25}$$

where $r_{ij} = |\mathbf{r}_0^{(i)} - \mathbf{r}_0^{(j)}|$ is the interpacket center separation and $\zeta = d/[2(M_{2,0}^{(i)} + M_{2,0}^{(j)})]$ is the reciprocal sum of the density packet widths (the mean square spread in position). The effective potential is easily evaluated when $V(r)$ is a Gaussian, exponential, or a polynomial. As we will discuss below, it is most common to represent the nonbonded potential energy as a sum of Gaussians or exponentials, and the intermolecular potential in the form of polynomials.

The general form for the empirical potential energy function most popular in studies of proteins is[81,82,85]

$$V(r^N) = V_{\text{bond}} + V_{\text{angle}} + V_{\text{torsion}} + V_{\text{LJ}} + V_{\text{Coulomb}} \,. \tag{26}$$

In evaluating the effective potential for the density distribution we can employ a very useful trick — we will write the total potential as a sum of pair potential terms. This is a simple matter for the bond potential which is typically modeled as a harmonic Hook's law form $V_{\text{bond}} = \kappa |r_i - r_{i+1}|^2$. The nonbonded interaction is typically a pair potential consisting of the Lennard-Jones interaction of a hard core repulsion (modeled as a $1/r^{12}$ form or just as well as an exponential) and a longer range attractive London dispersion energy (which varies as $1/r^6$). In some cases it is important to include ion-induced dipole terms (varying as $1/r^4$ and proportional to the atomic polarizability). Higher order effects such as self-consistent polarizability can be added but with a significant computational overhead. For all but the last interactions, the potential is a simple pairwise additive function.

For the Coulombic interaction (which varies as $1/r$), the average potential of Eq. (24) can be calculated exactly in a simple form. For the Lennard-Jones potential we find it easiest to fit the potential to a sum of Gaussians and/or exponentials for which the average potential is easily calculated.[74]

Other interactions are often treated using three or four atom interactions. The angle term is usually treated using internal angle coordinates defined by the three atoms forming the angle. However, this form may be replaced by a Urey-Bradley interaction of the form $V_{\text{angle}} = \kappa|r_i - r_{i+2}|^2$ (however, the force constants must be adjusted). For the case of rigid bonds, the usual angle term may be written as a harmonic interaction in $|r_i - r_{i+2}|^2$. The torsional potential is usually written in terms of four atom torsional angles but in the case where the bonds and angles are constant it may be rewritten using a simple polynomial as a sum over powers of $|r_i - r_{i+3}|^2$ making it a simple pair potential. The take home message is that the protein potential function can be readily written in terms of pairwise additive interactions. Once this is done, we can use Eq. (24) to evaluate the effective potential energy of interaction between the density of two atoms in the protein modeled using Gaussian phase packets.

An alternative is to treat the internal potential (bond, angle, torsions) as a force acting only on the packet centers. The effective potential is

$$\langle V \rangle = V_{\text{bond}} + V_{\text{angle}} + V_{\text{torsion}} + \langle V \rangle_{\text{LJ}} + \langle V \rangle_{\text{Coulomb}} \,, \tag{27}$$

where only the nonbonded potential is averaged over the density distribution. The centers of the packets define an underlying protein structure which is confined to a manifold of states of well defined bond lengths and angles.

4.1.2. *Temperature Control during Annealing*

In applying the GPP algorithm to an optimization problem we perform a simulated annealing of the continuous density distribution, which requires that we control the temperature during cooling. There are a variety of ways to do this. One method involves a rigid constraint on the temperature.

Other methods involve coupling the system to a heat bath such that the temperature of the system may fluctuate. We will discuss and compare these methods in the next two sections.

4.1.2.1. *A Rigid Constraint on the Temperature*

To apply the constant temperature constraint which truly holds the temperature constant, we appeal to Gauss's principle[86,87] and write the generalized Liouvillian[74]

$$\mathcal{L} = \mathcal{L}_0 + \mathcal{L}_c = \mathcal{L}_0 - \gamma \frac{\partial}{\partial p} \cdot p\,, \tag{28}$$

where $\mathcal{L}_0$ is the streaming operator of the system and $\mathcal{L}_c$ is the collision operator which couples the system to the bath. γ is determined by the external temperature constraint equation $dT(t)/dt = 0$ or equivalently $d\langle p^2\rangle/dt = 0$ which guarantees that the temperature remains constant at every instant in time.

Using the Liouville equation we find equations of motion for the average position and momentum[74]

$$\frac{d\langle r\rangle}{dt} = \frac{\langle p\rangle}{m}, \qquad \frac{d\langle p\rangle}{dt} = \langle F\rangle - \gamma\langle p\rangle \tag{29}$$

and for the higher-order moments of position and momentum

$$\frac{dM_{n,k}}{dt} = \frac{n}{m} M_{n-1,k+1} + k\Big(W_{n,k-1} - \gamma M_{n,k}\Big)\,. \tag{30}$$

The above equations are exact and completely general. For the Gaussian phase packet, the time evolution of the first and second-order moments is given by[74]

$$\dot{r}_0 = \frac{p_0}{m}, \qquad \dot{p}_0 = -\nabla_{r_0}\langle V\rangle - \gamma p_0 \tag{31}$$

$$\begin{aligned}
\dot{M}_{2,0} &= \frac{2}{m} M_{1,1} \\
\dot{M}_{1,1} &= \frac{1}{m} M_{0,2} - \frac{1}{d} M_{2,0} \nabla^2_{r_0}\langle V\rangle - \gamma M_{1,1} \\
\dot{M}_{0,2} &= -\frac{2}{d} M_{1,1} \nabla^2_{r_0}\langle V\rangle - 2\gamma M_{0,2}\,.
\end{aligned}$$

The temperature constraint equation which specifies the cooling schedule $dT(t)/dt$ determines the value of γ. For an exponential cooling schedule $dT(t)/dt = -\gamma T(t)$. Note that $\langle V \rangle$ is the bare potential energy averaged over the phase space distribution

$$\langle V \rangle \equiv \int dr \int dp\ \rho(r,p) V(r)\,. \tag{32}$$

Averaging over the density distribution $\rho(r,p)$ effectively coarse-grains the potential surface. In Fig. 6 we show how the effective potential is smoothed as the width of the density distribution increases.

Note that in general

$$m\ddot{r}_0 = -\nabla_{r_0} \langle V \rangle - \gamma p_0 \tag{33}$$

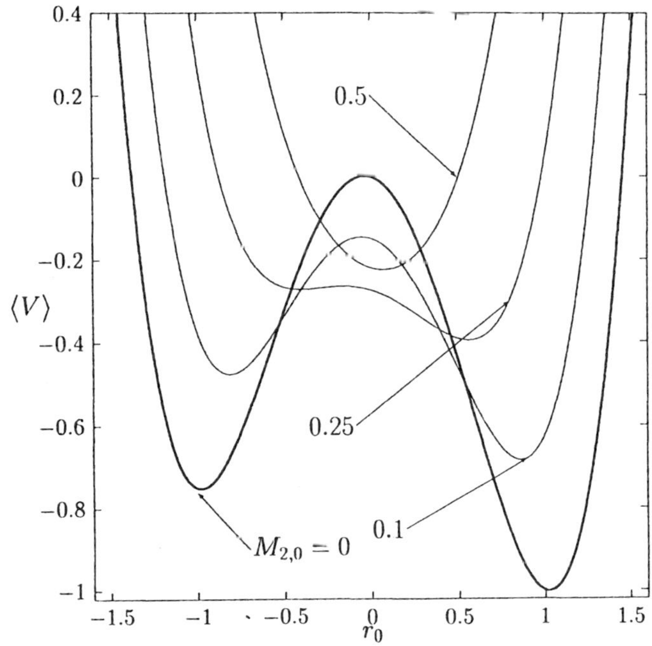

Fig. 6. The effective potential defined by Eq. (32) is plotted for an asymmetric double well for a series of squared widths $M_{2,0}$. The bare potential used in molecular dynamics corresponds to $M_{2,0} = 0$. The units are arbitrary. The potential is approximately what would be seen by a Lennard-Jones atom moving along r between two atoms (of different types) centered at $r = \pm 2$.

which has the form of a Langevin equation (minus the explicit random force) for the Gaussian center r_0 moving on a coarse-grained effective potential. In the limit that $M_{2,0} = 0$ the equations of motion reduce to the constant temperature algorithm of Evans and Hoover[87] for classical point particles.

4.1.2.2. *Coupling to a Bath through Weak and Strong Collisions*

The rigorous constraint on the temperature leads to a set of equations which are easily integrated and which in practice lead to a continually fluctuating $\rho(r,p,t)$. In a more realistic treatment of coupling a system to a heat bath we expect that the density distribution will relax to the static equilibrium density distribution where $\dot{\rho}_{\rm eq} = -\mathcal{L}\rho_{\rm eq} = 0$. This can be achieved, even in an approximate representation of $\rho(r,p,t)$, by coupling the system to the heat bath through a collision operator $\mathcal{L}_c$ of the Fokker–Planck[88] or BGK form.[89]

Consider a "weak collision" Fokker–Planck operator[88]

$$\mathcal{L}_c = -\gamma \frac{\partial}{\partial p} \cdot \left[p + mk_BT \frac{\partial}{\partial p} \right] . \tag{34}$$

The final equations of motion for the Gaussian density distribution of a particle coupled to a heat bath through the Fokker–Planck collision operator are identical to Eq. (31) with the exception that the equation of motion for the second moment of the momentum $M_{0,2}$ is

$$\dot{M}_{0,2} = -\frac{2}{d} M_{1,1} \nabla^2_{r_0} \langle V \rangle - 2\gamma \left[M_{0,2} - dmk_BT \right] .$$

A related result has been presented where the momenta are adiabatically eliminated leading to a Gaussian packet solution of the Smoluchowski equation.[153,154]

Following the hybrid constant temperature algorithm of Andersen,[55] one might choose a BGK-like collision operator[89,90]

$$\mathcal{L}_c = \gamma \left[1 - \left(\frac{m}{2\pi k_BT} \right)^{d/2} \exp(-p^2/2mk_BT) \int dp \right] \tag{35}$$

which thermalizes the momentum with a rate γ according to the Maxwell distribution. In our GPP dynamics algorithm this corresponds to integrating a deterministic dynamics specified by Eq. (31) where the equation of motion for $M_{0,2}$ is replaced by[74]

$$\dot{M}_{0,2} = -\frac{2}{d} M_{1,1} \nabla^2_{r_0} \langle V \rangle - \gamma \Big[M_{0,2} - (p_0^2 + dmk_B T) \Big] . \tag{36}$$

In each algorithm the value of the friction γ is fixed and the temperature is controlled by an external cooling schedule. If the equations of motion are integrated at a fixed T for a long enough time (long compared to $1/\gamma$) the distribution is damped and $\rho(r,p)$ approaches an estimate of the steady state distribution.

The simulated annealing using GPP dynamics and an exponential cooling schedule has been applied to the global energy minimization of atomic clusters ranging from 2 to 55 atoms with excellent results.[74,91] The results

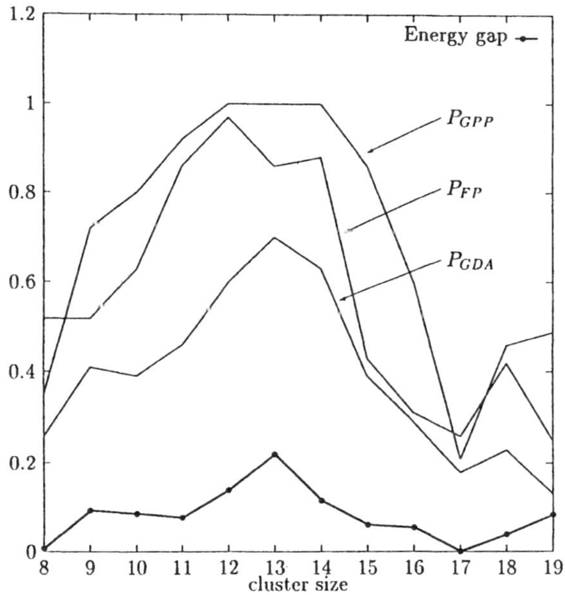

Fig. 7. The probability of locating the global energy minimum for a sample of one hundred independent configurations is plotted for a series of Lennard-Jones clusters for three algorithms: simulated annealing with Gaussian Phase Packets (GPP), simulated annealing using the Fokker–Planck GPP algorithm (FP) and simulated annealing in temperature using the Gaussian density annealing (GDA) algorithm. The energy gap ΔE between the global energy minimum and the next lowest energy state found is plotted (in Lennard-Jones reduced units per particle) to show the inverse correlation between the size of the energy gap and the probability for finding the global minimum.

for annealing using the (1) a rigid temperature control (GPP) and (2) coupling to a bath using the Fokker–Planck collision operator (FP) are summarized in Fig. 7. For comparison, the energy gap $\Delta E = E_1 - E_0$ between the global and next lowest lying energy minimum found in the study is plotted. There is a strong correlation between the probability of finding the global minimum and the magnitude of the energy gap. The hardest optimization problems are those for which the energy gap is small (for example, $n = 17$).

4.2. *Annealing using Locally Enhanced Sampling (LES)*

For conformational searches of peptides and small proteins, a powerful method which stays close to the spirit of sampling conformation space through molecular dynamics is the Locally Enhanced Sampling (LES) algorithm developed by Elber and coworkers.[11,84,92] This method relies on making a "mean-field" approximation where a set of atoms or amino acid residues are simulated by an averaged "field of copies" of those residues. To explain, assume we have a protein which binds small peptides. While the protein is large, we are mostly interested in the binding-site region and how the peptide binds to it. If we were to run Z simulations of the full system, for all these simulations much of the protein will be in the same conformation. That is, the time spent updating the unaffected portion of the system is wasted.

In the LES procedure we divide the atoms or residues into a "system," consisting of the peptide and those residues in the binding region, and a "bath" of residues which are not strongly affected by the binding process. Z trajectories of the "system" are run simultaneously feeling the full force of the single set of "bath" coordinates; atoms of the "bath" feel the averaged force of the Z systems. This is an approximate method which has been quite successful in exploring reaction pathways for carbon monoxide diffusing through the heme protein myoglobin where the ligand is treated as the "system" by Z copies (Z as large as 60).[11] For most problems of interest the "system" will be small compared to the "bath" and the LES algorithm provides an approximate method of dramatically increasing the sampling of peptide conformations (optimally by a factor of Z) while only marginally increasing the computer time required.

Using the formalism of the Liouville equation based algorithms discussed above, we express the density distribution of the LES algorithm as

$$\rho(r^N, p^N, t) = \rho_{\text{bath}}\rho_{\text{system}}$$
$$= \prod_{k=1}^{M} \delta(r - r_k)\delta(p - p_k) \prod_{k=M+1}^{N} \rho_{\text{system}}(r_k, p_k) \tag{37}$$

and

$$\rho_{\text{system}}(r_k, p_k) = \frac{1}{Z}\sum_{i=1}^{Z} \delta(r - r_{ki})\delta(p - p_{ki}) \tag{38}$$

where Z is the number of "copies" of the system particles. The density distribution of the peptide bath interacts with an approximate mean field distribution of the protein system.

While the LES method is a powerful technique, it is approximate. For example, as a result of the lack of equal-and-opposite forces the system does not equipartition energy as one would find in normal molecular dynamics — energy is collected in the "system" coordinates effectively heating the system to Z times the temperature of the bath.[93] However, there are approximate methods for correcting this problem[94] including holding the bath coordinates fixed when no sampling of the bath coordinates is desired.[95,96] An additional difficulty arises from the excluded volume of the system. In the limit that the system copies have little overlap, the excluded volume will be Z times that of a single system. Nevertheless, the LES algorithm is a very effective enhanced sampling algorithm which is easily applied using molecular dynamics or Monte Carlo.

Roitberg and Elber have applied the LES method to optimization problems using a dynamical simulated annealing strategy with impressive results for two tetrapeptides and aromatic side chains in BPTI.[97] Moreover, they provide a beautifully simple proof that the result of an ideal annealing run using the LES algorithm will result in the *exact* result for the global energy minimum which would be found using MD simulated annealing.

The strength of the LES algorithm is that is it very simple to implement. No effective potential is required, and the equations of motion are Newton's equations. For problems involving proteins, the empirical potential energy function is a combination of terms simply expressed in either Cartesian or internal coordinates, but not both. This makes the calculation of the effective potential required by the wave packet methods rather tedious. The LES algorithm maintains many of the desirable features of a mean field approximation (lower barriers, fewer minima) while remaining simple.

4.3. *Selectively Optimized Sampling*

We can imagine a combination of the Gaussian phase packet (GPP)[74] and LES[11] algorithms which provides enhanced sampling for a selected subset of coordinates (as in LES). The bath coordinates are represented by a delta-function distribution of single point particles while a mobile basis of Gaussian packets is used to represent the system coordinates where enhanced sampling is desired.[98] For example, the density distribution of the peptide "system" coordinates can be smeared out as a product of single Gaussian phase packets while the protein "bath" is represented using point particles as in standard molecular dynamics. Such a choice for the density distribution allows for optimized sampling of a *selected set of coordinates*, $i \in (M+1, N)$, while the remaining coordinates, $i \in (1, M)$, are treated as point particles in the standard way using Newton's equations of motion

$$m\ddot{r}_k = \sum_{j=1}^{M}{}' F(|r_k - r_j|) + \sum_{j=M+1}^{N} \langle F(|r_k - r_j|)\rangle_j \,. \tag{39}$$

The primed sum indicates the restriction $j \neq k$. The bracket $\langle\rangle_j$ indicates an average over the density distribution of the jth phase packet. We expect to find the correct energy equipartitioning and avoid any excluded volume problems using this hybrid representation.

5. Classical Simulated Annealing in Temperature

The nut to crack in applying simulated annealing to proteins is the definition of a cooling schedule which leads to a reasonable probability of finding the global minimum in a computationally realizable time. An appealing alternative is to replace the cooling schedule necessary when annealing in real or Monte Carlo time with direct integration in temperature. As we shall see later, this is closely related to quantum dynamics in imaginary time. There is no real time and therefore no cooling schedule that must be specified as part of the annealing protocol. There is only a set of equations describing the evolution of the ensemble of systems in temperature. In this section I discuss a method which directly anneals an approximation to the classical density distribution in temperature and then describe a few closely related algorithms.

5.1. *The Gaussian Density Annealing (GDA) Algorithm*

Ma and Straub[91] have recently explored the possibility of performing simulated annealing directly in temperature by approximately integrating the classical Bloch equation[99]

$$\frac{\partial \rho_{\text{eq}}}{\partial \beta} = -(\mathcal{H} - \langle \mathcal{H} \rangle)\rho_{\text{eq}} \tag{40}$$

to obtain the equilibrium classical density distribution for the canonical ensemble $\rho_{\text{eq}}(r, p, \beta), = \exp(-\beta\mathcal{H})/Q(\beta)$ at a given temperature. $\mathcal{H}(r, p)$ is the classical Hamiltonian energy function and $Q(\beta) = \int drdp \, \exp(-\beta\mathcal{H})$ is the canonical partition function. In fact, Eq. (40) is the classical analog of the imaginary time Schrödinger equation (discussed in the next section). For an N body system it is convenient to make the Hartree approximation to the many body density distribution as a product of the single body density distributions (employed in the GPP integration[91] of the Liouville equation)

$$\rho(r, \beta) = (2\pi M_2/d)^{-d/2} \exp\left[-\frac{d}{2M_2}(r - r_0)^2\right]. \tag{41}$$

The equations of motion in reciprocal *temperature* for the center r_0 and second moment M_2 of a single Gaussian packet in d-dimensions are

$$\begin{aligned} \frac{\partial r_0}{\partial \beta} &= -\frac{1}{d} M_2 \nabla_{r_0} \langle V \rangle \\ \frac{\partial M_2}{\partial \beta} &= -\frac{1}{d^2} M_2^2 \nabla_{r_0}^2 \langle V \rangle \, . \end{aligned} \tag{42}$$

$\langle V \rangle$ is the pair potential averaged over the density distribution

$$\langle V \rangle_{GDA}(r_0, \beta) = (2\pi M_2/d)^{-d/2} \int dr' V(r') e^{-d\|r_0 - r'\|^2/2M_2} \, . \tag{43}$$

This effective potential is of the form that appears in the approximate solution of the Liouville equation using Gaussian phase packets. The evaluation of this potential for a protein was discussed in the previous section.

The form of these equations is quite appealing. The centers move according to a steepest descent energy minimization equation on the effective potential energy surface while the widths of the density distribution adjust themselves to the curvature of the effective potential surface. Therefore, the method incorporates the general properties of potential smoothing algorithms. Moreover, the annealing minimization evolves directly in temperature with no real time dynamics. While the equations define a simulated annealing protocol the algorithm is independent of cooling schedule — if the equations of motion are integrated accurately we should have the optimal annealing protocol for that representation of the density distribution.

The strength of the GDA method is that the problematic cooling schedule has been replaced by a set of deterministic equations of motion for the annealing of the density distribution. It is a simple matter to interpret the evolution of the density distribution in temperature. Starting with the known infinite temperature distribution

$$\rho_{\mathrm{eq}}(\beta = 0) = \text{constant} \tag{44}$$

the equations are integrated in temperature to a final value β_{max} which must satisfy $1/\beta_{\mathrm{max}} \ll \Delta E$ where ΔE is the energy gap between the global and next lowest energy minimum (see Fig. 7). For small β, the Gaussian representation of the density distribution should be quite reasonable. In the intermediate β regime, we expect that many minima contribute to the exact equilibrium density distribution of the protein and the Gaussian density approximation will be at best an incomplete representation of the equilibrium ensemble of configurations. However at large β (low enough temperatures), we expect that the equilibrium density distribution will be localized in a single potential minimum and the Gaussian density approximation should be a good representation of the exact equilibrium distribution.

This method has been applied to the standard case of Lennard-Jones clusters ranging from 2 to 55 atoms with excellent results[91] (see Fig. 7). It has also been shown that the evolution of the density distribution in temperature according to the GDA algorithm leads to physically reasonable results for the system density and distribution functions. Initially, at small β (high temperature), the density corresponds to that of a gas. As the temperature is lowered there is a weak phase change to a liquid-like density distribution, followed by a stronger phase change to a solid-like density distribution localized in the global (or a low lying) energy minimum. During the periods of phase change the equations of motion can become stiff;

however, as long as they are integrated accurately there is an adequate amount of time spent in the transition region. This is in contrast to the standard simulated annealing algorithm where one must guess how long to spend in a region of phase change where there are long length scale correlations and relaxation times.

Amara and Straub[100] have applied the GDA algorithm to a model protein.[68] The model consists of 22 residues where each residue is a sphere of neutral (N), hydrophobic (B) or hydrophilic (L) nature. A pair of hydrophobic sites interact with a favorable Lennard-Jones 12-6 potential. A neutral site interacts with any other site through a purely repulsive $1/r^{12}$ interaction. A hydrophilic site interacts with a hydrophilic or hydrophobic residue through a 12-6 type potential where the $1/r^6$ term, as well as the core potential, is purely repulsive. The torsion potential for dihedral angles

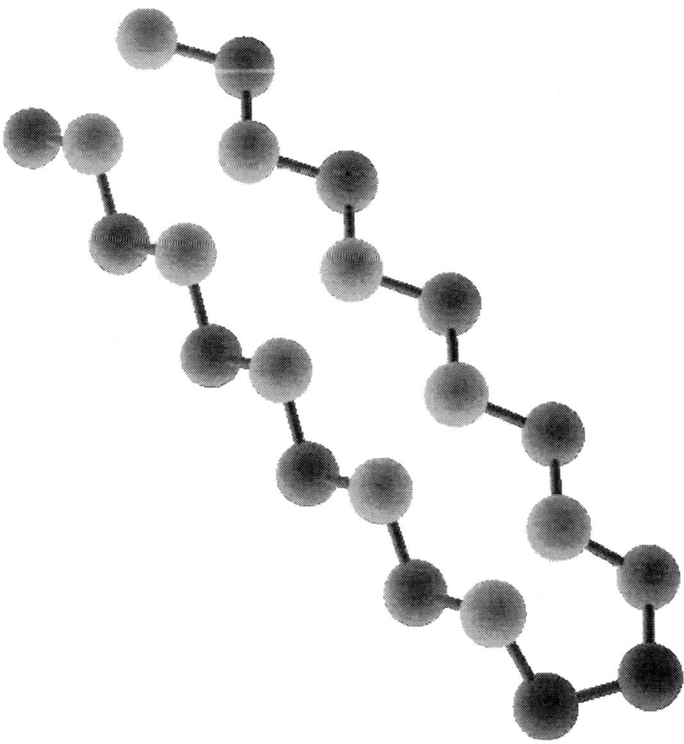

Fig. 8. The global energy minimum of the model protein studied by Amara and Straub using the GDA algorithm.

made-up of two or more neutral residues (in the center of the chain) are floppy, one fold sinusoidal functions while all other dihedrals are more rigid and made of a linear combination of one and three fold sinusoidal terms. The details of this potential can be found elsewhere.[68]

The sequence studied is $B(LB)_5\ N_2\ (LB)_5$ which leads to a global energy minimum configuration of a β-sheet (see Fig. 8). The role of the rigid dihedrals is to form the two linear chain regions while the more flexible central torsion angles allow for the bend to form a β-sheet. The global energy minimum configuration allows for a maximum number of energetically favorable hydrophobic pair contacts. This is a minimal model for protein folding

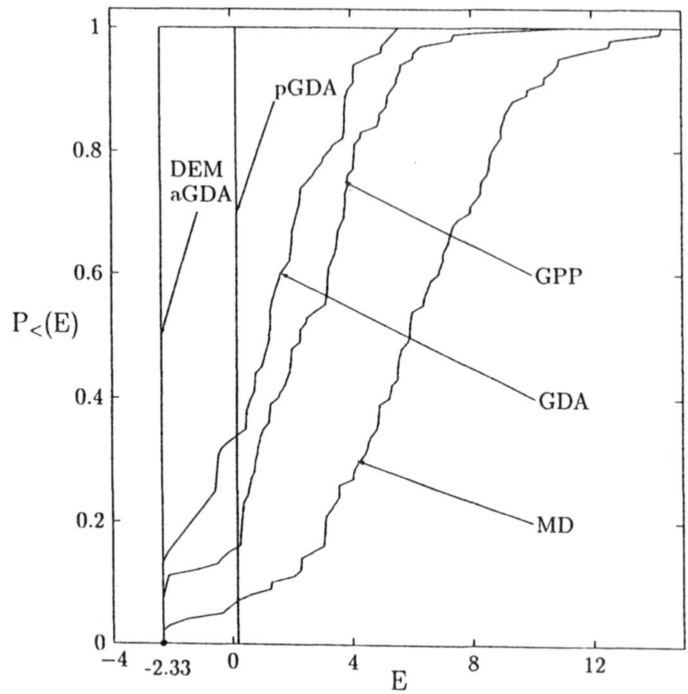

Fig. 9. The probability of locating a configuration with energy E or less from an initial distribution of one hundred independent configurations is plotted for a model protein for five algorithms: simulated annealing in time using molecular dynamics (MD) and Gaussian phase packet dynamics (GPP), simulated annealing in temperature using the Gaussian density annealing (GDA) algorithm, the "adiabatic" GDA algorithm (aGDA), the GDA algorithm using a single "preconditioning" step (pGDA) and the diffusion equation method (DEM) of Scheraga and coworkers.

and represents a serious test of the optimization algorithms. It is a short step from the potential function used for this model protein to the more general all atom empirical potential energy function such as ECEPP or CHARMM.

The results for the energy minimization of this model are presented in Fig. 9.[100] Simulated annealing using molecular dynamics (MD) was applied using an exponential cooling schedule. This schedule is not optimal and was chosen as a standard for the comparison of algorithms. The same cooling schedule was employed in a simulated annealing protocol using the Gaussian phase packet (GPP) dynamics with the rigid temperature constraint. The third method applied was the Gaussian Density Annealing (GDA) algorithm which is free of a cooling schedule. The initial mean squared width for the Gaussian density of each residue was $M_2 = 5$. In each case a weak pairwise harmonic boundary potential was used to encourage the molecule to explore compact configurations and the effective potential of Eq. (27) was used. The results are summarized in Fig. 9 in terms of the probability of finding a minimum energy configuration with energy E or less from a distribution of independent configurations chosen from a high temperature molecular dynamics trajectory. The data indicate that the GPP annealing algorithm performs significantly better than the standard MD simulated annealing for the same cooling schedule. The GPP results locate the global energy minimum more frequently and tend to find significantly lower local energy minima. The GDA algorithm outperforms both the GPP and MD simulated annealing algorithms by isolating the global minimum more frequently and tending to locate only very low energy local minima. However, while the GPP results can be improved by optimizing the cooling schedule (allowing for slower cooling), the GDA algorithm (which is free of any cooling schedule) can be improved only through a better choice of initial conditions and boundary potential or a more complete representation of the density distribution $\rho(r, \beta)$.

5.2. *The Diffusion Equation Method (DEM)*

I find it useful to think of the highly successful DEM of Piela, Kostrowicki and Scheraga[101] as a classical annealing algorithm. In the DEM, a steepest descent minimization is carried out on a transformed potential energy surface. The potential transformation is the convolution of the potential with a Gaussian function of a given width. This leads to a coarse graining of the potential function over a length scale defined by the width of the Gaussian. Initially, the Gaussian width is taken to be large and it is hoped that the transformed potential has one or a few minima. (This can be guaranteed by a clever adjustment of the boundary potential which primarily changes the initial conditions.) Energy minimization is used to find the minimum of the transformed surface. The width is then gradually decreased and the minimum is mapped to an energy minimum of the physical potential surface (where the Gaussian widths are zero).

The transformed potential energy function used in the DEM can be written[102]

$$\langle V \rangle_{DEM}(r,t) = (4\pi t)^{-d/2} \int dr' V(r') e^{-\|r-r'\|^2/4t}, \tag{45}$$

where t is the diffusion "time." The larger the value of t, the wider the Gaussian and the longer the length scale for the potential smoothing. This transformed interaction potential employed in the DEM is exactly the effective potential employed in the GDA algorithm where the squared width of the Gaussian $M_2 = 2td$ is proportional to the diffusion "time" of the DEM. In fact, the *ad hoc* potential deformation used in the DEM results from a fundamental treatment of the temperature evolution of the equilibrium density distribution for the special case of a Gaussian density approximation.

We can take this comparison an extra step. In the GDA algorithm one initially chooses large M_2's and then allows the centers to follow a steepest descent trajectory to the minimum potential energy conformation. In the DEM there is a single diffusion time meaning that all particles are constrained to have the same M_2, and as the diffusion time is reduced, corresponding to $\beta \to \infty$, the effective widths decrease monotonically. In the case of the GDA algorithm the widths are variationally optimized depending on the environment of each particle. The DEM can be thought of as a special case of the GDA algorithm where (1) the second-moments M_2 for all wave packets are identical, and (2) the magnitude of M_2 is initially set to something large and then monotonically decreased to zero while integrating the r_0 equation to converge at each value of M_2. For this reason, it is profitable to think of the DEM as a classical annealing algorithm.

The DEM has been applied to water clusters with mixed results.[103] As has been noted, a Gaussian coarse grained smoothing of the sort employed in the GDA and DEM might fill in a deep narrow well (the global energy minimum) before it fills a shallow broad well. Perhaps this "pathology" is present in the water–water interactions which makes certain cluster sizes difficult optimization problems. Of course, it might be that the higher lying broad *potential* minimum is a lower *free energy* minimum.[104]

In an application of the DEM to the Met-enkephalin pentapeptide (see Fig. 10), the torsional angle potential was set to zero and the bond lengths and angles were fixed. The solution found by the diffusion equation method was one on the manifold of peptide conformational states where the bonds and angles are fixed and the nonbonded (Lennard-Jones and Coulombic) potential is favorable. In this study, Kostrowicki and Scheraga showed that the initial boundary conditions are extremely important in the solution of the DEM. Depending on the form of the boundary potential and the initial diffusion time it is possible to find one or a number of initial conformations and some or none of these can lead to the global energy minimum. For example, a two Gaussian representation of the nonbonded and hydrogen bonding functions of the ECEPP potential was used. After the initial minimization on the transformed potential there was one minimum corresponding to a rather extended conformation. This minimum mapped onto

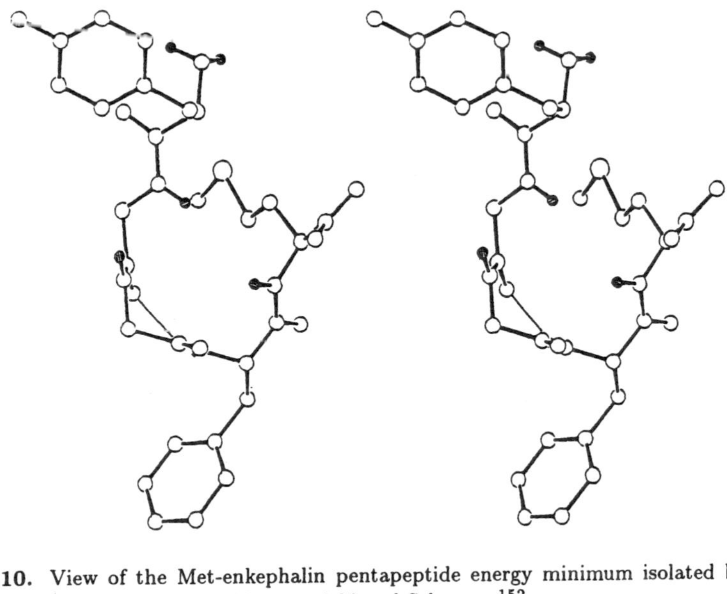

Fig. 10. View of the Met-enkephalin pentapeptide energy minimum isolated by the DEM taken from the study of Kostrowicki and Scheraga.[152]

a local (and relatively high lying) potential minimum for the pentapeptide. However, when a three Gaussian representation of the nonbonded and hydrogen bonding functions was used, two compact initial minima were found for the transformed potential and both of these minima mapped to a single structure which was very close to the form of the global energy minimum. The small differences between the DEM result and the global minimum are due to the absence of the torsional angle potential.

These results demonstrate the crucial importance of the initial conditions used for the DEM. Similar conclusions follow for the application of the GDA method. Their study also demonstrates that with a good choice of boundary conditions the DEM can be a very successful algorithm for energy minimization on the complex potential energy surface of peptides and proteins. This point has been discussed in detail by Scheraga and coworkers.[105] More research needs to be performed to understand the details of the influence of the initial (boundary) conditions for these methods. Another important point is that the DEM is a deterministic procedure, in that once the few minima are initially isolated on the transformed potential surface, there is an exact procedure for mapping these minima on to the physical potential surface. This is a positive aspect of the DEM and GDA methods which those methods built on simulated annealing in time lack.

To allow for comparison with other optimization methods discussed in this chapter, the DEM was applied to the model protein (discussed in the previous section) and the results are displayed in Fig. 9. (The comparison made here is used to demonstrate the general properties of the algorithms and not the absolute best results which can be obtained.) Note that the DEM isolates a single final configuration and that configuration is the global energy minimum. In contrast, the GDA algorithm finds a distribution of minima which includes the global minimum and all low lying "excited" states. In the DEM, the first step consisted of a local energy minimization on the transformed potential surface for a set of one hundred random configurations taken from a high temperature molecular dynamics run. With a large enough initial diffusion time (in this case 1000) and a proper choice of boundary conditions one can expect to find a few (in this case one) energy minima on the transformed potential surface. The DEM then consisted of a single trajectory which mapped the lone minimum energy configuration on the transformed surface on to the global energy minimum of the physical potential surface.

5.3. *Adiabatic Gaussian Density Annealing (aGDA)*

An important variation of the GDA algorithm is the "adiabatic GDA" (aGDA) method. As I noted above in the discussion of the simulated annealing algorithms, the hope is that at each temperature the density distribution simulated is close to the equilibrium distribution. In the primitive GDA algorithm there are errors that develop due to the fact that the basis set for $\rho(r)$ is incomplete. These errors can lead the density distribution away from the equilibrium value. Tsoo and Brooks recognized that such errors propagate and increase during integration.[106] They suggested that this problem could be minimized by occasionally fixing the values of the set of $\{M_2\}$ and running a steepest decent trajectory on the corresponding effective potential surface. This step allows the centers to adjust to the set of packet widths leading to a closer approximation to the "steady state" or hopefully equilibrium solution. The results for a series of Lennard-Jones clusters were excellent. They also applied the modified GDA algorithm to water clusters with interesting results.[106]

In earlier work on a related imaginary time optimization algorithm,[107] it was suggested that the equations of motion for r_0 and M_2 could be decoupled. For the primitive GDA, problems result when the M_2 values decrease too quickly. Therefore, a useful "adiabatic" approximation provides that for every set of $\{M_2\}$ the position of the density distribution center relaxes "instantaneously" to its equilibrium value. The algorithm is applied as follows. (1) The steepest descent equation

$$\dot{r}_o = -\nabla_{r_0}\langle V(r_0, M_2)\rangle \tag{46}$$

is solved with a fixed set of widths $\{M_2\}$ to find the minimum on the effective potential surface for a particular temperature $r_0^{ad}(\beta)$. This can be performed using a favorite local minimizer. (2) The widths of each packet are integrated out in β holding fixed the set of center positions $\{r_0^{ad}\}$ by solving

$$\frac{\partial M_2}{\partial \beta} = -\frac{1}{d^2} M_2^2 \nabla_{r0}^2 \langle V(r_0^{ad}, M_2)\rangle \,. \tag{47}$$

(3) Return to step (1) and repeat the cycle.

The resulting algorithm is very much like the DEM in that the positions of the centers follow changes in the widths adiabatically. We refer to this method as the "adiabatic" GDA algorithm (aGDA). An important difference between this algorithm and the DEM is that each width is allowed to evolve in an optimal way which depends on the curvature of the effective potential surface; in the DEM each particle "density" has the same width. This difference should be particularly important in studies of inhomogeneous systems such as proteins which lack the symmetry of the solutions for the atomic clusters. The results for the aGDA algorithm are shown in Fig. 9. As we found in the case of the DEM, the distribution of initial configurations mapped on to a single energy minimum on the transformed potential surface where the particle widths were $M_2 = 5$. Following the initial minimization, there was a single trajectory which led to the global energy minimum.

It is also possible to perform a "preconditioning" step by solving Eq. (46) for the initial set of $\{M_2\}$ and subsequently running a trajectory according to the usual GDA equations of motion Eq. (42). This preconditioned GDA (pGDA) was applied to the model protein and the results are shown in Fig. 9. Like the aGDA and DEM algorithms, the pGDA algorithm tends to isolate very few minima. However, for this particular run the aGDA and DEM isolated the global energy minimum while the pGDA algorithm found a higher lying local minimum. Similar results were found by Tsoo and Brooks[106] in their optimization study of water clusters. It is important to allow the packet centers to "keep up" with any rapid variation in the packet widths. This seems to explain the success of the aGDA method in finding the global energy minimum.

5.4. *Shalloway's Packet Annealing (PA)*

The strength of both the Gaussian Density Annealing algorithm[91] and the Diffusion Equation Method[101] rests on the use of a similar transformed potential energy function. In the former case, the use of the transformed potential energy follows from a solution of the classical Bloch equation. However, while the lowest potential energy is desirable, thermodynamically, it is the basin of lowest free energy that will predominate at low temperatures. Shalloway has developed an intriguing method which resembles the GDA and DEM algorithms, with the important difference that the optimization problem is solved on a transformed *free energy* surface.[108–110]

The GDA and DEM methods employ a Gaussian transform of the potential energy function. In Shalloway's Packet Annealing (PA) method, the transformed potential is replaced by a Gaussian transform of the Boltzmann factor[108]

$$\langle F\rangle(r) = -\frac{1}{\beta}\ln\left[\left(\pi\Lambda^2\right)^{-d/2}\int dr' e^{-\beta V(r')} e^{-\|r-r'\|^2/\Lambda^2}\right]. \qquad (48)$$

The Boltzmann factor is coarse grained over a region of length scale Λ to provide an effective free energy $\langle F\rangle$. This is simple to write, but difficult to evaluate (except for quadratic potential functions). Nevertheless, Shalloway and coworkers have developed an efficient quadrature method which has been applied to the Lennard-Jones potential in the optimization of atomic clusters.[110]

The PA method begins with the specification of an initial value of the temperature and the smoothing length scale Λ. A schedule for the reduction of the temperature and Λ is then specified. At each temperature, a Monte Carlo search of the configuration space is carried out with a careful choice of step size. The distribution generated by the Monte Carlo search on the effective free energy surface is annealed according to the cooling schedule until a low temperature is reached. The resulting distribution provides a guess at the global energy minimum. The number of Monte Carlo steps taken at a given temperature was varied as $n\log(n)(1 - \log T)$ as the temperature decreased, n being the number of atoms in the cluster. The cooling schedule was defined by decreasing the temperature by a factor of $\xi < 1$ from the temperature at the previous step.[110]

Reasonable results are obtained for the optimization of Lennard-Jones clusters using the strategy described above. Coleman, Shalloway and Wu[110] examined two variations: (1) where a local minimization algorithm was applied at the end of the packet annealing process described above, and (2) where local annealing was performed at the end of each cooling step. The latter algorithm which makes use of the information provided by the whole trajectory, rather than the endpoint above, was the more effective one in our finding the global minima of Lennard-Jones clusters from 3 to 27 atoms as well as larger clusters of 36, 54 and 100 atoms with reasonable probability.

The most appealing aspect of the PA method is that it employs an effective free energy function rather than the potential energy surface. Like the GDA and DEM, the PA method has had success for the optimization of atomic clusters. However, for the optimization of complex molecular systems, calculation of the effective free energy function is expected to be significantly more time consuming. As discussed above, the technical problems are easily solved for the calculation of $\langle V\rangle$, required by the GDA and DEM algorithms, from an empirical energy function of a protein. One expects to have a more difficult time in the calculation of $\langle F \rangle$. Recent application of the PA method to met-enkephalin has provided very encouraging results.[155]

6. Quantum Mechanical Annealing

The dynamics of classical and quantum mechanical systems are qualitatively different. To a classical point particle, energy barriers are impregnable. Energy barriers must be crossed over, and if the total energy for the system is less than the barrier energy, the barrier cannot be crossed. However, a quantum mechanical particle may tunnel through a barrier whose height is greater than the total energy of the system. As a result, quantum mechanical annealing methods are particularly attractive as methods for global optimization (see Fig. 1).

I will mention several approaches to this problem. Each method is an extension of a well developed method for finding the ground state wave function of a quantum mechanical system. The hope is that for a system with a nondegenerate ground state, the density $||\phi(r)||^2$ associated with the wave function $\phi(r)$ will be localized in the region of the classical global energy minimum. The optimization process is then approached in two steps, (1) an estimate for the ground state wave function is found, and (2) the classical limit is taken (by reducing $\hbar$ to zero, or increasing the mass, or simply reducing the ratio $\hbar^2/2m$ to zero).

6.1. *Variational Calculations*

The most successful methods for finding an estimate of the ground state wave function of an atomic or molecular system rely on the time-indepen -dent variational method.[111] Given a system where the ground state energy is $\mathcal{E}_0$, any guess $\phi(r)$ at the wave function of the system will have an energy

$$E = \langle \phi | \mathcal{H} | \phi \rangle \geq \mathcal{E}_0 \, , \tag{49}$$

where $\mathcal{H}$ is the Hamiltonian operator. That is, the energy of the guess is always greater than or equal to the exact ground state energy.

Somorjai proposed using a distributed fixed Gaussian basis set.[112] While this suggestion is a reasonable approach for one-dimensional problems,[113] there are reasons to doubt the efficacy of such an approach for larger systems of hundreds of atoms which can extend over large regions of configuration space. For a long chain molecule, extended states of the system may explore large volumes while occupying relatively little of it at any given time. The space fixed basis set is everywhere, all the time. As such, it can spend significant amounts of computational time monitoring empty space. This problem can be addressed by exploring mobile basis sets that are functions of atomic or internal molecular coordinates.

Olszewski, Piela and Scheraga[114] have developed a method for finding the global energy minimum of a protein by approximating the ground state wave function of the system. Their Self-Consistent Multi-Torsional Field (SCMTF) method is based on approximating the protein wave function by a Hartree product of normalized single torsion angle wave functions $\phi_k(\theta_k)$, leading to a set of coupled one-dimensional time-independent Schrödinger equations with the effective Hamiltonian

$$\mathcal{H}_k = -\frac{\hbar^2}{2I_k}\frac{\partial^2}{\partial\theta_k^2} + \langle V\rangle_k(\theta_k) \tag{50}$$

where I_k is an averaged moment of inertia for the kth torsion angle, and $\langle V\rangle_k(\theta_k)$ is the mean field potential energy for the kth torsion angle — the potential energy averaged over all degrees of freedom other than θ_k. The minimization procedure consists of (1) generating an initial configuration of the protein (a set of dihedral angles), (2) estimating the various

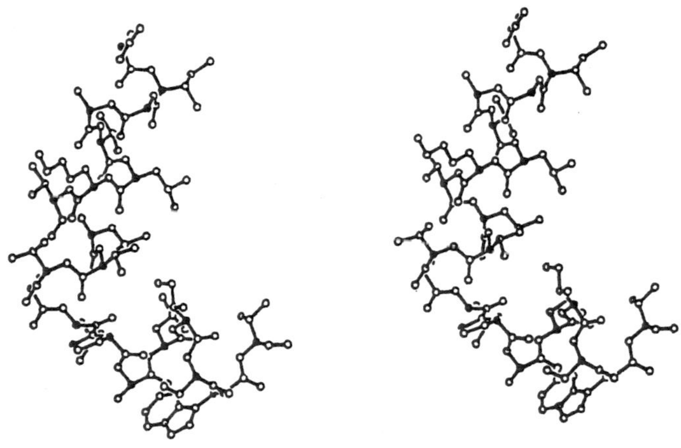

Fig. 11. View of the global energy minimum configuration of melittin isolated using the SCMTF method taken from the study of Olszewski, Piela and Scheraga.[116]

moments of inertia I_k from an ensemble of configurations, (3) calculation of the mean field effective potential $\langle V\rangle_k(\theta_k)$ where the required integrals are computed using a Monte Carlo procedure, and (4) solution of the set of coupled one-dimensional Schrödinger equations, iterating until the estimates of the $\phi_k(\theta_k)$ no longer change. The lowest energy configuration found in this procedure is the guess at the global minimum. The SCMTF method has been applied to Met-enkephalin,[114] decaglycine and icosalanine,[115] and a 20 residue portion of the protein melittin.[116] For melittin the SCMTF method effectively located the global potential energy minimum for the system from four independent starting configurations (see Fig. 11).

6.2. *Imaginary (Euclidean) Time Methods*

The time-dependent Schrödinger equation is[117]

$$i\hbar\frac{\partial}{\partial t}\phi(r,t) = -\frac{\hbar^2}{2m}\nabla^2\phi(r,t) + V(r)\phi(r,t) \tag{51}$$

where $\phi(r,t)$ is in general a complex function. $\hbar$ is Planck's constant, $V(r)$ is the potential energy, and m is the mass. The imaginary (Euclidean) time form of the time-dependent Schrödinger equation is defined by substituting the imaginary time variable $\tau = it/\hbar$ so that

$$\frac{\partial}{\partial \tau}\phi(r,\tau) = \frac{\hbar^2}{2m}\nabla^2\phi(r,\tau) - V(r)\phi(r,\tau)\,. \tag{52}$$

We arbitrarily take $\phi(r,\tau)$ to be real. The wonderful property of this equation is that the formal solution is of the form

$$\phi(r,\tau) = e^{-\mathcal{H}\tau}\phi(r,0)\,. \tag{53}$$

Suppose we know the eigenfunctions and energies for $\mathcal{H}$ which are $u_n(r)$ and $\mathcal{E}_n$ for the nth excited state. We can rewrite the wave function as

$$\phi(r,\tau) = \sum_n c_n u_n(r) e^{-\mathcal{E}_n\tau}\,. \tag{54}$$

At large enough values of τ the dominant contribution to the wave function comes from the ground state

$$\phi(r,\tau) \simeq u_0(r) e^{-\mathcal{E}_0\tau} \tag{55}$$

while the contribution from all excited states is exponentially smaller. Starting from a guess at $\phi(r,\tau)$, the imaginary time method consists of integrating Eq. (52) to find an estimate of the ground state wave function. This method is widely employed and can be used to find accurate ground and excited state energies of small molecular systems.

6.2.1. *Quantum Mechanical Annealing using Diffusion Monte Carlo (DMC)*

As Schrödinger originally noted, the imaginary time Schrödinger equation has the familiar form of the classical diffusion equation

$$\frac{\partial}{\partial\tau}C(r,\tau) = \mathcal{D}\nabla^2 C(r,\tau) - kC(r,\tau) \tag{56}$$

with concentration $C(r,\tau)$, diffusion constant $\mathcal{D}$, and rate constant k (for the first-order rate process).[118] As this analogy makes clear, one can simulate the imaginary time Schrödinger equation as a classical diffusion equation for a distribution of system replicas (a concentration profile or density distribution). The first-order rate process leads to the spawning and death of system replicas. In practice, one simulates an ensemble of replicas of the system. Each replica diffuses through a Monte Carlo walk in configuration space. At each step a decision is made whether to spawn or kill system replicas based on the rate constant $k = V(r)$.[119,118]

Doll and coworkers have recently adapted the DMC method[119,120] to treat the classical optimization problem. They apply the DMC to determine the ground state wave function and energy at a given value of $\hbar^2/2m$. The mass of the particles is then gradually increased, each time applying the DMC method, until the classical limit is reached. At this point, they arrive at an estimate of the global energy minimum of the system. Their method has been applied to Lennard-Jones atomic clusters with perfect results for clusters of up to 19 atoms.[121]

The use of DMC is particularly attractive since (1) no approximation to the wave function is required, and (2) no effective potential (or matrix elements) or derivatives are required, only evaluations of the potential energy function. It appears to be a very attractive method for applications to biomolecules where an empirical energy function is complicated by the mix of terms depending on Cartesian and internal coordinates (see Sec. 4.1.1).

There are two possible drawbacks. One is that DMC is exact and becomes more computationally intensive than an approximate solution. Moreover, there is the difficulty of applying Monte Carlo to proteins in an efficient manner. It is well-known that a simple Metropolis algorithm of random atomic moves distributed uniformly in a cube will typically lead to an increase in energy due to the stiff bond and angle constraints. However, more intelligent algorithms have been designed which make use of internal coordinates or normal modes[122,123] and these appear to be quite promising. Nevertheless, molecular dynamics remains the paradigm for the simulation of thermodynamic properties of proteins. For these reasons, it is important to develop an approximate method for solving the imaginary time Schrödinger equation based on the solution of deterministic equations of motion (akin to molecular dynamics). One such method is based on Gaussian wave packet dynamics in imaginary time.

6.2.2. *Quantum Mechanical Annealing using Gaussian Wave Packets*

The formal solution of the imaginary time Schrödinger equation is $\phi(r,\tau) = \exp(-\mathcal{H}\tau)\phi(r,0)$. The form of the equilibrium distribution function is $\exp(-\mathcal{H}/k_BT)$, where k_B is Boltzmann's constant and T is the temperature. Here, time plays the role of an inverse temperature. Lowering the temperature to anneal the system is equivalent to increasing the imaginary time. Amara, Hsu and Straub have exploited this idea in searching for the global energy minimum through approximate solution of the imaginary time Schrödinger equation for the ground state wave function using Gaussian wave packets.[107]

The starting point is an alternative normalized expression which is a variation of Eq. (52), namely

$$\frac{\partial}{\partial\tau}\phi(r,\tau) = -\left(\mathcal{H} - \langle\mathcal{H}\rangle\right)\phi(r,\tau)\,. \tag{57}$$

If we take the wave function of each particle to be a Gaussian

$$\phi(r,\tau) = (2\pi M_2/d)^{-d/4}\exp\Big[-\frac{d}{4M_2}(r-r_0)^2\Big] \tag{58}$$

we can fully define the Gaussian by its average position $r_0 = \langle r\rangle$, and its squared width or second moment, $M_2 = \langle(r-r_0)^2\rangle = d\sigma^2$. It is a simple matter to derive equations of motion for the evolution of the GWP in imaginary time[107]:

$$\dot{r}_0 = -\frac{2}{d} M_2 \nabla_{r_0} \langle V \rangle \,, \tag{59}$$

$$\dot{M}_2 = \frac{d\hbar^2}{2m} - \frac{2}{d^2} M_2^2 \nabla_{r_0}^2 \langle V \rangle \,.$$

These are the optimal equations of motion for the Gaussian which minimizes the square of the error between the approximate and exact solutions. (1) As has been noted by Heller, in real time the equations of motion for a GWP have no explicit dependence on $\hbar$.[124] However, the imaginary time equations do depend explicitly on $\hbar$. (2) The evolution of the packet center resembles that of a steepest descent[44] down the effective potential $\langle V \rangle$

$$\dot{r}_0 \propto -\nabla_{r_0} \langle V \rangle \,. \tag{60}$$

The term $d\hbar^2/2m$ acts to expand the wave packet and is a measure of the quantum nature of the particle. For an effectively classical system we expect $\hbar^2/2m$ to be small. If we choose, we can increase this parameter to encourage the system to delocalize and tunnel. (3) The packet will evolve to minimize the effective *total* energy, a sum of the potential $\langle V(r_0, M_2) \rangle$ and kinetic $d\hbar^2/8mM_2$ energies (see Fig. 12). Larger values of $\hbar$ increase the importance of the kinetic energy and raise the zero point energies of the well minima. The effective total energy surface has many of the features of quantum effective potentials which may be used in the simulation of quantum systems.[58,125] (4) For a many-body system we approximate the total wave function by a Hartree product of N single particle wave functions

$$\psi(r^N, \tau) = \prod_{k=1}^{N} \phi_k(r_k, \tau) \,. \tag{61}$$

The effective potential energy can be written as a sum over the individual pair interaction energies

$$\langle V \rangle = \int dr \prod_k \rho_k(r_k) V(r) \tag{62}$$

where the density of a given particle is $\rho_i(r_i) \equiv \phi_i^*(r_i)\phi_i(r_i)$ and $V(r)$ is averaged over the positions of all the other particles — a mean field potential.[126] This method has been very successful in locating the global energy minimum for a series of Lennard-Jones clusters.[107]

The difficulty in applying this method to biomolecules is that the effective potential $\langle V \rangle$ must be calculated. When the GWP is written in Cartesian coordinates, the effective potential for the Coulombic interaction may be calculated exactly. The Lennard-Jones potential (or an ion induced dipole interaction, etc.) must be fit to a series of Gaussians. The dihedral potential may be expressed in terms of a polynomial in the distance $r_{i,i+4}$ to perform the integrals in Eq. (62). So, it is possible to evaluate $\langle V \rangle$ for the standard empirical energy function for a protein, but with some added complexity when compared with the DMC method which requires only an evaluation of the empirical energy function itself (see Sec. 4.1.1).

The appeal of the imaginary time method is that the equations of motion are deterministic and, due to the small number of parameters which must be followed, the integration is readily performed for a biomolecule or polymer. One is relieved of the burden of finding clever Monte Carlo moves which efficiently explore configuration space. A natural and promising extension of the methods discussed in this section involves quantum mechanical annealing using a path integral representation of the protein.[127,128]

6.3. *"Tunneling" in General*

In an early study of this kind of quantum mechanical annealing, Ruján adapted the Green Function Monte Carlo (GFMC) technique to create an algorithm called "simulated tunneling."[129] In the GFMC method, an ensemble of systems (or population of configurations) is generated, and each member of the ensemble is allowed to diffuse until an equilibrium fluctuating population of configurations is found. The end result is an estimate of the energy and wave function of the quantum mechanical ground state.

Since the goal is to find the classical global energy minimum, and not the quantum mechanical ground state energy, Ruján made some simplifications of the GFMC method inspired by classical simulated annealing which led to his simulated tunneling algorithm. (1) An ensemble of systems is generated and a cooling schedule is chosen. (2) Each system in the ensemble is allowed to diffuse following a Monte Carlo procedure as the temperature is lowered according to a cooling schedule. (3) If a new minimum is located by the ensemble, return to (1) and generate a new ensemble of states; if not, stop.

Ruján applied his method to the traveling salesman problem, with apparently excellent results. He observed that this algorithm, which employs a diffusing population of systems, resembles not only imaginary time quantum mechanical evolution and classical reaction diffusion dynamics, but also evolutionary systems, and computational ecologies.

Reflecting on our discussion of the GDA method, we can recognize that while a single classical point particle cannot tunnel through an energy barrier, a distribution of classical particles can. This is evident in Fig. 12 where the barrier in the effective total energy is lopped off even in the classical $\hbar = 0$ limit. With a distribution of particles, even if the *average* energy of the distribution is less than the barrier energy, the distribution can cross the barrier. This is a form of "classical tunneling" which provides a mechanism for barrier crossing in simulated annealing optimization employing a density distribution (of the GPP or GDA sort) rather than a single point particle.

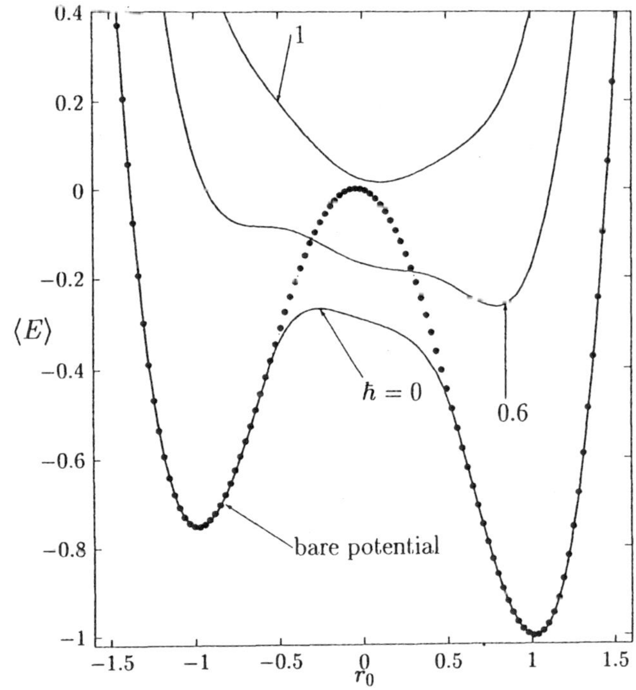

Fig. 12. The effective energy of the quartic double well potential is plotted for three values of $\hbar = 0$ (classical limit), 0.6 and 1 in Lennard-Jones reduced units. The effective energy is determined as the sum of the kinetic energy, $d\hbar^2/8mM_2$, and the effective potential, $\langle V(r_0, M_2^{ad})\rangle$, using the optimal value of M_2 at each value of r_0.

7. Sampling with Monte Carlo or Molecular Dynamics

In this review, I have tried to emphasize what a variety of successful and promising optimization strategies have in common. I have also stated what I think are the important differences. Occasionally, I have referred to the usefulness of having a deterministic set of equations of motion to integrate, as opposed to a function to sample using Monte Carlo. Also, some of the techniques presented involve some form of molecular dynamics or Monte Carlo calculation. As such, it seems worthwhile to mention some important developments in MC and MD methodology.

A great deal of work has been done since the first study of the effectiveness of Monte Carlo sampling for proteins.[130] Currently, there is some disagreement over the effectiveness of Brownian dynamics,[67] noisy molecular dynamics,[68] or molecular dynamics[131] when applied to large organic molecules and proteins.[132–135] Some evidence supports the use of improved Monte Carlo sampling using internal coordinate (torsion angle) moves[122,123] and Monte Carlo schemes coupled with energy minimization[136] which have led to significantly improved results over simulated annealing strategies[33] in applications to enkephalin.[136] However, most of this discussion has occurred before recent improvements in the numerical efficiency of MC and MD calculations on proteins.

A problem facing MC or MD simulations of complex systems is related to "broken ergodicity." The presence of high energy barriers separating regions of configuration space can cause a trajectory (Monte Carlo or molecular dynamics) to be locked into one phase of the system.[13] It may well be that there are phases of lower free energy meaning that the simulated phase is a supercooled state of the system.[137] A number of advances in Monte Carlo methodology which address the problem of broken ergodicity[138] have been reported in recent years. Cao and Berne have developed an "anti-force bias" method which helps to accelerate barrier crossing.[139] Ferrenberg and Swendsen,[140] and Frantz, Freeman and Doll[141] have proposed methods which use a high temperature MC run, where barriers may be crossed easily, overcoming problems of broken ergodicity, to compute averages at a lower temperature of interest. These latter methods have been combined by Tsai and Jordan to examine phase changes in small rare gas and water clusters.[142] Extended electrostatics techniques,[143,144] like potential smoothing methods, enhance both MD and MC calculation.

More recently, Andricioaei and Straub[156] have developed an enhanced sampling algorithm which uses a Monte Carlo protocol to explore a generalized statistical distribution of the form proposed by Tsallis

$$p(r) \propto [1 - (1 - q)\beta V(r)]^{\frac{1}{1-q}}$$

where $V(r)$ is the potential energy and "q" is a parameter. When $q \to 1$ the standard Gibbs–Boltzmann statistics are recovered. Otherwise, it has been shown that for $q > 1$ the sampling is greatly enhanced as is the probability of locating the global energy minimum.[156] The mechanism underlying the enhanced sampling is a dramatic increase in the probability of being in a barrier region relative to being in a minimum. The result is an increase in the rate of barrier crossing and sampling phase space. This is similar to the effect of tunneling in quantum annealing methods and barrier lowering due to potential smoothing.

There have also been significant advances in molecular dynamics methods. Multiple time step algorithms[145] can enhance molecular dynamics calculations on proteins by at least a factor of 4 to 5 over traditional integration methods.[146] This should be important not only for extending simulation times in the study of dynamics, but also in the use of MD for configurational space sampling.

Furthermore, molecular dynamics in higher dimensions, which provides a mechanism for lowering barriers, is akin to potential smoothing.[147] The transform from three to four dimensions appears as

$$r^2 \to r^2 + w^2 = x^2 + y^2 + z^2 + w^2 \tag{63}$$

and is easily implemented in a standard molecular dynamics calculation. A boundary potential may be placed on w to control its fluctuation. Fluctuations of the parameter w in the fourth dimension can help to lower barriers in the system by acting as an effective "random smoothing" of the potential in three dimensions. This method looks a great deal like the potential shift algorithm of Piela and coworkers mentioned previously, and has some conceptual similarity to distance geometry[28,148] and energy embedding procedures.[149,150] Similar effects have been achieved through coupling of the system to an "energy bath" which has a well-defined reference potential energy.[151]

Clearly, there are many possibilities involving both MD and MC techniques, combined with some form of potential smoothing and modern algorithms for integration and updating of nonbonded interaction lists, which will lead to improved algorithms for enhanced sampling of proteins.

8. Future Directions

There are many important directions for future research into optimization algorithms for proteins. This chapter was written to give an overview, however incomplete, of a number of areas where progress has been made and where improvements are likely to follow. In addition to many areas in the field of optimization not discussed in this chapter, I believe there is no method discussed here which can be described as finished or complete. As such, there remain many interesting unanswered questions.

The work discussed in this chapter points to the importance of potential smoothing techniques in molecular optimization algorithms. While smoothing methods may be applied *ad hoc* to take advantage of the special properties of a given problem, it is also possible to develop algorithms which represent the system using continuous classical density distributions or quantum mechanical wave functions which display the positive properties of smoothing methods. The quantum mechanical annealing methods have the further advantage of being able to tunnel to a lower energy minimum, although algorithms based on the purely classical density distribution display generalized "tunneling" as well — an ensemble of states may pass over a barrier whose energy is higher than the mean energy of the ensemble.

The GDA algorithm, which is essentially an annealing algorithm but which evolves in temperature rather than time, has the desirable property of being free of a cooling schedule. This property is shared by the quantum mechanical algorithms which evolve in imaginary time. It is easy to imagine a variety of quantum mechanical annealing methods beyond the wave packet and DMC methods discussed here and these methods seem well worth exploring.

Some of the algorithms discussed are deterministic and their results are strongly dependent on the choice of boundary conditions. An important area for future work on methods such as the DEM and aGDA algorithms is in specifying improved boundary conditions and understanding in detail their link to the final minimum energy structures isolated. Other methods such as the GDA and LES algorithms consist of deterministic equations of motion which evolve structures from a random set of initial conditions. In this way these algorithms have a non-deterministic component — the choice of initial conditions. This randomness can be seen as a complication, requiring more effort to solve the set of equations, but can also be of some value in providing a distribution of low energy structures.

An important question concerns the relative importance of new potentials for proteins and optimization algorithms.[15] Clearly, the development of coarse grained or minimal models of proteins is an important direction of research. These model potentials have the benefit of being considerably simplified when compared to all atom models. At the same time, the optimization problem remains, even when using model potentials. Developments in each area will represent important contributions to the effort to identify effective algorithms to explore protein folding.

Acknowledgments

I have enjoyed discussions with my students Patricia Amara and Jianpeng Ma who have made many contributions to the work summarized in this chapter. Special thanks to David Hsu for his numerous insightful observations and helpful comments. Thanks to Bruce Berne, Anders Wallqvist and Dave Thirumalai for comments on this manuscript. I thank Harold Scheraga for providing me with copies of Figs. 10 and 11 and the American Chemical Society for permission to use them. This work was supported in part by a grant from the donors of the Petroleum Research Fund, administered by the American Chemical Society, and a grant from the National Science Foundation (CHE-9306375).

References

1. C. Levinthal, in *Mossbauer Spectroscopy in Biological Systems* eds. P. Debrunner, J. C. M. Tsibris and E. Münck (Univ. of Illinois Press, Urbana-Champaign, 1969) p. 22.
2. G. D. Fasman, *Prediction of Protein Structure and the Principles of Protein Conformation* (Plenum, New York, 1989).
3. T. E. Creighton, *Protein Folding* (Freeman, New York, 1992).
4. N. Gō, *Ann. Rev. Biophys. Bioeng.* **12**, 183 (1983), N. Gō and H. Abe, *Adv. Biophys.* **18**, 149 (1984).
5. J. D. Bryngelson and P. G. Wolynes, *J. Phys. Chem.* **93**, 6902 (1989); J. D. Bryngelson and P. G. Wolynes, *Biopolymers* **30**, 177 (1990), *Proc. Natl. Acad. Sci. USA* **84**, 7524 (1987).
6. Z. Guo, D. Thirumalai and J. D. Honeycutt, *J. Chem. Phys.* **97**, 525 (1992).
7. M. Karplus and E. Shakhovich, in *Protein Folding*, ed. T. E. Creighton (Freeman, New York, 1992).

8. I will not attempt to provide a summary of Scheraga's many contributions to this problem. Fortunately, he has written a number of clear and recent reviews which I can recommend as a summary of the many interesting approaches to the global energy minimization of proteins which do not fit the scope of this chapter. See H. A. Scheraga, *Rev. Comp. Chem. III*, 73 (1992), eds. K. B. Lipkowitz and D. B. Boyd; *Theor. Biochem. Mol. Biophys.* **2**, 231 (1991); *Chemica Scripta* **29A**, 3 (1989); *Ann. N.Y. Acad. Sci.* **439**, 170 (1985).
9. M. B. Smith, *Aust. J. Biol. Sci.* **17**, 261 (1964); M. B. Smith and J. F. Back, *ibid.* **18**, 365 (1965); **21**, 539 (1968); **21**, 549 (1968).
10. H. Frauenfelder, S. G. Sligar and P. G. Wolynes, *Science* **254**, 1598 (1991); H. Frauenfelder, F. Parak and R. D. Young, *Ann. Rev. Biophys. Chem.* **17**, 451 (1988).
11. R. Elber and M. Karplus, *J. Am. Chem. Soc.* **112**, 9161 (1990); R. Czerminski and R. Elber, *Proteins* **10**, 70 (1991).
12. T. Noguti and N. Gō, *Proteins* **5**, 97 (1989); *ibid.* **5**, 104 (1989); *ibid.* **5**, 113 (1989); *ibid.* **5**, 125 (1989); *ibid.* **5**, 132 (1989).
13. J. E. Straub, A. Rashkin and D. Thirumalai, *J. Am. Chem. Soc.* **116**, 2049 (1994); J. E. Straub and D. Thirumalai, *Proc. Natl. Acad. Sci. USA* **90**, 809 (1993); J. E. Straub and D. Thirumalai, *Proteins* **15**, 360 (1993).
14. C. Camacho and D. Thirumalai, *Proc. Natl. Acad. Sci. USA* **90**, 6369 (1993).
15. A. Sali, E. Shakhnovich and M. Karplus, *Nature* **369**, 248 (1994).
16. H. Frauenfelder, S. G. Sligar and P. G. Wolynes, *Science* **254**, 1598 (1991).
17. E. I. Shakhnovich and A. M. Gutin, *Biophys. Chem.* **34**, 187 (1989).
18. E. Aarts and J. Korst, *Simulated Annealing and Boltzmann Machines* (Wiley, New York, 1990).
19. The term $\mathcal{NP}$-complete stands for "non-deterministic polynomial time" referring to the fact that problems in this class do not have solutions which can be found using a deterministic search algorithm in a time which scales as a polynomial of N, where N is a measure of the size of the problem.
20. P. G. Wolynes, in *Biologically Inspired Physics*, ed. L. Peliti (Plenum Press, New York, 1991) p. 15.
21. E. L. Lawler, J. K. Leustra, A. H. G. Rinnooy Kan and K. B. Shmoys, *The Travelling Salesman Problem* (Wiley, New York, 1985).
22. G. Baskaran, Y. Fu and P. W. Anderson, *J. Stat. Phys.* **45**, 1 (1986).
23. Nevertheless, as Scheraga has stated "It is an article of faith that we are looking for the global minimum." See H. A. Scheraga, *Alfred Benzon Symp.* **28**, 343 (1989).
24. S. Kirkpatrick, Jr., C. D. Gelatt and M. P. Vecchi, *Science* **220**, 671 (1983).
25. R. Czerminski and R. Elber, *Int. J. Quant. Chem.* **24**, 167 (1990); R. Elber and M. Karplus, *Chem. Phys. Lett.* **139**, 375 (1987).
26. R. Czerminski and R. Elber, *Proc. Natl. Acad. Sci. USA* **86**, 6963 (1989); R. Czerminski and R. Elber, *J. Chem. Phys.* **92**, 5580 (1990).
27. S. Fischer and M. Karplus, *Chem. Phys. Lett.* **96**, 5272 (1992).

28. T. F. Havel and K. Wütrich, *Bull. Math. Biol.* **46**, 673 (1984); *J. Mol. Biol.* **182**, 281 (1985).
29. G. M. Clore, A. M. Gronenborn, A. T. Brünger and M. Karplus, *J. Mol. Biol.* **186**, 435 (1985).
30. L. Nilsson, G. M. Clore, A. M. Gronenborn, A. T. Brünger and M. Karplus, *J. Mol. Biol.* **188**, 455 (1986).
31. A. T. Brünger, G. M. Clore, A. M. Gronenborn and M. Karplus, *Proc. Natl. Acad. Sci. USA* **83**, 3801 (1986).
32. H. J. Kushner, *SIAM J. Appl. Math.* **47**, 169 (1987).
33. A. Nayeem, J. Vila and H. A. Scheraga, *J. Comput. Chem.* **12**, 594 (1991).
34. R. Brower, G. Vasmatzis, M.. Silevrman and C. Delisi, *Biopolymers* **33**, 329 (1993).
35. M. R. Hoare and J. A. McInnes, *Adv. in Phys.* **32**, 791 (1983).
36. F. H. Stillinger and T. Weber, *J. Stat. Phys.* **52**, 1429 (1988). The effect on the number of minima of coarse graining the potential hypersurface was discussed earlier in F. H. Stillinger, *Phys. Rev.* B**32**, 3134 (195).
37. A recent study of C. J. Camacho and D. Thirumalai, *Phys. Rev. Lett.* **71** 2505 (1993), suggests that while the number of compact states of a protein grows exponentially with the number of amino acids, the number of *minimum energy* compact states is far smaller. They suggest that these minimum energy states act as basins of attraction in the intermediate stages of protein folding, and that the native state is reached from one of these minimum energy compact states.
38. F. H. Stillinger and D. K. Stillinger, *J. Chem. Phys.* **93**, 6106 (1990).
39. P. A. Braier, R. S. Berry and D. J. Wales, *J. Chem. Phys.* **93**, 8745 (1990).
40. R. S. Berry, *Chem. Rev.* **93**, 2379 (1993).
41. P. E. Gill, W. Murray and M. H. Wright, *Practical Optimization* (Academic Press, San Diego, 1981).
42. D. A. Pierre, *Optimization Theory with Applications* (Dover, New York, 1986).
43. T. Schlick, in *Rev. Comp. Chem.*, Vol. III, eds. K. B. Lipkowitz and D. B. Boyd, (VCH, New York, 1992) p. 1.
44. S. A. Teukolsky, W. H. Press, B. P. Flannery and W. T. Vetterling, *Numerical Recipes: The Art of Scientific Computing* (Cambridge Univ. Press, Cambridge, 1986).
45. M. R. Hoare and P. Pal, *Adv. Phys.* **20**, 161 (1971).
46. J. A. Northby, *J. Chem. Phys.* **87**, 6166 (1987).
47. C. D. Maranas and C. A. Floudas, *J. Chem. Phys.* **97**, 7667 (1992).
48. The focus of this chapter, which is methods for global energy minimization of macromolecules, is a subset of the large class of hard optimization problems and as such the "general" methods discussed in the remaining sections may appear to many to be tailored to problems of molecular optimization. However, all of those methods are in fact easily generalized to problems such as that of the traveling salesman.

49. J. Pillardy, K. A. Olszewski and L. Piela, *J. Phys. Chem.* **96**, 4337 (1992).
50. M. Dygert. N. Gō and H. A. Scheraga, *Macromol.* **8**, 750 (1975); K. A. Palmer and H. A. Scheraga, *J. Comp. Chem.* **12**, 505 (1991); *ibid.* **13**, 329 (1992).
51. Q. Zheng, R. Rosenfeld, S. Vajda and C. DeLisi, *J. Comput. Chem.* **14**, 556 (1993).
52. Q. Zheng, R. Rosenfeld, S. Vajda and C. DeLisi, *Protein Science* **2**, 1242 (1993).
53. R. Rosenfeld, Q. Zheng, S. Vajda and C. DeLisi, *J. Mol. Biol.* **234**, 515 (1993).
54. Z. Liu and B. J. Berne, *J. Chem. Phys.* **99**, 6071 (1993).
55. H. C. Andersen, *J. Chem. Phys.* **72**, 2384 (1980).
56. S. Nosé, *J. Chem. Phys.* **81**, 511 (1984).
57. W. G. Hoover, *Phys. Rev.* A**31**, 1695 (1985).
58. Similar effects are seen in the difference in structure between classical and quantum mechanical representations of atomic fluids using path integral methods or quantum effective potentials of the Feynman-Hibbs or Wigner-Kirkwood form. See D. Thirumalai, R. Hall and B. J. Berne, *J. Chem. Phys.* **81**, 2523 (1984). As we will see in Sec. 5, the use of a quantum effective potential which allows for tunneling can lead to effective global optimization routines.
59. T. Head-Gordon, F. H. Stillinger and J. Arrecis, *Proc. Natl. Acad. Sci.* **88**, 11076 (1991).
60. T. Head-Gordon and F. H. Stillinger, *Biopolymers* **33**, 293 (1993).
61. J. R. Gunn, A. Monge, R. A. Friesner and C. H. Marshall, *J. Phys. Chem.* **98**, 702 (1994).
62. M. Karplus and D. Weaver, *Biopolymers* **18**, 1421 (1979); D. Bashford, D. L. Weaver and M. Karplus, *J. Biomol. Struct. Dyn.* **1**, 1243; D. Bashford, F. E. Cohen, M. Karplus, I. D. Kuntz and D. L. Weaver, *Proteins* **4**, 211 (1988).
63. D. Vanderbilt and S. G. Louie, *J. Comput. Phys.* **56**, 259 (1984).
64. L. T. Wille, *Letters to Nature* **324**, 46 (1986).
65. L. T. Wille, *Chem. Phys. Lett.* **133**, 405 (1987).
66. J. D. Honeycutt and D. Thirumalai, *Proc. Natl. Acad. Sci. USA* **87**, 3526 (1990).
67. J. Rey and J. Skolnick, *J. Chem. Phys.* **158**, 199 (1991).
68. J. D. Honeycutt and D. Thirumalai, *Biopolymers* **32**, 695 (1992).
69. R. C. Tolman, *The Principles of Statistical Mechanics* (Dover, New York, 1979).
70. D. A. McQuarrie, *Statistical Mechanics* (Harper and Row, New York, 1976).
71. M. P. Allen and D. J. Tildesley, *Computer Simulation of Liquids* (Oxford, Bristol, 1990).
72. B. J. Berne and R. Pecora, *Dynamic Light Scattering* (Wiley-Interscience, New York, 1976).

73. R. W. Zwanzig, in *Lectures in Theoretical Physics*, Vol. III, eds. W. E. Britton, B. W. Downs and J. Downs (Wiley Interscience, New York, 1961) p. 106; see also R. Zwanzig, *J. Chem. Phys.* **33**, 1338 (1960).
74. J. Ma, D. Hsu and J. E. Straub, *J. Chem. Phys.* **99**, 4024 (1993).
75. J. Grad, J. Y. Yan and S. Mukamel, *Chem. Phys. Lett.* **134**, 219 (1987); J. Grad, Y. J. Yan, A. Haque and S. Mukamel, *J. Chem. Phys.* **86**, 3441 (1987).
76. S. Mukamel, *J. Phys. Chem.* **88**, 3185 (1984).
77. R. D. Coalson and M. Karplus, *J. Chem. Phys.* **79**, 6150 (1983); *ibid.* **81**, 2891 (1984); *ibid.* **93**, 3919 (1990). See also A. D. McLachlan, *Mol. Phys.* **8**, 39 (1964); E. J. Heller, *J. Chem. Phys.* **64**, 63 (1976).
78. J. P. Hansen and I. R. McDonald, *Theory of Simple Liquids* (Academic Press, London, 1986).
79. R. A. Laviolette and F. H. Stillinger, *J. Chem. Phys.* **83**, 4079 (1985).
80. G. Némethy, K. D. Gibson, K. A. Palmer, C. N. Yoon, G. Paterlini, A. Zagari, S. Rumsey and H. A. Scheraga, *J. Phys. Chem.* **96**, 6472 (1992); I. K. Roterman, M. H. Lambert, K. D. Gibson and H. A. Scheraga, *J. Biomol. Struct. Dyn.* **7**, 421 (1989).
81. F. A. Momany, R. F. McGuire, A. W. Burgess and H. A. Scheraga; *J. Phys. Chem.* **79**, 2361 (1975); G. Nementhy, M. S. Pottle and H. A. Scheraga, *ibid.* **87**, 1883 (1983); M. J. Sippl, G. Nementhly and H. A. Scheraga, *ibid.* **88**, 6231 (1984).
82. C. L. Brooks, M. Karplus and M. Pettitt, *Proteins: A Theoretical Perspective of Dynamics, Structure, and Thermodynamics* (John Wiley and Sons, New York, 1988).
83. J. A. McCammon and S. C. Harvey, *Dynamics of Proteins and Nucleic Acids* (Cambridge Univ. Press, New York, 1987).
84. The classical trajectory-bundle TDSCF method provides greater flexibility than Gaussian wavepackets at the cost of introducing more degrees of freedom. See R. B. Gerber and M. A. Ratner, *Adv. Chem. Phys.* **70**, 97 (1988); G. C. Schatz, V. Buch, M. A. Ratner and R. B. Gerber, *J. Chem. Phys.* **79**, 1808 (1983); V. Buch, R. B. Gerber and M. A. Ratner, *Chem. Phys. Lett.* **101**, 44 (1983); R. B. Gerber, V. Buch and M. A. Ratner, *J. Chem. Phys.* **77**, 3022 (1982); *Chem. Phys. Lett.* **91**, 173 (1982).
85. H. A. Scheraga, *Adv. Phys. Org. Chem.* **6**, 103 (1968).
86. D. J. Evans, W. G. Hoover, B. H. Failor, B. Moran and A. J. C. Ladd, *Phys. Rev.* A**28**, 1016 (1983).
87. W. G. Hoover, A. J. C. Ladd and B. Moran, *Phys. Rev. Lett.* **48**, 1818 (1982); D. J. Evans, *J. Chem. Phys.* **78**, 3297 (1983).
88. H. Risken, *The Fokker-Planck Equation* (Springer-Verlag, Berlin, 1989).
89. J. L. Skinner and P. G. Wolynes, *J. Chem. Phys.* **69**, 2143 (1978); *J. Chem. Phys.* **72**, 4913 (1980).
90. B. J. Berne, J. L. Skinner and P. G. Wolynes, *J. Chem. Phys.* **73**, 4314 (1980).

91. J. Ma and J. E. Straub, *J. Chem. Phys.* **101**, 533 (1994): *ibid.* **103**, 9113 (1995).
92. G. Verkhiver, R. Elber and W. Nowak, *J. Chem. Phys.* **97**, 7838 (1992).
93. J. E. Straub and M. Karplus, *J. Chem. Phys.* **94**, 6737 (1991).
94. A. Ulitsky and R. Elber, *J. Chem. Phys.* **98**, 1034 (1994); *ibid.* **98**, 3380 (1994).
95. A. Miranker and M. Karplus, *Proteins: Structure, Function, and Genetics* **11**, 29 (1991).
96. A. Caflisch, A. Miranker and M. Karplus, *J. Med. Chem.* **36**, 2142 (1993).
97. A. Roitberg and R. Elber, *J. Chem. Phys.* **95**, 9277 (1991).
98. N. Corbin and K. Singer, *Mol. Phys.* **46**, 671 (1982); K. Singer and W. Smith, *Mol. Phys.* **57**, 761 (1986) for Gaussian wavepackets applied to the quantum simulation of neon atoms in real time.
99. The unnormalized equilibrium density distribution for the canonical ensemble $\hat{\rho}(r, p, \beta) = \exp(-\beta H)$ satisfies the Bloch equation

$$\frac{\partial \hat{\rho}}{\partial \beta} = -H\hat{\rho}\,. \tag{64}$$

[See R. W. Zwanzig, in *Lectures in Theoretical Physics*, Vol. III, eds. W. E. Britton, B. W. Downs and J. Downs (Wiley Interscience, New York, 1961) p. 106.] Integrating the normalized distribution avoids the numerical difficulties associated with integration of unnormalized quantities, such as $\hat{\rho}(r; \beta)$, which become vanishingly small for large values of β.
100. P. Amara and J. E. Straub, **99**, 14840 (1995).
101. L. Piela, J. Kostrowicki and H. A. Scheraga, *J. Phys. Chem.* **93**, 3339 (1989).
102. J. Kostrowicki, L. Piela, B. J. Cherayil and H. Scheraga, *J. Phys. Chem.* **95**, 4113 (1991).
103. R. J. Wawak, M. M. Wimmer and H. A. Scheraga, *J. Phys. Chem.* **96**, 5138 (1992).
104. H. A. Scheraga, *Rev. Comp. Chem*, **III**, 73 (1992).
105. J. Kostrowicki, L. Piela, B. J. Cherayil and H. A. Scheraga, *J. Phys. Chem.* **95**, 4113 (1991); J. Kostrowicki and H. A. Scheraga, *J. Phys. Chem.* **96**, 7442 (1992).
106. C. Tsoo and III C. L. Brooks, *J. Chem. Phys.* **101**, 6405 (1994).
107. P. Amara, D. Hsu and J. E. Straub, *J. Phys. Chem.* **97**, 6715 (1993).
108. D. Shalloway, in *Recent Advances in Global Optimization*, eds. C. A. Floudas and P. M. Pardalos, (Princeton Univ. Press, Princeton, 1992).
109. D. Shalloway, *Global Optimization* **2**, 281 (1992); M. Orešič and D. Shalloway, *J. Chem. Phys.* **101**, 6405 (1994).
110. T. Coleman, D. Shalloway and Z. Wu, *Technical Report: Cornell Theory Center* (November 1992).
111. L. I. Schiff, *Quantum Mechanics* (McGraw-Hill, New York, 1968).
112. R. L. Somorjai, *J. Phys. Chem.* **95**, 4141 (1991).
113. M. Sylvain and R. L. Somorjai, *J. Phys. Chem.* **95**, 4147 (1991).

114. K. A. Olszewski and H. A. Scheraga, *J. Phys. Chem.* **96**, 4672 (1992).
115. K. A. Olszewski, L. Piela and H. A. Scheraga, *J. Phys. Chem.* **97**, 260 (1993).
116. K. A. Olszewski, L. Piela and H. A. Scheraga, *J. Phys. Chem.* **97**, 267 (1993).
117. A. Messiah, *Quantum Mechanics* (Wiley and Sons, New York, 1976).
118. J. B. Anderson, *J. Chem. Phys.* **63**, 1499 (1975); *ibid.* **65**, 4121 (1976); *Int. J. Quantum Chem.* **15**, 109 (1979).
119. D. Ceperley and B. Alder, *Science* **231**, 555 (1986).
120. D. L. Freeman and J. D. Doll, *J. Chem. Phys.* **82**, 462 (1985).
121. A. B. Finnila, M. A. Gomez, C. Sebenik, C. Stenson and J. D. Doll, *Chem. Phys. Lett.* **219**, 343 (1994).
122. T. Noguti and N. Gō, *Biopolymers* **24**, 527 (1985).
123. A. Kitao and N. Gō, *J. Comp. Chem.* **12**, 359 (1991).
124. E. J. Heller, *J. Chem. Phys.* **62**, 1544 (1975); E. J. Heller, *Acc. Chem. Res.* **14**, 368 (1981).
125. R. P. Feynman and H. Kleinert, *Phys. Rev.* A**34**, 5080 (1986).
126. A. D. McLachlan, *Mol. Phys.* **8**, 39 (1964).
127. R. P. Feynman, *Quantum Mechanics and Path Integrals* (Wiley, New York, 1965).
128. C. Zheng, C. F. Wong, J. A. McCammon and P. G. Wolynes, *Nature* **334**, 726 (1988).
129. P. Ruján, *Z. Phys. B - Condensed Matter* **73**, 391 (1988).
130. S. H. Northrup and J. A. McCammon, *Biopolymers* **19**, 1001 (1980).
131. Y. Sun and P. A. Kollman, *J. Comp. Chem.* **13**, 33 (1992).
132. M. Saunders, K. N. Houk, Y. -D. Wu, W. C. Still, M. Lipton, G. Chang and W. C. Guida, *J. Am. Chem. Soc.* **112**, 1419 (1990).
133. T. Schaumann, W. Braun and K. Wüthrich, *Biopolymers* **29**, 679 (1990).
134. B. Von Freyberg and W. Braun, *J. Comp. Chem.* **12**, 1065 (1991).
135. D. G. Garrett, K. Kastella and D. M. Ferguson, *J. Am. Chem. Soc.* **114**, 6555 (1992).
136. Z. Li and H. A. Scheraga, *Proc. Natl. Acad. Sci. USA* **84**, 6611 (1987).
137. J. P. Valleau and S. G. Whittington, in *Statistical Mechanics, Part A*, ed. B. J. Berne, (Plenum Press, New York, 1976).
138. J. I. Siepmann and M. Sprik, *Chem. Phys. Lett.* **199**, 220 (1992).
139. J. Cao and B. J. Berne, *J. Chem. Phys.* **92**, 1980 (1990).
140. A. M. Ferremberg and R. H. Swendsen, *Phys. Rev. Lett.* **61**, 2635 (1988).
141. D. D. Frantz, D. L. Freeman and J. D. Doll, *J. Chem. Phys.* **93**, 2769 (1990).
142. C. J. Tsai and K. D. Jordan, *J. Chem. Phys.* **99**, 6957 (1993).
143. R. H. Stote, D. J. States and M. Karplus, *J. Chim. Phys.* **88**, 2419 (1991).
144. H. -Q. Ding, N. Karasawa and W. A. Goddard III, *J. Chem. Phys.* **97**, 4309 (1992).
145. M. E. Tuckerman, G. J. Martyna and B. J. Berne, *J. Chem. Phys.* **93**, 1287 (1990); M. E. Tuckerman, B. J. Berne and A. Rossi, *J. Chem. Phys.* **94**,

1465 (1991); M. E. Tuckerman, B. J. Berne and G. J. Martyna, *J. Chem. Phys.* **94**, 6811 (1991).

146. D. D. Humphreys, R. A. Friesner and B. J. Berne, *J. Phys. Chem.* **98**, 6885 (1994).
147. R. C. van Schaik, H. J. C. Berendsen, A. E. Torda and W. F. van Gunsteren, *J. Mol. Biol.* **234**, 751 (1993).
148. L. M. Blumenthal, in *Theory and Applications of Distance Geometry* (Chelsea, New York, 1970) p. 97; G. M. Crippen, *Distance Geometry and Conformational Calculations* (Research Studies Press, England, 1981).
149. G. M. Crippen, *J. Comp. Chem.* **3**, 471 (1982); *J. Comp. Chem.* **5** 548 (1984).
150. E. O. Purisima and H. A. Scheraga, *Proc. Natl. Acad. Sci. USA* **84**, 2782 (1986); *J. Mol. Biol.* **196**, 697 (1987).
151. R. C. van Schaik, W. F. van Gunsteren and H. J. C. Berendsen, *J. Computer-Aided Mol. Design* **6**, 97 (1992).
152. J. Kostrowicki and H. A. Scheraga, *J. Phys. Chem.* **96**, 7442 (1992).
153. J. E. Straub, J. Ma and P. Amara, *J. Chem. Phys.* **103**, 1574 (1995).
154. P. Amara, J. Ma and J. E. Straub, in *Global Minimization of Nonconvex Energy Functions: Molecular Conformation and Protein Folding*, eds. P. M. Pardalos, D. Shalloway and G. Xue (American Mathematical Society, 1996), p. 41.
155. B. Chorch, M. Orešič and D. Shalloway, in *Global Minimization of Nonconvex Energy Functions: Molecular Conformation and Protein Folding*, eds. P. M. Pardalos, D. Shalloway and G. Xue (American Mathematical Society, 1996), p. 41.
156. I. Andricioaei and J. E. Straub, *Phys. Rev.* E**53**, 3055 (1996).

CHAPTER 4

ANALYTICAL THEORIES OF PROTEIN FOLDING

T. GAREL, H. ORLAND*

Service de Physique Théorique, CEA-Saclay
F-91191 Gif-sur-Yvette Cedex, France

and

D. THIRUMALAI

Institute for Physical Science and Technology
University of Maryland, College Park, Maryland 20742, USA

Contents

*Also at Groupe de Physique Statistique, Université de Cergy-Pontoise, 95806 Cergy-Pontoise Cedex, France.

1. Introduction

Proteins are (weakly branched) heteropolymers made out of twenty species of monomers (amino acids). Natural proteins have the property of folding into an (almost) unique compact native structure, which is of biological interest.[1,2] The compactness of this unique native state is largely due to the existence of an optimal amount of hydrophobic amino acid residues,[3] since these biological objects are usually designed to work in water. The task of predicting the conformation of the three-dimensional structure for the linear primary sequence is often referred to as the protein folding problem. Already at this stage, a rigorous analytical theory appears difficult, since it amounts to study a mesoscopic system (most proteins have between 100 and 500 residues), notwithstanding the solvent's properties.

This mesoscopic system is of a classical nature; the quantum mechanical valence electrons of the atoms induce interactions between the heavy "nuclei" which can then be treated as classical objects interacting through classical many-body interactions. This observation forms the basis of Molecular Dynamics, Monte Carlo, or Statistical Mechanical models.

To further complicate the matter, both the compactness and the chemical heterogeneity of a given protein tend to slow down dynamical processes: the question of the kinetic control of the folding (as opposed to a thermodynamic control) is therefore periodically asked. This remark suggests that the folding process may have something in common with the physics of glassy systems, where competing interactions (frustration) and/or disorder lead to a very rugged phase space, resulting in slow dynamical processes. This analogy is further strengthened, if we consider the following simple-minded approaches to proteins: (i) the total energy of a protein is the sum of bonded (geometrical) and non-bonded (Coulomb, Van Der Waals) terms which cannot be simultaneously satisfied. The commonly used CHARMM form of the energy function[4,5] is given in Appendix 2; (ii) experimentally (Crystallography, Nuclear Magnetic Resonance, Circular Dichroism), folded proteins have a local order due to hydrogen bonds (represented in the CHARMM energy function by the appropriate non-bonded terms). This order[6,7] is, roughly speaking, one-dimensional (α-helix) or two-dimensional (β-sheet) in nature, and is therefore incompatible with the requirement of global compactness.

In polymer physics,[8,9] the compact phase of a polymer in a bad solvent is reached through a phase transition, called the θ point. This compact phase has a sizeable conformational entropy, and thus no definite three-dimensional shape: a polymer globule (compact phase) is a statistical mixture of an exponentially large number of conformations.

In contradistinction, a folded protein has a unique (or quasi-unique) low temperature shape. This folded structure is reached in times of the order of a second. Although this time is considered small in the literature, it is to be compared with molecular time scales, which are of the order of a femtosecond. This discrepancy between the microscopic and mesoscopic time scales shows that the dynamics of protein folding cannot be accounted for by a model of relaxation in a single well, and thus confirms the existence of a very rugged phase space. In these facts lies the heart of the folding puzzle.[10,11]

In this review, we adopt a thermodynamic (vs dynamic) approach to, and present a physics-oriented view of, the protein folding problem. This simplified approach is restricted on (at least) two different levels. We will consider statistical mechanical models of the folding (or related) transitions. Furthermore, only approximate solutions (mean-field, saddle point method, etc.) of these models can in general be worked out. In this respect, we will work "in the thermodynamic limit" (number of residues going to infinity), neglecting non-extensive terms. The relation between these physical models and real molecules (a compact protein of one hundred residues has something like half of its atoms on the surface) may thus be quite remote. Nevertheless, we believe that the different models we will study shed an interesting, if partial, light on the difficult and fundamental folding puzzle. Following the energy frustration appeal (point (i) above) leads one to study spin glass-like (and more generally random heteropolymer) models of the folding transition. The spin-glass point of view[12,13] is of great interest for various (non-independent) reasons (dominance of a few ground states, coding *à la* Hopfield, insight into some Evolution models ...), even though its application to real three-dimensional proteins seems questionable. The very existence of secondary structures in proteins is more easily included along the lines of point (ii) above, and this approach suggests, for instance, an interpretation of the "molten globule" state. This geometrical competition between local order and global compactness may be called the "glassy-glass" point of view; it is linked with the Flory–Gibbs–Di Marzio[14,15] theory of the glass transition in polymers. These two approaches must be thought of as complementary: a more realistic theory should include both a local tendency to form, say, a helix and the existence of a hierarchy in amino acids concerning helix formation (alanine is a good helix former, as opposed to glycine). A review by M. Karplus and E. Shakhnovich[16] has recently appeared, where more general topics are included (dynamics of the folding, numerical simulations, etc.). The present work is more polymer-oriented and thus complements this review.

We also emphasize that the physical models considered in the following are appropriate for the description of protein folding *in vitro*. Indeed, these models describe the folding of a single heteropolymer chain. In recent years, it has become clear that *in vivo* folding may require the presence of chaperone molecules and other cofactors.[17] The folding in cells, where the protein concentration is much higher, can also be treated using ideas from polymer physics.[18]

The layout of this paper is as follows. Section 2 is devoted to a quick review of polymers through discrete or continuum models. Both the cases of good and bad solvent will be considered. A simple theoretical model which mimicks this behaviour is the Edwards model.[19] An approximate solution can be expressed in terms of the Schrödinger equation for the monomer concentration. On the other hand, since Flory's pioneering work, some properties of long chains have been linked with the field of critical phenomena, mostly with the $n = 0$ trick.[20] This fact will be used in the study of the collapsed phase, modelled through the statistics of Hamiltonian paths (Sec. 3): the inclusion of secondary structures in a compact homopolymer structure is then presented through the formalism of weighted Hamiltonian paths.[21–23] This approach is also of interest in the field of semiflexible polymer crystallization. Indeed, the Hamiltonian path approach yields a crystalline (almost unique) ground state. Moreover, the molten globule phase[24,25] receives a natural interpretation here. Section 4 presents the random heteropolymer approach (including spin glass-like models) to the folding transition. In the magnetic spin glass case, the problem of quenched disorder stems from the random positions of the magnetic impurities. Here, one models a protein through a (quenched) disordered primary sequence.[26–28] Various spin glass models (Sherrington-Kirkpatrick, Mattis, etc.) have been studied, from which one can imagine, *mutatis mutandis*, several protein folding scenarios. The quasi-uniqueness of the folded structure may therefore receive different interpretations. Most prominent among these models is the Random Energy Model of B. Derrida,[29,30] which although very simple, may well contain the essence of the folding transition of some random heteropolymers models, in high enough dimensions. Most spin glass technicalities, including the replica approach, are deferred to Appendix 3. There, we sketch the study of the "simplest" spin glass[31] and its connection to the Random Energy Model. The coding problem *à la* Hopfield (Refs. 32 and 33; see Appendix 4) is briefly examined, in the framework of a random sequence Mattis-like model.

The analytical approach clearly has a (serious) drawback, which makes connections with the three-dimensional biologists' proteins tenuous. Possible advantages include a deeper physical insight in the folding transition, a more global (less atomistic) view of proteins, and a possible bridge between proteins and glasses.

2. Basic Polymer Theory

2.1. *Introduction*

In this section, we introduce simple concepts of polymer physics that will be used in the following sections. General references, with a larger scope, include the books by Flory,[8,34] de Gennes,[9] Doi and Edwards,[35] Wiegel,[36] des Cloizeaux and Jannink,[37] and Freed.[38]

A polymer chain in a solvent is modeled as a sequence of N links representing the monomers. Typical values of the polymerization index N are 10^5 in polymer chemistry, $10^2 - 10^3$ in proteins, and up to 10^{10} for DNA molecules, which are the longest known polymers. Homopolymers are built from identical monomers, whereas heteropolymers may contain different units (e.g. 20 amino acids for proteins or 4 nucleotides for DNA).

In the following, we will consider only the single chain problem; indeed, the (*in vitro*) protein folding problem is *a priori* a one chain problem (we leave aside the question of chaperone assisted folding). The partition function for the chain can be written as:

$$Z = \int \prod_{i=1,\ldots,N+1} d\mathbf{r}_i \prod_{i=1,\ldots,N} g(\mathbf{r}_i, \mathbf{r}_{i+1}) \, \exp(-\beta\mathcal{H}) \tag{2.1}$$

where β is the inverse of the temperature T (in this article, we set Boltzmann's constant $k_B = 1$). The vectors $\mathbf{r}_i$ denote the position of the nodes. For example, in polyethylene, the $\mathbf{r}_i$ would represent the positions of the carbon atoms. Henceforth, we will assume that the polymer is embedded in a d-dimensional space.

The chain constraint is expressed by the factors $g(\mathbf{r}, \mathbf{r}')$. Possible choices for instance can be:

(i) $g(\mathbf{r}, \mathbf{r}') = \delta(|\mathbf{r} - \mathbf{r}'| - a)$, for freely hinged monomers of fixed length a. One can further modify g by restricting the nodes to belong to a regular lattice, which is a quite common simplification in Statistical Mechanics.

(ii) $g(\mathbf{r}, \mathbf{r}') = (\frac{d}{2\pi a^2})^{\frac{d}{2}} \, \exp(-d \, \frac{(\mathbf{r}-\mathbf{r}')^2}{2a^2})$, if the monomer length is allowed to fluctuate around a.

If one wishes to include curvature or torsion effects, one should use a function $g(\mathbf{r}_i, \mathbf{r}_{i+1}, \mathbf{r}_{i+2})$ or $g(\mathbf{r}_i, \mathbf{r}_{i+1}, \mathbf{r}_{i+2}, \mathbf{r}_{i+3})$ respectively. An example will be presented in Sec. 3.

The "Hamiltonian" $\beta\mathcal{H}$ may be expanded as:

$$\beta\mathcal{H} = \frac{1}{2} \sum_{i \neq j} v(\mathbf{r}_i, \mathbf{r}_j) + \frac{1}{6} \sum_{i \neq j \neq k} w(\mathbf{r}_i, \mathbf{r}_j, \mathbf{r}_k) + \ldots \tag{2.2}$$

and represents the direct monomer–monomer interactions (such as Coulomb or Van der Waals interactions) as well as the solvent-induced interactions.

The particular case of a chain with no interactions ($\mathcal{H} = 0$) is called a Brownian chain.

In many polymer applications, one is interested in quantities which do not depend on the microscopic scale, hence the need for a continuum description. Loosely speaking, this continuous limit is obtained (in a field theoretic approach) by taking $N \to +\infty$ and $a \to 0$ in such a way that the product $S = Na^2$, named the Brownian surface of the chain, remains constant. More rigorous arguments are given by des Cloizeaux and Jannink.[37]

From a practical point of view, it is sufficient to replace everywhere the discrete indices $i, j, k, \ldots = 1, \ldots, N$ of Eqs. (2.1) and (2.2) by continuous curvilinear abcissae $s, s', s'', \ldots$. Neglecting curvature and torsion effects, the partition function of the chain can be written as

$$Z = \int \mathcal{D}\mathbf{r}(s) \exp\left(-\frac{d}{2a^2} \int_0^N ds \left(\frac{d\mathbf{r}(s)}{ds}\right)^2 - \beta\mathcal{H}\right) \tag{2.3a}$$

with

$$\beta\mathcal{H} = \frac{1}{2} \int_0^N ds \int_0^N ds' v(\mathbf{r}(s), \mathbf{r}(s'))$$
$$+ \frac{1}{6} \int_0^N ds \int_0^N ds' \int_0^N ds'' w(\mathbf{r}(s), \mathbf{r}(s'), \mathbf{r}(s'')) \,. \tag{2.3b}$$

In the following, we shall use indistinctively the continuous or discrete versions of the partition function. However, as we shall see in Appendix 2, the continuous version is easier to handle for analytic calculations, since it can be mapped onto an imaginary time Schrödinger equation.

2.2. *Solvent-Induced Interactions*

It is physically intuitive that, in a good solvent, a polymer chain will swell so as to maximize contacts between monomers and solvent molecules. On the contrary, a bad solvent will lead to a collapse of the chain. One may equally well say that a good or bad solvent generates an effective repulsive or attractive monomer–monomer interaction respectively. In the context of protein folding, it is widely believed that the interactions between the hydrophobic residues (Trp, Ile, Tyr, etc.) are the main driving forces for the collapse of proteins: hydrophobic residues in water can be thought of as being in a bad solvent, and thus have an attractive effective interaction.

Let us consider a lattice model, with N monomers $\mathbf{r}_i$ and $\mathcal{N}$ solvent molecules $\mathbf{R}_\alpha$, where each site is occupied either by a monomer or a solvent molecule or a vacancy. The Hamiltonian of this model reads:

$$\mathcal{H}_{ms} = \sum_{i=1}^{N+1} \sum_{\alpha=1}^{\mathcal{N}} v_0(\mathbf{r}_i - \mathbf{R}_\alpha) \tag{2.4}$$

where v_0 denotes a short-ranged monomer-solvent molecule interaction.

Introducing the monomer and solvent volume concentrations:

$$\rho_m(\mathbf{r}) = \sum_{i=1}^{N+1} \delta(\mathbf{r} - \mathbf{r}_i) \tag{2.5a}$$

and

$$\rho_s(\mathbf{r}) = \sum_{\alpha=1}^{\mathcal{N}} \delta(\mathbf{r} - \mathbf{R}_\alpha) \tag{2.5b}$$

the Hamitonian (2.4) can be rewritten as:

$$\mathcal{H}_{ms} = \sum_{\mathbf{r}} \sum_{\mathbf{r}'} \rho_m(\mathbf{r}) v_0(\mathbf{r} - \mathbf{r}') \rho_s(\mathbf{r}'). \tag{2.6}$$

Assuming that the system is incompressible (no vacancies), we have, at each site $\mathbf{r}$:

$$\rho_s(\mathbf{r}) + \rho_m(\mathbf{r}) = 1. \tag{2.7}$$

Replacing (2.7) in (2.6) we obtain:

$$\mathcal{H}_{ms} = -\sum_{\mathbf{r}} \sum_{\mathbf{r}'} \rho_m(\mathbf{r}) v_0(\mathbf{r} - \mathbf{r}') \rho_m(\mathbf{r}') + \sum_{\mathbf{r}} \sum_{\mathbf{r}'} \rho_m(\mathbf{r}) v_0(\mathbf{r} - \mathbf{r}'). \tag{2.8}$$

The second term in (2.8) is a constant, equal to NV_0 where by definition $V_0 = \sum_{\mathbf{r}} v_0(\mathbf{r})$. It will therefore be omitted.

Note the change of sign of the first term between (2.6) and (2.8). It generates, as announced above, a contribution to the two-body interaction $v(\mathbf{r} - \mathbf{r}')$ of (2.2). The good (or bad) solvent is characterized by a negative (or positive) $v_0(\mathbf{r})$, and indeed generates a repulsive (or attractive) interaction $-\, v_0(\mathbf{r})$ between monomers. In order to apply these considerations to heteropolymers or proteins, it is necessary to allow for sequence dependent interactions v_0 in Eq. (2.4). This will be discussed in Sec. 5.

2.3. *Physical Quantities and the Edwards Model*

2.3.1. *Physical Quantities*

Several quantities can be measured experimentally, which characterize the scaling properties of long chains. For a detailed review, the reader is referred to the books mentioned in the beginning of this section.

We restrict ourselves to the following quantities:

(i) *the radius of gyration* is defined by:

$$R_G^2 = \left\langle \frac{1}{N} \sum_i \left(\mathbf{r}_i - \frac{1}{N} \sum_j \mathbf{r}_j\right)^2 \right\rangle \tag{2.9}$$

where $\langle \ldots \rangle$ denotes thermal averaging. In the Brownian case, it can easily be seen that R_G satisfies the scaling relation:

$$R_G \simeq N^{\frac{1}{2}}. \tag{2.10a}$$

For interacting monomers, a similar scaling relation holds for long chains with an index ν:

$$R_G \simeq N^{\nu}. \tag{2.10b}$$

Typical values in three dimensions[37] are $\nu \simeq 0.6$ for a chain in a good solvent, and $\nu = 1/3$ in a bad solvent. A chain is said to be swollen if $\nu > 0.5$ and collapsed if $\nu < 0.5$.

(ii) *the scattering function* is defined as the monomer–monomer density correlation function. It can be measured by various scattering techniques (light, X-rays, neutrons, etc.) and is given by:

$$g(\mathbf{q}) = \frac{1}{N} \int_0^N \mathrm{d}s \int_0^N \mathrm{d}s' \, \langle \exp(i\mathbf{q}(\mathbf{r}(s) - \mathbf{r}(s'))) \rangle \tag{2.11a}$$

in Fourier space. In three dimensions, the scattering function $g(\mathbf{q})$ for a Brownian chain has been calculated by Debye and reads:

$$g(\mathbf{q}) = \frac{2}{x}\left(1 - \frac{1}{x}(1 - \exp(-x))\right) \tag{2.11b}$$

where $x = Nq^2a^2/6$.

(iii) by analogy with the Brownian case, *the entropy S* satisfies the scaling relation:

$$\exp(S) \simeq \mu^N N^{\gamma-1} \tag{2.12}$$

where μ is a constant that depends on microscopic details of the system (lattice connectivity, continuous versus discrete, etc.), whereas γ is a universal critical exponent, independent of microscopic details. Note that in the case of a collapsed chain, there may be an additional surface contribution to (2.12).[39]

2.3.2. *Definition of the Edwards Model*

The Edwards model[19] is defined by a specific choice of the two-body interaction, (in the continuous limit); it assumes all interactions to be zero-ranged. Equations (2.3a) and (2.3b) can accordingly be rewritten as:

$$Z = \int \mathcal{D}\mathbf{r}(s) \exp(-\frac{d}{2a^2} \int_0^N ds \left(\frac{d\mathbf{r}(s)}{ds}\right)^2 - \beta\mathcal{H}) \tag{2.13a}$$

where now:

$$\beta\mathcal{H} = \frac{\lambda_2}{2} \int_0^N ds \int_0^N ds' \delta(\mathbf{r}(\mathbf{s}) - \mathbf{r}(s'))$$
$$+ \frac{\lambda_3}{6} \int_0^N ds \int_0^N ds' \int_0^N ds'' \delta(\mathbf{r}(s) - \mathbf{r}(s'))\delta(\mathbf{r}(s') - \mathbf{r}(s'')) . \tag{2.13b}$$

In (2.13b), λ_2 and λ_3 have the dimensions of volume and (volume)^2 respectively. This complicated model can be studied by various approximations, among which we quote:

(i) the renormalization group,[9,38] which is appropriate to describe large scale properties of the chain, such as critical exponents $(\nu, \gamma, \ldots)$, small q behavior of the scattering function, scaling functions, etc. This approach has been linked to the general theory of critical phenomena, by mapping the statistical mechanics of a polymer chain onto the critical behavior of an n-vector spin model, in the limit $n \to 0$.

(ii) the Flory approximation,[34] which will be described in the case of a chain in a good solvent.

(iii) the self-consistent field (SCF) approximation,[19] derived in Appendix 3, and used in the case of a bad solvent.

2.3.3. *A Chain in a Good Solvent: Flory Approximation*

In this case, the contribution of the solvent-induced interaction to λ_2 is positive, and thus adds up to the direct monomer–monomer "zero-ranged" interaction (van der Waals like), so that overall, λ_2, called the excluded volume, is positive. The net effect is a swelling of the chain, to avoid self-intersections, and the three-body term λ_3 can be set to zero. Such a chain is called a self-avoiding walk (SAW).

Flory[34] has devised a simple argument to compute the exponent ν of a chain in a good solvent, which gives amazingly good results for all dimensions d. The argument, as presented by de Gennes,[9] goes as follows: Consider a chain of N monomers of length a, with local repulsive interaction characterized by an excluded volume parameter λ_2. The swollen chain will acquire a radius of gyration R; the monomer concentration $c = N/R^d$, and the repulsive energy is thus

$$\begin{aligned} E_{\text{rep}} &= \frac{T}{2}\lambda_2\, c^2\, R^d \\ &= \frac{T}{2}\lambda_2 \frac{N^2}{R^d} \end{aligned} \tag{2.14a}$$

where T is the temperature.

In Appendix 3, we show how to calculate the entropy (or elastic energy) of a Brownian chain. Approximating the self-interacting chain by a Brownian one, we get an elastic energy of the chain, given by:

$$E_{\text{el}} = T\frac{dR^2}{2Na^2}. \tag{2.14b}$$

The total free energy is:

$$F = T\frac{dR^2}{2Na^2} + \frac{T\lambda_2}{2}\frac{N^2}{R^d} \tag{2.15}$$

and minimization with respect to R yields the celebrated Flory exponent:

$$\nu_F = \frac{3}{d+2} \tag{2.16}$$

which is exact for $d = 1$, 2, 4, and almost exact for $d = 3$ ($\nu_F = 0.6$ whereas best numerical estimates (based on renormalization theory) give $\nu = 0.588$. See Ref. 38). This value is in excellent agreement with light scattering experiments.[40]

The Flory exponent (for $d = 3$) can also be obtained by Edwards' self-consistent field theory. For the case of a chain in a good solvent though, this theory is delicate to implement, and we refer the reader to the literature. Furthermore, other derivations (e.g. see Ref. 41 and references therein) of the Flory theory allow one to calculate the exponent γ of Eq. (2.12) and the asymptotic form of the scattering function.

Finally, replacing R by its Flory approximation in (2.14a), we obtain, up to a multiplicative constant:

$$E_{\text{rep}} \sim \frac{1}{2}\lambda_2 N^{\frac{4-d}{d+2}} . \tag{2.17}$$

We see that this repulsive energy vanishes for a long enough chain, if $d > 4$, which implies that the chain is Brownian. The physical interpretation of this phenomenon is that the probability of self-intersection of a Brownian chain vanishes in dimensions larger than four. This dimension four is called the upper critical dimension.

2.3.4. *A Chain in a Bad Solvent: Self-Consistent Field Theory*

In this case, the contribution of the solvent induced interaction to λ_2 is negative, and thus competes with the direct monomer–monomer interaction. This competition leads to the existence of a change of the sign of λ_2 when temperature decreases. The temperature at which λ_2 vanishes is called the θ point. Below this temperature, λ_2 is negative, and thus tends to collapse the chain. To avoid complete collapse of the chain onto a microscopic volume, one must take into account the three-body repulsive term $\lambda_3 > 0$.

Within the Flory approximation, the free energy of the model now reads:

$$F = T\frac{dR^2}{2Na^2} + \frac{T\lambda_2}{2}\frac{N^2}{R^d} + \frac{T\lambda_3}{6}\frac{N^3}{R^{2d}} . \tag{2.18}$$

Below the θ point, $\lambda_2 < 0$, and the minimization of F yields:

$$R \simeq N^{\frac{1}{d}} . \tag{2.19}$$

This means that the chain is collapsed into a compact globule, with a finite monomer concentration. At the θ point, we have $\lambda_2 = 0$, and thus:

$$\nu = \frac{2}{d+1} . \tag{2.20}$$

We see that in this case, the upper critical dimension is three.

As we shall use it in the context of proteins, we now present the SCF approximation, within ground state dominance (see Appendix 2). Since we expect the system to collapse at low temperature, this corresponds to the existence of a bound state (in a Schrödinger-like equation). We may thus restrict ourselves to a ground state dominance approximation in the corresponding Schrödinger equation[42–44] as detailed in Eqs. (A3.9)–(A3.14). Within this framework, the variational free energy of the chain reads (in $d = 3$):

$$\beta\frac{F}{N} = \int d\mathbf{r}\left[\frac{a^2}{6}(\nabla\Psi)^2 + N\frac{\lambda_2}{2}\Psi^4 + N^2\frac{\lambda_3}{6}\Psi^6 - E\Psi^2\right]. \tag{2.21a}$$

Minimization with respect to Ψ yields the nonlinear equation:

$$-\frac{a^2}{6}\nabla^2\Psi + N\lambda_2\Psi^3 + N^2\frac{\lambda_3}{2}\Psi^5 = E\Psi \tag{2.21b}$$

where Ψ^2 should be normalized to unity.

Exact solutions of this equation cannot be obtained analytically. It is then useful to make a variational approximation to (2.21), with trial wave functions of the form:

$$\Psi = \left(\frac{1}{2\pi R^2}\right)^{\frac{d}{4}} \exp\left(-\frac{\mathbf{r}^2}{4R^2}\right) \tag{2.22a}$$

where R is a variational parameter, simply related to the radius of gyration of the chain. With this ansatz, the free energy is equal to:

$$\beta\frac{F}{N} = \frac{a^2}{8R^2} + \frac{\lambda_2}{16\pi^{3/2}}\frac{N}{R^3} + \frac{\lambda_3}{144\sqrt{3}\pi^3}\left(\frac{N}{R^3}\right)^2. \tag{2.22b}$$

Minimization of this equation with respect to R yields essentially the same results as in Flory theory, namely, the existence of a collapsed phase with finite monomer concentration for $\lambda_2 < 0$, and a θ point for $\lambda_2 < 0$.

Due to the importance of the globular phase of polymers to the physics of proteins, we will return to this subject, using different methods, in Chapter 3, devoted to the formation of secondary structures in compact globules.

2.4. *Conclusions*

We have considered very simplified models of interactions for the monomers. Many other polymer problems are relevant to the physics of biopolymers.

To name but a few, let us mention

(i) charged polymers (polyelectrolytes, polyampholytes) in saline solutions[45–49] or subject to an electric field (Ref. 50 and references therein);
(ii) polymers at interfaces (in relation to membrane proteins)[51–53];
(iii) polymers in random media, (Refs. 54–58, and references therein); and
(iv) heteropolymers (which will be discussed in Sec. 4).

3. Secondary Structures in a Compact Phase

We see in Appendix 1 that a commonly used form of the total energy of a protein is the sum of a bonded (geometric) part and of a non-bonded (Coulomb, Van der Waals) part. In particular, the Coulomb part is responsible, in this formalism, for the formation of hydrogen bonds,[6] which tend to locally stabilize one-dimensional (α-helix) or two-dimensional (β-sheets) structures (see Fig. 1). Since we know that the biologically active protein is compact, we are typically faced with the problem of geometrical frustration, where local and global orders are weakly incompatible. This approach is familiar in glasses where one tries to solve this contradiction by a mapping onto a curved space.[59–62] In this section, we shall describe how one can study, in a very simplified and schematic way, the thermodynamics of secondary structure formation in a compact phase. For earlier works,

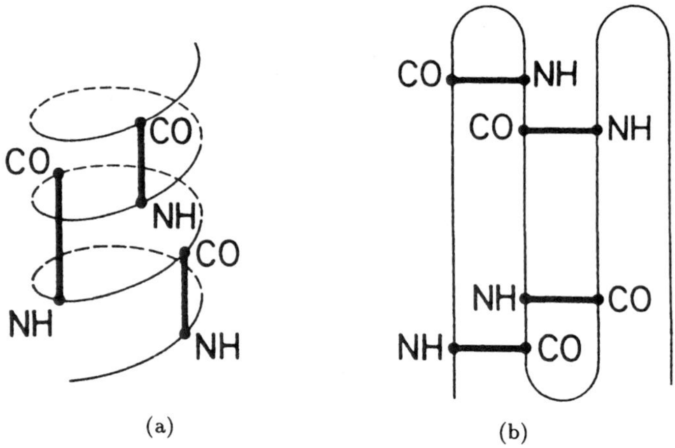

Fig. 1. Schematic representation of hydrogen bonds in (a) α-helix (b) β-sheet.

the reader is referred to Lauritzen and Zwanzig.[63] However, before doing so, let us reconsider the problem of the collapsed chain on a lattice.

3.1. *Hamiltonian Paths and the Compact Phase*

We consider a d-dimensional hypercubic lattice of $N = L^d$ sites, and its associated Hamiltonian paths. We recall that a Hamiltonian path (HP) is a path which visits all sites of the lattice once and only once (Fig. 2). Hamiltonian paths have been often used to model collapsed polymer globules. In this approach, the entropy of collapsed phase can be obtained by evaluating the number of Hamiltonian paths on the lattice. For simplicity, we consider closed paths, but, as is well-known in polymer theory[37] boundary conditions play a role only in subdominant terms of the free energy.

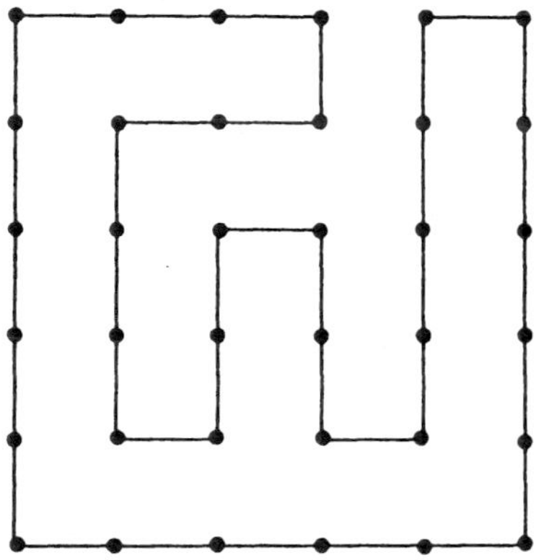

Fig. 2. A Hamiltonian path in $d = 2$ with $N_c = 14$ corners.

Thus we define the partition function $\mathcal{Z}$:

$$\mathcal{Z} = \sum_{\{HP\}} 1. \tag{3.1}$$

Note that $\mathcal{Z}$ is nothing but the total number of HP. Following Orland *et al.*[64] one may rewrite (3.1) as:

$$\mathcal{Z} = \lim_{n \to 0} \frac{1}{n} \frac{\int \prod_{\{\mathbf{r}\}} \mathrm{d}\varphi(\mathbf{r})\, e^{-A_G} \prod_{\{\mathbf{r}\}} \frac{1}{2}\varphi^2(\mathbf{r})}{\int \prod_{\{\mathbf{r}\}} \mathrm{d}\varphi(\mathbf{r})\, e^{-A_G}} \tag{3.2a}$$

where

$$A_G = \frac{1}{2} \sum_{rr'} \varphi(\mathbf{r}) \left(\Delta_{\mathbf{rr}'}\right)^{-1} \varphi(\mathbf{r}') . \tag{3.2b}$$

In Eq. (3.2), $\varphi(\mathbf{r})$ is a n-component field attached to each point $\mathbf{r}$ of the lattice, and the operator $\Delta_{\mathbf{rr}'}$ is 1 if $\mathbf{r}$ and $\mathbf{r}'$ are nearest-neighbour, and 0 otherwise. The denominator is a simple Gaussian integral, equal to $\exp(-\frac{n}{2} \log \det \Delta_{\mathbf{rr}'})$. It is equal to one in the limit $n \to 0$, and will be omitted henceforth.

To prove the identity between (3.2) and (3.1), namely that (3.2) indeed counts all HP, we use Wick's theorem (see Appendix 1). At each site $\mathbf{r}$, one attaches a factor $\frac{1}{2}\varphi^2(\mathbf{r})$, which can be viewed graphically as one line coming into, and one line coming out of site $\mathbf{r}$. The elementary contraction (see Eq. (3.2b)) reads:

$$\overline{\varphi_\alpha(\mathbf{r}) \cdot \varphi_\beta(\mathbf{r}')} = \delta_{\alpha\beta} \Delta_{\mathbf{rr}'} \tag{3.3}$$

where α and β denote any of the n-components of the field $\varphi(\mathbf{r})$. Since Eq. (3.3) conserves the index α, each connected part of the path yields a factor n through the summation over α. To deal with the one chain problem (or with a single connected path), one must therefore extract the term proportionnal to n, hence the factor $\lim_{n \to 0} \frac{1}{n}$ in Eq. (3.2a). By doing so, paths with more than one connected part, which have higher powers of n, are cancelled, and this proves the identity of (3.1) and (3.2). In the field of critical phenomena, and in particular in polymer physics, this limit has been first considered by P.G. de Gennes.[20]

The evaluation of (3.2) is out of reach, and we restrict ourselves to a simple approximation, namely the Saddle-Point Method (SPM). Minimizing the exponent of the numerator of (3.2a) with respect to $\varphi(\mathbf{r})$ yields:

$$\sum_{\mathbf{r}'} \Delta^{-1}_{\mathbf{rr}'} \varphi(\mathbf{r}') = \frac{2\varphi(\mathbf{r})}{\varphi(\mathbf{r})^2} . \tag{3.4}$$

If we assume periodic boundary conditions on the lattice, it is legitimate to look for a homogeneous solution to (3.4). We thus take a solution of the form:

$$\varphi(\mathbf{r}) = (\varphi, 0, \ldots) \tag{3.5a}$$

where φ is a constant, independent of $\mathbf{r}$. Using (3.4) we get $\varphi^2 = 2q$, where $q = 2d$ is the coordination number of the lattice. In the SPM, the partition function (3.4) reads:

$$Z = \left(\frac{q}{e}\right)^N \tag{3.5b}$$

where $e = \exp(1) = 2.718\cdots$. This result is in excellent agreement with numerical simulations and exact results, on various lattices, even in low dimensions (Derrida unpublished, Camacho and Thirumalai[65]). This is due to the fact that fluctuations around the SPM are indeed small.[21,64]

A different approach by Flory[14] yields a form similar to (3.5), with q replaced by $q-1$. This replacement spoils the agreement mentioned above. Equation (3.5b) shows that the compact phase has a sizeable conformational entropy $S/N = \log(q/e)$, and thus no definite three-dimensional shape: a polymer globule (compact phase) is a statistical mixture of an exponentially large number of conformations. (We remind the reader that these results are strongly dependent on the boundary conditions (see e.g. Gordon *et al.*[66]), assumed here to be periodic. We do not consider here the question of dynamics in the compact phase. This exponential number of conformations (and the energy barriers between them) will dramatically slow down all processes. In a related problem, the presence of an exponentially large number of metastable states has been shown to induce glassy dynamics.[67]

To sum up, we have modeled the collapsed phase of a single polymer chain below the θ point by lattice Hamiltonian paths. In this approach, one loses track of the relation between the spatial degrees of freedom and the curvilinear abcissa along the chain; thus, it cannot be easily generalized to heteropolymers. Note that the representation (3.2) is also of interest in other problems of statistical physics, in the case where $n \neq 0$.

3.2. *Alpha Helices in the Compact Phase*

Following Flory,[14] we take each link of the Hamiltonian path to represent a helical turn. Since hydrogen bonds have a tendency to favor long helices, that is to align the links of our model, we attribute an energy penalty ε to the breaking of a helix, that is whenever the Hamiltonian path makes a turn (corner), (see Fig. 3). This model has attracted a lot of attention in the theory of polymer melting.[14,68] The partition function of the system, at inverse temperature $\beta = \frac{1}{T}$, reads

$$\mathcal{Z} = \sum_{\{HP\}} e^{-\beta\varepsilon N_c\{\mathcal{H}\}} \tag{3.6}$$

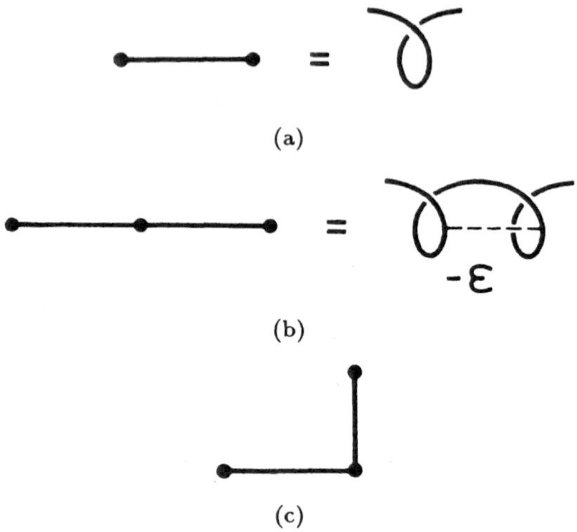

Fig. 3. (a) Definition of the model: one lattice link represents a helical turn. (b) Two aligned links allow for the formation of a hydrogen bond (of energy $-\varepsilon$). (c) Two unaligned links have zero energy.

where $\{HP\}$ denotes the ensemble of all Hamiltonian paths, and $N_c\{\mathcal{H}\}$ denotes the number of corners present in path $\mathcal{H}$. Following the previous section, we introduce on each site $\mathbf{r}$ and for each direction $\alpha = 1, \ldots, d$ an n-component real field $\varphi_\alpha(\mathbf{r})$. The partition function Z can be rewritten:

$$Z = \lim_{n \longrightarrow 0} \frac{1}{n} \int \prod_{\mathbf{r}} \prod_{\alpha=1}^{d} d\varphi_\alpha(\mathbf{r})\, e^{-A_G}$$

$$\times \prod_{\mathbf{r}} \left[\frac{1}{2} \sum_{\alpha=1}^{d} \varphi_\alpha^2(\mathbf{r}) + e^{-\beta\varepsilon} \sum_{1 \le \alpha < \gamma \le d} \varphi_\alpha(\mathbf{r}) \cdot \varphi_\gamma(\mathbf{r}) \right] \tag{3.7a}$$

with

$$A_G = \frac{1}{2} \sum_{\alpha=1}^{d} \sum_{\mathbf{r},\mathbf{r}'} \varphi_\alpha(\mathbf{r}) \cdot (\Delta_{\mathbf{r}\mathbf{r}'}^{\alpha})^{-1} \cdot \varphi_\alpha(\mathbf{r}') \tag{3.7b}$$

where the operator $\Delta_{\mathbf{r}\mathbf{r}'}^{\alpha}$ is 1 if $\mathbf{r}$ and $\mathbf{r}'$ are nearest-neighbour in direction α, and 0 otherwise. Note again that we have omitted the denominator in (3.7a), since it goes to unity when $n \to 0$.

In order to prove the equivalence of (3.6) and (3.7), we will use again Wick's theorem: we define, from (3.7a), the elementary contraction:

$$\overline{\varphi_{\alpha}^{(u)}(\mathbf{r}) \cdot \varphi_{\gamma}^{(v)}(\mathbf{r}')} = \delta_{uv}\delta_{\alpha\gamma}\Delta_{\mathbf{rr}'}^{\alpha} \tag{3.8}$$

where u and v are component indices of $\varphi_{\alpha}(\mathbf{r})$, running from 1 to n.

Expanding the product over $\{\mathbf{r}\}$ in (3.7a), we must choose at each site $\mathbf{r}$ either a term of the form $\frac{1}{2}\varphi_{\alpha}^{2}(\mathbf{r})$ or one of the form $e^{-\beta\varepsilon}\varphi_{\alpha}(\mathbf{r}) \cdot \varphi_{\gamma}(\mathbf{r})$, corresponding respectively to a path going straight through $\mathbf{r}$ in direction α, or to a path making a turn at $\mathbf{r}$, from direction α to direction γ. By contracting the fields according to (3.8), we construct a sum over all self-avoiding compact closed paths with appropriate weights, with an additional factor n (due to the summation over component index u) for each closed loop. As above, we extract the single chain contribution by taking the limit $n = 0$. This concludes the proof.

From this proof, it is clear that vacancies (empty sites) can easily be included in the model, by adding to the terms in brackets in (3.7a) a term of the form $e^{\beta\mu}$, where μ is the chemical potential for the vacancies.

In order to evaluate (3.7a), we will again resort to a mean-field theory, that is, a saddle point method (SPM).

The ground state of the model consists of straight paths, making turns on the surface of the lattice (their free energy being non-extensive, of the order L^{d-1}). A correct mean-field treatment should, *a priori*, take this one-dimensional character into account, but it has been shown elsewhere[21] that excellent results are obtained by using a straightforward isotropic homogeneous mean-field. We have also shown in this article that fluctuations around this isotropic mean field should not be included, since it spoils the quality of the results.

The mean field equations (saddle point on (3.7a)) read:

$$\sum_{\mathbf{r}'} \left(\Delta_{\mathbf{rr}'}^{\alpha}\right)^{-1} \varphi_{\alpha}(\mathbf{r}') = \frac{\varphi_{\alpha}(\mathbf{r}) + e^{-\beta\varepsilon}\sum_{\gamma(\neq\alpha)}\varphi_{\gamma}(\mathbf{r})}{\frac{1}{2}\sum_{\alpha'}\varphi_{\alpha'}^{2}(\mathbf{r}) + \frac{1}{2}e^{-\beta\varepsilon}\sum_{\gamma'\neq\alpha'}\varphi_{\alpha'}(\mathbf{r}) \cdot \varphi_{\gamma'}(\mathbf{r})} . \tag{3.9}$$

Assuming periodic boundary conditions, the isotropic homogenous mean-field solution takes the form $\varphi_{\alpha}(\mathbf{r}) = \varphi$ for any $\alpha = 1, \ldots, d$ and $\mathbf{r}$. We further break the $O(n)$ symmetry by choosing φ in a given direction, say 1. From (3.9), we obtain:

$$\varphi^{(1)} = \sqrt{\frac{4}{d}}\,. \tag{3.10a}$$

At this mean-field level, the free energy per site reads:

$$f = -T \log\left(\frac{q(\beta)}{e}\right) \tag{3.10b}$$

with

$$q(\beta) = 2 + 2(d-1)e^{-\beta\varepsilon} \tag{3.10c}$$

and $e = 2.71828\cdots$. Note that $q(\beta)$ plays the role of an effective coordination number.

On the other hand, the free energy is a decreasing function of temperature (since the entropy is positive), and as mentioned above, the ground state free energy per site vanishes (it is of order $1/L$). Thus, the free energy per site remains negative (or zero) at all temperatures.

There is a temperature T_F for which the effective coordination number $q(\beta)$ is equal to e, and f vanishes. For $d = 3$, $T_F = 0.58\ \varepsilon$. Below this temperature, the free energy remains at zero (Fig. 4).

Physically, there is a competition between the entropy gain of making turns, and the corresponding energy loss. At high temperature, the corners are mobile in the bulk, leading to a liquid-like structure, whereas at low temperature, the system is frozen, in stretched walks, with the corners

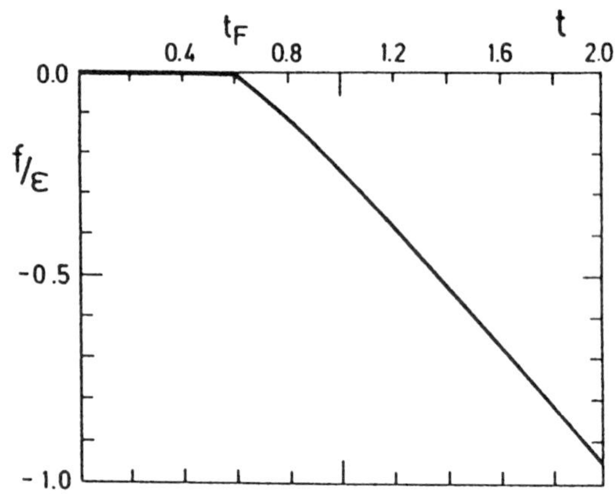

Fig. 4. Plot of the free energy per site (Eq. (3.10b)) versus temperature, for $d = 3$. t is the reduced temperature: $t = T/\varepsilon$.

expelled on the surface. The two regions are separated by a first order phase transition at T_F. The average length of a helix is given by

$$\ell = \frac{N\varepsilon}{U\left(\beta_F\right)} \tag{3.11}$$

where $U(\beta) = -\frac{\partial}{\partial\beta}\log Z$ is the internal energy.

At the freezing point, in $d = 3$, the length is equal to $\ell_F = 3.78$, and it is infinite ($\ell = L$) in the low temperature phase.

Note that ℓ_F corresponds to a typical number of amino acids per α-helix of the order of 15, since one link corresponds to a helical turn, that is 3.6 amino acids.

In a more elaborate treatment,[21] one finds a very weak temperature dependence, of the low temperature phase, but the overall freezing picture remains correct. Indeed, in the frozen phase, the typical length of a helix (in $d = 3$) is

$$\ell \simeq \frac{e^{6\beta\varepsilon}}{12}$$

yielding $\ell = 2592$ just below T_F (corresponding to 9330 residues). This length scale is clearly out of reach in any realistic protein, or polymer system. The (first order) complete freezing picture is thus adequate for any practical purpose, which corresponds to a crystal-like ground state, with non-extensive entropy. There is however a residual surface entropy (non-extensive), essentially equal to the number of HP in $d-1$ $(=2)$ dimensions.

For completeness, we now sketch the derivation of the situation with vacancies. Let us denote by Ω the number of lattice sites and N the number of monomers. If μ is the chemical potential for the vacancies, the associated fugacity is $K = \exp(\beta\mu)$. The partition function reads

$$Z = \lim_{n \longrightarrow 0} \frac{1}{n} \int \prod_{\mathbf{r}} \prod_{\alpha=1}^{d} \mathrm{d}\varphi_\alpha\left(\mathbf{r}\right) e^{-A_G}$$

$$\times \prod_{\mathbf{r}} \left[K + \frac{1}{2} \sum_{\alpha=1}^{d} \varphi_\alpha^2\left(\mathbf{r}\right) + e^{-\beta\varepsilon} \sum_{1 \leq \alpha < \gamma \leq d} \varphi_\alpha\left(\mathbf{r}\right) \cdot \varphi_\gamma\left(\mathbf{r}\right) \right] . \tag{3.12}$$

Denoting by ρ the vacancy concentration, the saddle-point method for φ yields:

$$\varphi = 4\varphi \frac{q(\beta)}{4K + dq(\beta)\varphi^2} . \tag{3.13}$$

There are two solutions: (i) $\varphi = 0$, corresponding to a pure vacancy phase, with free energy per site $f = -T\log(K)$; and (ii) $\varphi = (4(q(\beta) - K)/dq(\beta))^{1/2}$, corresponding to a mixture of polymer and vacancies, with free energy per site $f = -T\log(\frac{q(\beta)}{e}\exp(K/q(\beta)))$. If the vacancy concentration is small enough, the latter solution is the correct one. Since there are N monomers and $\Omega - N$ vacancies, according to (3.12), we must extract the coefficient of $K^{\Omega-N}$ in the expression of the partition function $Z = (\frac{q(\beta)}{e}\exp(K/q(\beta)))^{\Omega}$. The free energy per monomer, as a function of ρ reads:

$$f = -T\log(\frac{q(\beta)}{e}\rho^{-\frac{\rho}{1-\rho}}). \tag{3.14}$$

There exists a critical concentration $\rho_c = 0.13$ above which the first order freezing transition ceases to exist. This is due to the fact that mobile vacancies decrease the frustration: the entropy of mixing of the vacancies compensate the energy of corners, and may thus forbid the occurrence of freezing.

3.3. β-Sheets

We again use the Hamiltonian path formalism, with the slight modification that a node of a path is now to be interpreted as an amino acid.

The model[23] can be described in the following way. Consider a Hamiltonian path. To mimic the formation of CO–HN, (i.e. a hydrogen-bond (H-bond)), in β-sheets, we allow a H-bond (energy gain ε), whenever two pairs of aligned links belong to two non-intersecting neighboring strands (see Fig. 5)). We of course do not make any distinction between parallel and antiparallel sheets. The partition function reads:

$$Z = \sum_{\{HP\}} \sum_{\{\text{H-bonds}\}} e^{-\beta\varepsilon\, N_H(\mathcal{H})}. \tag{3.15}$$

The summation runs over all possible Hamiltonian paths $\mathcal{H}$, and over all possible sets of H-bonds compatible with the path (see below).

In order to give an integral representation of Z, in addition to the previous fields $\varphi_\alpha(\mathbf{r})$ which generate the paths, we need to introduce two scalar fields $\psi_\alpha^+(\mathbf{r})$ and $\psi_\alpha(\mathbf{r})$ which respectively initiate and terminate, and H-bond at site $\mathbf{r}$ in direction α. We obtain

$$Z = \lim_{n\longrightarrow 0}\frac{1}{n}\frac{\int\prod_{\alpha,\mathbf{r}} d\varphi_\alpha(\mathbf{r})\, d\psi_\alpha(\mathbf{r})\, d\psi_\alpha^+(\mathbf{r})\, e^{-A_G}\prod_{\mathbf{r}} D(\mathbf{r})}{\int\prod_{\alpha,\mathbf{r}} d\varphi_\alpha(\mathbf{r})\, d\psi_\alpha(\mathbf{r})\, d\psi_\alpha^+(\mathbf{r})\, e^{-A_G}} \tag{3.16a}$$

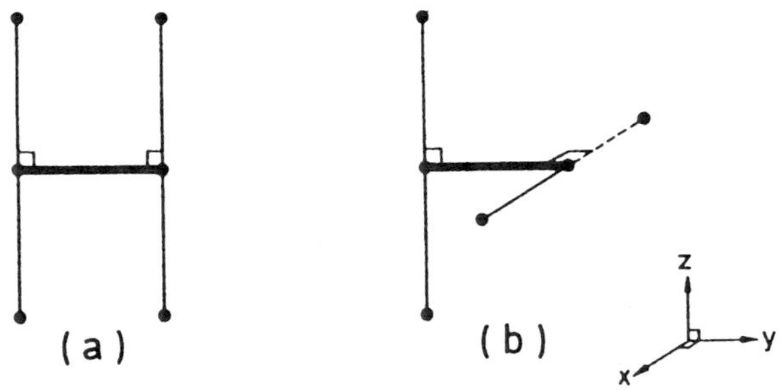

Fig. 5. The two different possible types of H-bonds in the β-sheet model.

with

$$A_G = \sum_{\alpha=1}^{d} \sum_{\mathbf{rr}'} \left[\frac{1}{2} \varphi_\alpha(\mathbf{r}) \left(\Delta_{\mathbf{rr}'}^{\alpha}\right)^{-1} \varphi_\alpha(\mathbf{r}') + \psi_\alpha^+(\mathbf{r}) \left(\Delta_{\mathbf{rr}'}^{\alpha+}\right)^{-1} \psi_\alpha(\mathbf{r}') \right] \tag{3.16b}$$

and

$$D(\mathbf{r}) = \sum_{\alpha} \frac{1}{2} \varphi_\alpha^2(\mathbf{r}) \; G_\alpha(\mathbf{r}) + \sum_{\alpha<\delta} \varphi_\alpha(\mathbf{r}) \cdot \varphi_\delta(\mathbf{r}) \tag{3.16c}$$

and

$$G_\alpha(\mathbf{r}) = 1 + e^{\beta\varepsilon/2} \sum_{\gamma(\neq\alpha)} \left(\psi_\gamma^+(\mathbf{r}) + \psi_\gamma(\mathbf{r})\right) + e^{\beta\varepsilon} \sum_{\gamma(\neq\alpha)} \psi_\gamma^+(\mathbf{r}) \psi_\gamma(\mathbf{r}) \,. \tag{3.16d}$$

The operator $\Delta_{\mathbf{rr}'}^{\alpha}$ is the one defined in Sec. 3.2, whereas $\Delta_{\mathbf{rr}'}^{\alpha+} = \delta(\mathbf{r}' - (\mathbf{r} + \mathbf{e}_\alpha))$, where $\mathbf{e}_\alpha$ is the unit vector in direction α.

Expanding the product over $\vec{r}$ in (3.16a) at each site $\mathbf{r}$, we must choose either (i) a term of the form $\frac{1}{2}\varphi_\alpha^2(\mathbf{r})G_\alpha(\mathbf{r})$, or (ii) a term of the form $\varphi_\alpha(\mathbf{r}) \cdot \varphi_\delta(\mathbf{r})$. The latter (ii) represents a corner and does not allow for a H-bond. The former (i) represents a path going straight through $\mathbf{r}$ in direction α, and according to (3.16d), allows for four possibilities, namely, no bond with weight 1, one bond entering or leaving $\mathbf{r}$ in direction $\gamma(\neq \alpha)$ with weight $e^{\beta\varepsilon/2}$, and, one bond entering and one leaving at $\mathbf{r}$ in direction $\gamma(\neq \alpha)$ with weight $e^{\beta\varepsilon}$.

As in Sec. 3.2, the identification of (3.15) and (3.16a) proceeds through the use of the Wick's theorem, with the following elementary contractions:

$$\begin{aligned}
\overline{\varphi_\alpha^{(u)}(\mathbf{r})\varphi_\beta^{(v)}(\mathbf{r}')} &= \delta_{\alpha\beta}\delta_{uv}\delta^\alpha_{\mathbf{rr}'} \\
\overline{\psi_\alpha^+(\mathbf{r})\psi_\beta(\mathbf{r}')} &= \delta_{\alpha\beta}\Delta^{\alpha+}_{\mathbf{rr}'} \\
\overline{\psi_\alpha(\mathbf{r})\psi_\beta(\mathbf{r}')} &= \overline{\psi_\alpha^+(\mathbf{r})\psi_\beta^+(\mathbf{r}')} = 0\,.
\end{aligned} \tag{3.17}$$

These contractions indeed generate the required partition function. As mentioned in the previous section, vacancies can be included in the model by a slight modification of $D(\mathbf{r})$.

As in the previous section, we study this model in a mean-field approximation. The mean-field equations read:

$$\sum_{\mathbf{r}'}(\Delta^\alpha_{\mathbf{rr}'})^{-1}\varphi_\alpha(\mathbf{r}') = \frac{\varphi_\alpha(\mathbf{r})G_\alpha(\mathbf{r}) + \sum_{\delta(\neq\alpha)}\varphi_\delta(\mathbf{r})}{D(\mathbf{r})}, \tag{3.18a}$$

$$\sum_{\mathbf{r}'}(\Delta^{\alpha+}_{\mathbf{rr}'})^{-1}\psi_\alpha(\mathbf{r}') = \frac{\frac{1}{2}\sum_{\gamma(\neq\alpha)}\varphi_\gamma^2[e^{\beta\varepsilon/2} + e^{\beta\varepsilon}\psi_\alpha(\mathbf{r})]}{D(\mathbf{r})}, \tag{3.18b}$$

$$\sum_{\mathbf{r}'}(\Delta^{\alpha+}_{\mathbf{r}'\mathbf{r}})^{-1}\psi_\alpha^+(\mathbf{r}') = \frac{\frac{1}{2}\sum_{\gamma(\neq\alpha)}\varphi_\gamma^2[e^{\beta\varepsilon/2} + e^{\beta\varepsilon}\psi_\alpha^+(\mathbf{r})]}{D(\mathbf{r})}. \tag{3.18c}$$

Restricting ourselves to an isotropic homogeneous solution, $[\varphi_\alpha(\mathbf{r}) = \varphi,\ \psi_\alpha^+(\mathbf{r}) = \psi_\alpha(\mathbf{r}) = \psi]$, we get:

$$\varphi^2 = \frac{4}{d} \tag{3.19a}$$

and ψ satisfies the equation:

$$\psi = \frac{2(d-1)}{d}\,\frac{e^{\beta\varepsilon/2}\left(1 + e^{\beta\varepsilon/2}\psi\right)}{D} \tag{3.19b}$$

with

$$D = 2d + 4(d-1)e^{\beta\varepsilon/2}\psi + 2(d-1)e^{\beta\varepsilon}\psi^2\,. \tag{3.19c}$$

The free energy per site reads:

$$f = T\left\{1 + \mathrm{d}\psi^2 - \log D\right\}\,. \tag{3.20}$$

As in the previous model, the ground state configuration corresponds to frozen fully stretched configurations saturated with transverse H-bonds (stacked β-sheets), with energy per site equal to $-\varepsilon$. Note that, compared to Eq. (3.7), the origin of the energies was shifted by $-\varepsilon$. Therefore, we have

$$f \leq -\varepsilon \,. \tag{3.21}$$

Solving numerically Eq. (3.19b), we find again a first order freezing transition at a temperature T_F where $f = -\varepsilon$. In $d = 3$, we get $T_F = 0.86\,\varepsilon$.

The physics of this model is very similar to that of the model of Sec. 3.2, namely, a liquid-like high temperature phase with no definite β-sheet structure, and a low temperature frozen phase, consisting of stacks of parallel β-sheets.

A non-isotropic type of mean-field, as in Bascle *et al.*,[21] would probably again induce a very weak temperature dependence in the low temperature phase, but the biological robustness of the two-dimensional β-sheets should make the temperature dependence even weaker than in the previous case. Furthermore, due to the higher dimensionality of the β-sheets, we believe the isotropic mean-field approximation to be even better than in the previous case. Finally, vacancies are easily included along the lines of the previous section.

3.4. *Conclusion*

We have studied in a mean-field approximation two models for the formation of secondary structures in globular proteins (in the thermodynamic limit).

These models exhibit a first order transition between a high-temperature liquid-like compact phase and a quasi-frozen low-temperature phase. In this quasi-frozen phase, secondary structures span the whole system. In our models, the denaturation proceeds in two steps: a first step, where secondary structures unfreeze and become mobile. This phase seems closely related to the molten globule[24,25]: In a second step (not described by the above models) the compact globule would unfold, into a swollen coil.

A naive transposition from the melting of crystals suggests to study the scattering function $g(q)$. Indeed, the existence of order (related to secondary structures) in proteins has a signature in the scattering function. As noted by Ptitsyn,[24,25] the molten globule phase and the native protein have a slightly different $g(q)$. The smallness of the difference is due to the fact that, as we have seen in Sec. 2.2, the scattering function tests essentially fluctuations of monomer concentration. These fluctuations are similar in both phases, since they are both compact. In the lattice models depicted above, the monomer concentration is constant, and in order to see differences in the scattering function, it is necessary to decorate the backbone chain, by inserting side-links, which mimick the side-chains. In doing so, it is clearly seen that the scattering function of the molten globule (unfrozen phase) has much less structure than that of the crystal-like native protein (frozen phase).[69]

As far as comparison with proteins is concerned, several caveats are in order:

(i) The lattice approach, where one link is taken as a helix turn or an amino acid, may be too crude.

(ii) The models are studied in the thermodynamic limit ($N = L^d$ going to infinity). Choosing arbitrarily a size of $N = 10^2 - 10^3$ amino acids for a protein, the quasi-frozen phase is in fact completely frozen in the bulk, since thermal fluctuations would be present only in much larger systems, $N \sim 10^8$ amino acids (see Sec. 2). However, surface modes should not be frozen, and a more detailed study of finite size effects is in order.

(iii) Since we have considered only compact structures, we have implicitly assumed the total collapse (van der Waals) energy E_c to be much larger than the total intramolecular H-bond energy E_H. In real systems, the ratio $\frac{E_c}{E_H}$ seems rather large, of order 20 (J. Garnier, private communication).

The opposite limit, where $\frac{E_c}{E_H} < 1$ is closely related to the standard "helix-coil" transition, and can be modeled through the use of a one-dimensional Ising model.[70] If one neglects the couplings between the different helical portions of the chain, the typical length of a helix is governed by a balance between energy and entropy, and is given by:

$$l_H \sim \exp(\beta \varepsilon_H) \tag{3.22a}$$

where ε_H is the microscopic hydrogen bond energy (called ε in the previous subsections). The compact phase will occur below a critical temperature T_c given by:

$$T_c \sim l_H(T_c)\epsilon_c \tag{3.22b}$$

where ϵ_c denotes the microscopic attractive part of the Van der Waals energy. The physical image emerging from this approach is completely different from the one depicted in the previous sections, since it corresponds to a θ point of helical portions.

(iv) Finally, let us note that amino acids have given probabilities to be present in α-helices, β-sheets or turns (i.e. no secondary structure). This is reflected in the Ramachandran plots[1]; a more elaborate approach should include these weights.

4. Protein Folding and Random Heteropolymers

4.1. *Introduction*

To view proteins as random polymers may seem very unrealistic to the biologist. The usual sentence "evolution has resulted in ..." is often opposed to the physicist. Nevertheless, we believe that the study of various random polymers (or heteropolymers) may be helpful in catching some aspects of protein folding (quasi-uniqueness of the ground-state coding, of the structure in the primary sequence, etc.). Here, by protein folding, we mean the description of global tertiary folding, with no mention of secondary structures. This approach is complementary to the Hamiltonian path description of Sec. 3.

A (small) problem with this approach is that one is faced with spin glass like technicalities (replica symmetry breaking, utrametricity, etc.). In this review, we have tried to adopt a middle course. Some useful facts about spin glasses are gathered in Appendices 4 and 5.

The random polymers we have in mind are described by a partition function, which is similar to Eqs. (2.1) and (2.2) of Sec. 2. The randomness is described by the two-body Hamiltonian: the interaction $v(\mathbf{r}_i, \mathbf{r}_j)$ is replaced by a quenched random interaction $v_{ij}(\mathbf{r}_i, \mathbf{r}_j)$, which depends on the chemical nature of links i and j. These links may be here viewed as coarse-grained representations of amino acids. We therefore have:

$$Z\left(\{v_{ij}\}\right) = \int \prod_{i=1,\ldots,N+1} d\mathbf{r}_i \prod_{i=1,\ldots,N} g(\mathbf{r}_i, \mathbf{r}_{i+1}) \exp\left(-\beta\mathcal{H}(\{v_{ij}\})\right) \qquad (4.1)$$

where

$$\beta\mathcal{H}\left(\{v_{ij}\}\right) = \beta\mathcal{H} = \frac{1}{2}\sum_{i\neq j} v_{ij}(\mathbf{r}_i, \mathbf{r}_j) + \frac{1}{6}\sum_{i\neq j\neq k} w(\mathbf{r}_i, \mathbf{r}_j, \mathbf{r}_k) + \ldots \qquad (4.2)$$

and $g(\mathbf{r}_i, \mathbf{r}_{i+1})$ describes the chain constraint.

In Eqs. (4.1) and (4.2), we have emphasized the fact that the partition function Z depends on the quenched variables v_{ij}. In other words, the thermodynamics variables are the amino acids positions $\mathbf{r}_i$, and the quenched nature of the v_{ij} means that the chain is not allowed to break up and recombine (the primary sequence is fixed and random). At this stage, the analogy with spin glasses should be clear ($\mathbf{S}_i \to \mathbf{r}_i$, $J_{ij} \to v_{ij}$); see Appendix 4. We have illustrated the heteropolymer approach to protein folding in the following way. Section 4.2 discusses various kinds of randomness, and their "biological" interpretation. In Sec. 4.3, we study a particular kind of randomness, namely, the randomly hydrophobic-hydrophilic chain, which is certainly relevant to folding in water. We use for this model a (replica) variational function such as the one discussed in Sec. 2.2. The "real" spin glass-like random chain, where the v_{ij} are independent random variables, will be treated in Sec. 4.4. We will consider first a mean-field like approach, which is somewhat academic, but seems to capture the essence of the problem, namely the equivalence with a Random Energy Model (REM). Then, we will sketch the variational approach to the finite dimensional case. Finally, Sec. 4.5 adumbrates the neural network approach to protein folding, in the framework of a Mattis-like random sequence model.

4.2. *Quenched Disorder in Polymers*

Let us first mention a small difference between the case of spin glasses and random heteropolymers. We are interested in the statistical mechanics of a long chain, with disordered v_{ij} interactions (we do not consider the possibility of disordered three (and more)-body interactions). The domain vs. domain wall of Eq. (A4.3) has to be modified as follows: We consider a very dilute "soup" of M random chains, each of length N, and each representing a different random primary sequence (RPS). The free energy per chain of the soup is given by:

$$F = \frac{1}{M} \sum_{\text{RPS}} F(\text{RPS})\,. \tag{4.3a}$$

This expression neglects the interchain interactions, and is thus valid provided (i) the soup is dilute enough and (ii) the interactions are short-ranged. For a large M, one may interpret F as an average overall possible choices of RPS, that is, over all possible realizations of the v_{ij}. Denoting this average by $\overline{F}$, we have:

$$F = \overline{F} = \int \prod P(\{v_{ij}\}) F(\{v_{ij}\}) d(\{v_{ij}\})\,. \tag{4.3b}$$

This point of view clearly leads to the use of replicas, in order to perform the quenched average.

If one wants to study a specific primary sequence, i.e. a given set of v_{ij} (without any averaging procedure), one must resort to the self-consistent field method, described in Appendix 3. Equations (A3.17) must be modified, to account for the sequence dependence of the v_{ij},: this would require an involved numerical treatment, which has not been undertaken so far.

We now consider various possible choices of the two-body Hamiltonian $v_{ij}(\mathbf{r}_i, \mathbf{r}_j)$.

(i) The random hydrophilic-hydrophobic (LB) chain

$$v_{ij}(\mathbf{r}_i, \mathbf{r}_j) = v_0(\mathbf{r}_i, \mathbf{r}_j) + \beta(\lambda_i + \lambda_j) f(\mathbf{r}_i, \mathbf{r}_j) \tag{4.4}$$

where v_0 is a repulsive (excluded volume) short range interaction, λ_i is a quenched random variable describing the hydrophilicity degree of link i, and f is a short range repulsive function. This particular choice[71,72] will be studied in Sec. 4.3.

(ii) The "random-bond" chain
The interactions $v_{ij}(\mathbf{r}_i, \mathbf{r}_j)$, $i < j$ are random independent variables, whose positive mean describe excluded volume effects. Since two couples (i, j) and (i', j') of the same chemical nature and at different positions of the primary sequence are characterized by independent values of the interactions $v_{ij}(\mathbf{r}_i, \mathbf{r}_j)$ and $v_{i'j'}(\mathbf{r}_{i'}, \mathbf{r}_{j'})$, this model assumes a very strong influence of the environment. This case[26–28,73] will be studied in Sec. 4.4.

(iii) The "random sequence" chain

$$v_{ij}(\mathbf{r}_i, \mathbf{r}_j) = v_0(\mathbf{r}_i, \mathbf{r}_j) + q_i q_j f(\mathbf{r}_i, \mathbf{r}_j) \tag{4.5}$$

where v_0 refers to excluded volume effects, and the second term describes the interaction of randomly charged monomers. Each monomer i is here characterized[27,74] by a charge q_i, regardless of the biochemical environment. This model can be generalized *à la* Hopfield,[75–78] by assigning M "charges" $q_i^{(p)}$ to monomer i, yielding an interaction:

$$v_{ij}(\mathbf{r}_i, \mathbf{r}_j) = v_0(\mathbf{r}_i, \mathbf{r}_j) + \sum_{p=1}^{M} q_i^{(p)} q_j^{(p)} f_p(\mathbf{r}_i, \mathbf{r}_j) \,. \tag{4.6}$$

These models, which are considered in Sec. 4.5, illustrate how the low temperature phase may be linked with (or coded in) the primary sequence "charges" $q_i^{(p)}$. The corresponding spin models are described in Appendix 5.

4.3. *The Random Hydrophilic-Hydrophobic Chain*

We consider the same lattice model as in Sec. 2.1, with a monomer-solvent molecule interaction given by (see Eq. (2.4)):

$$\mathcal{H}_{ms} = -\sum_{i=1}^{N+1}\sum_{\alpha=1}^{\mathcal{N}} \lambda_i \delta(\mathbf{r}_i - \mathbf{R}_\alpha) \tag{4.7}$$

where the R_α are the positions of the solvent molecules. Here, the λ_i are independent quenched random variables. A positive or negative λ corresponds to a hydrophilic or hydrophobic monomer respectively. Following the procedure of Sec. 2.1, we can eliminate the solvent molecules (using the incompressibility condition (2.7)), and we obtain:

$$\mathcal{H}_{ms} = \sum_{i,j} \lambda_i \; \delta(\mathbf{r}_i - \mathbf{r}_j)\,. \tag{4.8}$$

The full Hamiltonian, including the direct monomer–monomer (van der Waals, etc.) interaction ($v_0(\mathbf{r}_i, \mathbf{r}_j) = v_0\delta(\mathbf{r}_i - \mathbf{r}_j)$) reads:

$$\begin{aligned}\beta\mathcal{H}(\lambda_i) = &\frac{1}{2}\sum_{i\neq j} \left[v_0 + \beta(\lambda_i + \lambda_j)\right] \delta(\mathbf{r}_i - \mathbf{r}_j)\\ &+ \frac{1}{6}\sum_{i\neq j\neq k} w_3 \; \delta(\mathbf{r}_i - \mathbf{r}_j)\; \delta(\mathbf{r}_j - \mathbf{r}_k)\\ &+ \frac{1}{24}\sum_{i\neq j\neq k\neq l} w_4 \; \delta(\mathbf{r}_i - \mathbf{r}_j)\; \delta(\mathbf{r}_j - \mathbf{r}_k)\; \delta(\mathbf{r}_k - \mathbf{r}_l)\,.\end{aligned} \tag{4.9}$$

We have also included in (4.9) a repulsive four-body term, for reasons that will become clear further on. This model is studied in details by Garel *et al.*[72]

The partition function reads:

$$Z(\lambda_i) = \int \prod_{i=1,\ldots,N+1} d\mathbf{r}_i \prod_{i=1,\ldots,N} g(\mathbf{r}_i, \mathbf{r}_{i+1}) \; \exp(-\beta\mathcal{H}(\lambda_i)) \tag{4.10}$$

leading to an average free energy:

$$\overline{F} = \int \prod d\lambda_i \; P(\lambda_i) \; F(\lambda_i) \tag{4.11a}$$

where

$$F(\lambda_i) = -T \; \log Z(\lambda_i)\,. \tag{4.11b}$$

Our calculations will be performed with the particular choice:

$$P(\lambda_i) = \frac{1}{\sqrt{2\pi\lambda^2}} \exp\left(-\frac{(\lambda_i - \lambda_0)^2}{2\lambda^2}\right)\,. \tag{4.12}$$

With our conventions, a positive λ_0 corresponds to a majority of hydrophilic links.

Rewriting (4.9) and (4.10) in a continuous form, we have:

$$Z(\lambda(s)) = \int \mathcal{D}\mathbf{r}(s) \exp\left(-\frac{d}{2a^2}\int_0^N ds \left(\frac{d\mathbf{r}(s)}{ds}\right)^2 - \beta\mathcal{H}(\lambda(s))\right) \tag{4.13a}$$

with

$$\begin{aligned}
\beta\mathcal{H}(\lambda(s)) &= \frac{1}{2}\int_0^N ds \int_0^N ds' \,(v_0 + \beta(\lambda(s) + \lambda(s')))\,\delta(\mathbf{r}(s) - \mathbf{r}(s')) \\
&+ \frac{w_3}{6}\int_0^N ds \int_0^N ds' \int_0^N ds'' \;\delta(\mathbf{r}(s) - \mathbf{r}(s'))\;\delta(\mathbf{r}(s') - \mathbf{r}(s'')) \\
&+ \frac{w_4}{24}\int_0^N ds \int_0^N ds' \int_0^N ds'' \int_0^N ds''' \;\delta(\mathbf{r}(s) - \mathbf{r}(s')) \\
&\quad \delta(\mathbf{r}(s') - \mathbf{r}(s''))\;\delta(\mathbf{r}(s'') - \mathbf{r}(s'''))
\end{aligned} \tag{4.13b}$$

where we have implicitly assumed that the origin of the chain is fixed at $\mathbf{0}$ and that the extremity is free.

To perform the quenched average (4.11), we introduce replicas (see Appendix 4): denoting the replica indices by $a, b, \ldots$ we get:

$$\begin{aligned}
\overline{Z^n} &= \int \prod_{a=1}^{n} \mathcal{D}\mathbf{r}_a(s) \exp\left(-\frac{d}{2a^2}\int_0^N ds \sum_a \left(\frac{d\mathbf{r}_a(s)}{ds}\right)^2 - A\right) \\
&\cdot \exp\left(\frac{\beta^2\lambda^2}{2}\int_0^N ds \int_0^N ds' \int_0^N ds'' \sum_{a,b} \delta(\mathbf{r}_a(s) - \mathbf{r}_a(s'))\delta(\mathbf{r}_b(s) - \mathbf{r}_b(s''))\right)
\end{aligned} \tag{4.14a}$$

with

$$A = \frac{1}{2}(v_0 + 2\beta\lambda_0) \int_0^N ds \int_0^N ds' \sum_a \delta(\mathbf{r}_a(s) - \mathbf{r}_a(s'))$$

$$+\frac{w_3}{6} \int_0^N ds \int_0^N ds' \int_0^N ds'' \sum_a \delta(\mathbf{r}_a(s) - \mathbf{r}_a(s'))\, \delta(\mathbf{r}_a(s') - \mathbf{r}_a(s''))$$

$$+\frac{w_4}{24} \int_0^N ds \int_0^N ds' \int_0^N ds'' \int_0^N ds'''$$

$$\cdot \sum_a \delta(\mathbf{r}_a(s) - \mathbf{r}_a(s'))\, \delta(\mathbf{r}_a(s') - \mathbf{r}_a(s''))\delta(\mathbf{r}_a(s'') - \mathbf{r}_a(s'''))\,. \tag{4.14b}$$

Introducing the parameters $q_{ab}(\mathbf{r}, \mathbf{r}')$, with $a < b$, and $\rho_a(\mathbf{r})$ by:

$$\begin{aligned} q_{ab}(\mathbf{r}, \mathbf{r}') &= \int_0^N ds\ \delta(\mathbf{r}_a(s) - \mathbf{r})\delta(\mathbf{r}_b(s) - \mathbf{r}') \\ \rho_a(\mathbf{r}) &= \int_0^N ds\ \delta(\mathbf{r}_a(s) - \mathbf{r}) \end{aligned} \tag{4.15}$$

we may write Eqs. (4.14) as:

$$\begin{aligned} \overline{Z^n} = \int \mathcal{D}q_{ab}(\mathbf{r}, \mathbf{r}')\mathcal{D}\hat{q}_{ab}(\mathbf{r}, \mathbf{r}')\mathcal{D}\rho_a(\mathbf{r})\mathcal{D}\phi_a(\mathbf{r}) \\ \cdot \exp\left(G(q_{ab}, \hat{q}_{ab}, \rho_a, \phi_a) + \log\zeta(\hat{q}_{ab}, \phi_a)\right) \end{aligned} \tag{4.16a}$$

where $\hat{q}_{ab}(\mathbf{r}, \mathbf{r}')$ and $\phi_a(\mathbf{r})$ are the Lagrange multipliers associated with (4.15), and:

$$\begin{aligned} &G(q_{ab}, \hat{q}_{ab}, \rho_a, \phi_a) \\ &= \int d\mathbf{r} \sum_a \left(i\rho_a(\mathbf{r})\phi_a(\mathbf{r}) - (v_0 + 2\beta\lambda_0)\, \frac{\rho_a^2(\mathbf{r})}{2} - \frac{w_3'}{6}\rho_a^3(\mathbf{r}) - \frac{w_4}{24}\rho_a^4(\mathbf{r}) \right) \\ &+ \int d\mathbf{r} \int d\mathbf{r}' \sum_{a<b} \left(iq_{ab}(\mathbf{r}, \mathbf{r}')\hat{q}_{ab}(\mathbf{r}, \mathbf{r}') + \beta^2\lambda^2\, q_{ab}(\mathbf{r}, \mathbf{r}')\rho_a(\mathbf{r})\rho_b(\mathbf{r}') \right) \end{aligned} \tag{4.16b}$$

and

$$\zeta(\hat{q}_{ab}, \phi_a) = \int \prod_a \mathcal{D}\mathbf{r}_a(s)$$

$$\times \exp\left(- \frac{d}{2a^2} \int_0^N ds \sum_a \dot{\mathbf{r}}_a^2 - i \int_0^N ds \sum_a \phi_a(\mathbf{r}_a(s)) \right.$$
$$\left. - i \int_0^N ds \sum_{a<b} \hat{q}_{ab}(\mathbf{r}_a(s), \mathbf{r}_b(s)) \right). \tag{4.16c}$$

In Eq. (4.16b), we have defined:

$$w_3' = w_3 - 3\beta^2 \lambda^2 . \tag{4.16d}$$

According to Appendix 3, we can rewrite:

$$\zeta(\hat{q}_{ab}, \phi_a) = \int \prod_a d\mathbf{r}_a \langle \mathbf{r}_1 \cdots \mathbf{r}_n | e^{-NH_n(\hat{q}_{ab}, \phi_a)} | \mathbf{0} \cdots \mathbf{0} \rangle \tag{4.17a}$$

where the H_n is a "quantum-like" $n \to 0$ Hamiltonian, given by:

$$H_n = -\frac{a^2}{2d} \sum_a \nabla_a^2 + \sum_a i\phi_a(\mathbf{r}_a) + \sum_{a<b} i\hat{q}_{ab}(\mathbf{r}_a, \mathbf{r}_b) . \tag{4.17b}$$

Anticipating some kind of (hydrophobically-driven) collapse, we assume that we can use ground-state dominance to evaluate (4.17), and write, omitting the non-extensive prefactors of (A3.11a):

$$\zeta(\hat{q}_{ab}, \phi_a) \simeq e^{-NE_0(\hat{q}_{ab}, \phi_a)}$$
$$= \exp\left(-N \min_{\{\Psi(\mathbf{r})\}} \{ \langle \Psi | H_n | \Psi \rangle - E_0 \left(\langle \Psi | \Psi \rangle - 1 \right) \} \right) \tag{4.18}$$

where E_0 is the ground state energy of H_n. At this point, the problem is still untractable, and we make the extra approximation of saddle-point method (SPM). The extremization with respect to q_{ab} reads:

$$i\hat{q}_{ab}(\mathbf{r}, \mathbf{r}')$$
$$= -\beta^2 \lambda^2 \rho_a(\mathbf{r}) \, \rho_b(\mathbf{r}') .$$

This equation shows that replica symmetry is not broken (at least in the mean field approximation), since $\hat{q}_{ab}$ is a product of two single-replica quantities ρ_a. To get more analytic information, we follow Appendix 3, and use the Rayleigh–Ritz variational principle (see Eq. (A3.8a)). Due to the absence of replica symmetry breaking (RSB), we further restrict the variational wave-function space to Hartree-like replica-symmetric wave-functions, and write:

$$\Psi(\mathbf{r}_1, \ldots, \mathbf{r}_n) = \prod_{a=1}^{n} \varphi(\mathbf{r}_a)\,. \tag{4.19}$$

Because of replica symmetry, we can omit replica indices and take the $n \to 0$ limit easily. The variational free energy now reads:

$$\begin{aligned} &-\beta\overline{F}(q, \hat{q}, \rho, \phi, \varphi) \\ &= \int d\mathbf{r} \left(i\rho(\mathbf{r})\phi(\mathbf{r}) - (v_0 + 2\beta\lambda_0)\,\frac{\rho^2(\mathbf{r})}{2} - \frac{w_3'}{6}\rho^3(\mathbf{r}) - \frac{w_4}{24}\rho^4(\mathbf{r}) \right) \\ &- \frac{1}{2}\int d\mathbf{r} \int d\mathbf{r}' \left(iq(\mathbf{r},\mathbf{r}')\hat{q}(\mathbf{r},\mathbf{r}') + \beta^2\lambda^2\, q(\mathbf{r},\mathbf{r}')\rho(\mathbf{r})\rho(\mathbf{r}') \right) \\ &- N\left\{ \int d\mathbf{r}\ \varphi(\mathbf{r}) \left(-\frac{a^2}{2d}\nabla^2 + i\,\phi(\mathbf{r}) \right) \varphi(\mathbf{r}) - \frac{i}{2}\int d\mathbf{r} \right. \\ &\left. \cdot \int d\mathbf{r}'\hat{q}(\mathbf{r},\mathbf{r}')\varphi^2(\mathbf{r})\varphi^2(\mathbf{r}') \right\} + NE_0 \left(\int d^d r \varphi^2(\mathbf{r}) - 1 \right). \end{aligned} \tag{4.20}$$

The SPM equations read:

$$\begin{aligned} \rho(\mathbf{r}) &= N\varphi^2(\mathbf{r})\,, \\ q(\mathbf{r},\mathbf{r}') &= N\varphi^2(\mathbf{r})\varphi^2(\mathbf{r}')\,, \\ i\phi(\mathbf{r}) &= (v_0 + 2\beta\lambda_0)\rho(\mathbf{r}) + \frac{w_3'}{2}\rho^2(\mathbf{r}) + \frac{w_4}{6}\rho^3(\mathbf{r}) + \beta^2\lambda^2 \int d\mathbf{r}' q(\mathbf{r},\mathbf{r}')\rho(\mathbf{r}')\,, \\ i\hat{q}(\mathbf{r},\mathbf{r}') &= -\beta^2\lambda^2\rho(\mathbf{r})\rho(\mathbf{r}')\,. \end{aligned} \tag{4.21}$$

We still have to minimize with respect to the normalized wave-function $\varphi(\mathbf{r})$. This leads to a very complicated nonlinear Schrödinger equation, and we shall restrict ourselves to a one-parameter family of Gaussian wave-functions of the form:

$$\varphi(\mathbf{r}) = \left(\frac{1}{2\pi R^2} \right)^{d/4} \exp(-\frac{\mathbf{r}^2}{4R^2}) \tag{4.22}$$

where R is the only variational parameter.

Using Eqs. (4.20) and (4.21), the variational free energy per monomer ($\overline{f} = \overline{F}/N$) reads:

$$\beta\overline{f} = \frac{a^2}{2d}\int d\mathbf{r}\,(\nabla\varphi(\mathbf{r}))^2 + \frac{(v_0+2\beta\lambda_0)}{2}\,N\int d\mathbf{r}\varphi^4(\mathbf{r})\,. \tag{4.23}$$
$$+N^2\left\{\frac{w_3'}{6}\int d\mathbf{r}\varphi^6(\mathbf{r}) + \frac{\beta^2\lambda^2}{2}\left(\int d\mathbf{r}\varphi^4(\mathbf{r})\right)^2\right\} + \frac{w_4}{24}N^3\int d\mathbf{r}\varphi^8(\mathbf{r})\,.$$

Using the Gaussian integrals of Appendix 1, and the value of w_3' given in (4.16d), the free energy reads:

$$\beta\overline{f} = \frac{a^2}{8R^2} + \frac{1}{(2\sqrt{\pi})^d}\frac{(v_0+2\beta\lambda_0)}{2}\frac{N}{R^d}$$
$$+\left(\frac{1}{(2\pi\sqrt{3})^d}\frac{w_3}{6} - \frac{1}{(2\pi)^d}\frac{\beta^2\lambda^2}{2}(3^{-d/2}-2^{-d})\right)\left(\frac{N}{R^d}\right)^2$$
$$+\left(\frac{1}{(32\pi^3)}\right)^{d/2}\frac{w_4}{24}\left(\frac{N}{R^d}\right)^3\,. \tag{4.24}$$

At low temperatures, one has to study the sign of the third term of Eq. (4.24). The repulsive four-body term is necessary to yield a stable theory at low temperatures, when the coefficient of $(\frac{N}{R^d})^2$ changes sign, due to disorder fluctuations. A detailed study yields the following results:

(i) $\lambda_0 > 0$: The Hydrophilic Case

In our approximation, we find that there is a first order transition towards a collapsed phase ($R \sim N^{1/d}$) induced by a negative three-body term (and stabilized by a positive four-body term). This transition is neither an ordinary θ point (since the two-body term is positive) nor a freezing point (the replica-symmetry is not broken). Note that since our approach is variational, the true free energy of the system is lower than the variational one, and we thus expect the real transition to occur at an even higher temperature. Since the transition is first order, we expect metastability and retardation effects to be important near the transition; we also expect that (due to the latent heat) there will be a reduction of entropy in the low temperature phase compared with an ordinary second order θ point.

(ii) $\lambda_0 < 0$: The Hydrophobic Case

In this case, defining the two temperatures:

$$T_0 = 2|\lambda_0|/v_0$$
$$T_1 = \lambda\frac{\sqrt{3}}{\sqrt{w_3(1-3^{d/2}/2^d)}} \tag{4.25}$$

leads to two possible scenarios:

(a) $T_1 > T_0$: the collapse transition is again driven by the disorder fluctuations of the three-body interactions. The resulting first-order transition is very similar to the hydrophilic case (i).
(b) $T_0 > T_1$: the collapse transition is now driven by the strong two-body λ_0 term. The resulting phase transition is very similar to an ordinary θ point, and is therefore second-order.

4.4. *The "Random Bond" Chain*

There are various ways to attack this difficult case, and we will present our point of view along two lines. The simple high dimension approach[27] leads directly to a Potts glass[79] with a large number of states, which is known to be equivalent to a Random Energy Model (REM). Already at this level, as we can see in Appendix 4, there is a freezing transition, characterized by a "one-step replica symmetry breaking". The low dimensional space dependent approach utilizes a variational approach similar to the one of the preceding section, *except* that the symmetry between replicas is broken in a more involved way.[80,81]

4.4.1. *High Dimension Approach*

Let us consider the Hamiltonian:

$$H = \sum_{i<j} (v_0 + v_{ij})\, \delta(\mathbf{r}_i - \mathbf{r}_j) + \frac{w_3}{6} \sum_{i \neq j \neq k} \delta(\mathbf{r}_i - \mathbf{r}_j)\, \delta(\mathbf{r}_j - \mathbf{r}) \qquad (4.26)$$

where the couplings $\{v_{ij}\}$ are random independent couplings, and v_0 represents the overall effect of the solvent, as well as the direct non-random pair interactions. Note that, compared with Eq. (4.2), we have rescaled the Hamiltonian by a factor β^{-1}. This rescaling makes the temperature dependence of the phase diagram more transparent.

For the sake of simplicity, we use a Gaussian probability distribution for the couplings:

$$P(v_{ij}) = \frac{1}{\sqrt{2\pi v^2}} \exp\left(-\frac{v_{ij}^2}{2v^2}\right). \qquad (4.27)$$

The partition function is:

$$Z(\{v_{ij}\}) = \int \mathcal{D}\mathbf{r}_i \prod_i g(\mathbf{r}_i, \mathbf{r}_{i+1}) e^{-\beta\mathcal{H}} \tag{4.28}$$

where the function g enforces the chain constraint.

Replicating and averaging (4.27) yields:

$$\begin{aligned}\overline{Z^n} = &\int \prod_a \mathcal{D}\mathbf{r}_i^a \prod_{i,a} g(\mathbf{r}_i^a, \mathbf{r}_{i+1}^a)\\ &\times \exp\left(-\beta\tilde{v}_0 \sum_{i<j}\sum_a \delta(\mathbf{r}_i^a - \mathbf{r}_j^a) + \frac{\beta^2 v^2}{2}\sum_{a\neq b}\sum_{i<j} \delta(\mathbf{r}_i^a - \mathbf{r}_j^a)\delta(\mathbf{r}_i^b - \mathbf{r}_j^b)\right)\\ &\times \exp\left(-\frac{\beta w_3}{6}\sum_{i\neq j\neq k}\sum_a \delta(\mathbf{r}_i^a - \mathbf{r}_j^a)\,\delta(\mathbf{r}_j^a - \mathbf{r}_k^a)\right)\end{aligned} \tag{4.29}$$

with $\tilde{v}_0 = v_0 - \beta\frac{v^2}{2}$.

The above replicated Hamiltonian has three characteristics:

(i) the chain constraint term;
(ii) a possible θ point if $\tilde{v}_0 < 0$; and
(iii) a possible freezing transition due to the $a \neq b$ term of (4.29).

Since we wish to emphasize the freezing transition here, we will assume that $\tilde{v}_0$ is indeed negative, so that the system is in the collapsed phase (see Sec. 2). To get a tractable model, we further consider a simple but unrealistic geometry, namely a collapsed chain on a fully connected lattice. On such a lattice, by definition, each point is neighbour to all the other points of the lattice. Examples are provided by a triangle (three points, two dimensions), and a tetrahedron (four points, three dimensions), so that for a large number Ω of points, it is a high dimensional $(\Omega - 1)$ polyhedron. On such a lattice, the chain constraint (i) is automatically satisfied, since a site has $\Omega - 1$ neighbors, so that the chain constraint is a $1/\Omega$ effect.

Let us show that in the collapsed phase, the model is equivalent to a Random Energy Model (REM), as studied in Appendix 4. For that purpose, we will show that the energies of this system are random independent Gaussian variables. First, we note that in the collapsed phase, the monomer density ρ is finite and constant in space; only a number N/ρ of sites are occupied. This implies that the only conformation dependent term of Eq. (4.26) is the random two-body term $\sum_{i<j} v_{ij}\delta(\mathbf{r}_i - \mathbf{r}_j)$. Since this term is a linear combination of Gaussian variables, it is also a Gaussian variable, and thus, its distribution is entirely characterized by its correlation functions. Therefore, instead of computing the joint probability $P(E_1, E_2)$ for two copies of the chain as in Eq. (A4.21), we will calculate directly the correlation $\overline{E_1 E_2}$ between the energies E_1 and E_2 of two conformations $\{\mathbf{r}_i^{(1)}\}$ and $\{\mathbf{r}_i^{(2)}\}$ of the chain in the collapsed phase. We have:

$$\begin{aligned}\overline{E_1^2} &= \frac{v^2}{2}\sum_{i,j}\delta(\mathbf{r}_i^{(1)} - \mathbf{r}_j^{(1)}) \\ &= \frac{v^2}{2}\sum_{\mathbf{r}}\rho_1^2(\mathbf{r}) \\ &= \frac{v^2}{2}N\rho\end{aligned} \tag{4.30}$$

where $\rho_1(\mathbf{r}) = \sum_i \delta(\mathbf{r} - \mathbf{r}_i^{(1)})$. Therefore, $\overline{E_1^2}$ is independent of the (collapsed) conformation. Similarly, we have:

$$\begin{aligned}\overline{E_1 E_2} &= \frac{v^2}{2}\sum_{i,j}\delta(\mathbf{r}_i^{(1)} - \mathbf{r}_j^{(1)})\delta(\mathbf{r}_i^{(2)} - \mathbf{r}_j^{(2)}) \\ &= \frac{v^2}{2}\sum_{\mathbf{r},\mathbf{r}'} q_{12}^2(\mathbf{r}, \mathbf{r}')\end{aligned} \tag{4.31a}$$

where $q_{12}(\mathbf{r}, \mathbf{r}') = \sum_i \delta(\mathbf{r} - \mathbf{r}_i^{(1)})\ \delta(\mathbf{r}' - \mathbf{r}_i^{(2)})$. We have:

$$\sum_{\mathbf{r},\mathbf{r}'} q_{12}(\mathbf{r}, \mathbf{r}') = N \tag{4.31b}$$

and since all occupied points are equivalent (constant density), we have:

$$q_{12}(\mathbf{r}, \mathbf{r}') = \frac{\rho^2}{N} \tag{4.31c}$$

which in turn implies:

$$\overline{E_1 E_2} = \frac{v^2}{2}\rho^2 \tag{4.31d}$$

from which we get the joint probability:

$$\lim_{N\to\infty} P(E_1, E_2) \sim \exp\left(-\frac{E_1^2 + E_2^2}{N\rho v^2}\right)$$

leading to a REM behavior. This behavior, in the context of protein folding, was first put forward, on phenomenological grounds, by Bryngelson and Wolynes.[26] Thus, in retrospect, it comes as no surprise that the random heteropolymer model behaves very much like the REM.

It is of course possible to obtain these results using replicas. In this case, one shows that the chain model as we have stated it, is equivalent to an infinite-range Potts-glass, with $N/\rho \to \infty$ states. This model can then be solved by a one-step replica symmetry breaking scheme. The physics of the REM is discussed in Appendix 3: the chain undergoes a freezing transition at a temperature T_c. Above T_c, the system has a finite entropy, whereas below T_c, it vanishes. The system is then frozen in a small number of dominant states, determined by subtle effects.

4.4.2. *Finite Dimension Approach*

To go beyond the REM freezing behavior, one has to take into account the chain constraint $g(\mathbf{r}_i^a, \mathbf{r}_{i+1}^a)$ in Eq. (4.29), as was already done in Sec. 4.2. We have

$$\overline{Z^n} = \int \prod_a \mathcal{D}\mathbf{r}_i^a \prod_{i,a} g(\mathbf{r}_i^a, \mathbf{r}_{i+1}^a)$$

$$\times \exp\left(-\beta\tilde{v}_0 \sum_{i<j}\sum_a \delta(\mathbf{r}_i^a - \mathbf{r}_j^a) + \frac{\beta^2 v^2}{2} \sum_{a\neq b}\sum_{i<j} \delta(\mathbf{r}_i^a - \mathbf{r}_j^a)\delta(\mathbf{r}_i^b - \mathbf{r}_j^b)\right)$$

$$\times \exp\left(-\frac{\beta w_3}{6} \sum_{i\neq j\neq k}\sum_a \delta(\mathbf{r}_i^a - \mathbf{r}_j^a)\, \delta(\mathbf{r}_j^a - \mathbf{r}_k^a)\right) \qquad (4.29)$$

that we now rewrite as

$$\overline{Z^n} = \int \prod_{a=1}^{n} \mathcal{D}\mathbf{r}_a(s)\, \exp\left(-\frac{d}{2a^2}\int_0^N ds \sum_a \left(\frac{d\mathbf{r}_a(s)}{ds}\right)^2 - A\right)$$

$$\times \exp\left(\frac{\beta^2 v^2}{2}\int_0^N ds \int_0^N ds' \sum_{a<b} \delta(\mathbf{r}_a(s) - \mathbf{r}_a(s'))\, \delta(\mathbf{r}_b(s) - \mathbf{r}_b(s'))\right) \qquad (4.32a)$$

with

$$A = \frac{1}{2}(\beta\tilde{v}_0) \int_0^N ds \int_0^N ds' \sum_a \delta(\mathbf{r}_a(s) - \mathbf{r}_a(s'))$$

$$+\frac{\beta w_3}{6} \int_0^N ds \int_0^N ds' \int_0^N ds'' \sum_a \delta(\mathbf{r}_a(s) - \mathbf{r}_a(s'))\, \delta(\mathbf{r}_a(s') - \mathbf{r}_a(s'')) . \tag{4.32b}$$

Introducing the parameters $q_{ab}(\mathbf{r}, \mathbf{r}')$, with $a < b$, and $\rho_a(\mathbf{r})$ by:

$$q_{ab}(\mathbf{r}, \mathbf{r}') = \int_0^N ds\, \delta(\mathbf{r}_a(s) - \mathbf{r})\, \delta(\mathbf{r}_b(s) - \mathbf{r}')$$

$$\rho_a(\mathbf{r}) = \int_0^N ds\, \delta(\mathbf{r}_a(s) - \mathbf{r}) \tag{4.33}$$

we may write Eqs. (4.32) as:

$$\overline{Z^n} = \int \mathcal{D}q_{ab}(\mathbf{r}, \mathbf{r}')\mathcal{D}\hat{q}_{ab}(\mathbf{r}, \mathbf{r}')\mathcal{D}\rho_a(\mathbf{r})\mathcal{D}\phi_a(\mathbf{r})$$
$$\times \exp\left(G(q_{ab}, \hat{q}_{ab}, \rho_a, \phi_a) + \log\zeta(\hat{q}_{ab}, \phi_a)\right) \tag{4.34a}$$

where $\hat{q}_{ab}(\mathbf{r}, \mathbf{r}')$ and $\phi_a(\mathbf{r})$ are the Lagrange multipliers associated with (4.33), and:

$$G(q_{ab}, \hat{q}_{ab}, \rho_a, \phi_a) = \int d\mathbf{r} \sum_a \left(i\rho_a(\mathbf{r})\phi_a(\mathbf{r}) - (\beta\tilde{v}_0)\, \frac{\rho_a^2(\mathbf{r})}{2} - \frac{\beta w_3}{6}\rho_a^3(\mathbf{r}) \right) \tag{4.34b}$$

$$+ \int d\mathbf{r} \int d\mathbf{r}' \sum_{a<b} \left(i q_{ab}(\mathbf{r}, \mathbf{r}')\hat{q}_{ab}(\mathbf{r}, \mathbf{r}') + \frac{\beta^2 v^2}{2}\, q_{ab}^2(\mathbf{r}, \mathbf{r}') \right)$$

and

$$\zeta(\hat{q}_{ab}, \phi_a) = \int \mathcal{D}\mathbf{r}_a(s)$$

$$\times \exp\left(-\frac{d}{2a^2} \int_0^N ds\, \dot{\mathbf{r}}_a^2 - i \int_0^N ds \sum_a \phi_a(\mathbf{r}_a(s)) \right.$$
$$\left. - i \int_0^N ds \sum_{a<b} \hat{q}_{ab}(\mathbf{r}_a(s), \mathbf{r}_b(s)) \right) . \tag{4.34c}$$

To go further, one follows the approximations of Sec. 4.2:

(i) one assumes ground state dominance in the "quantum" Hamiltonian associated with Eq. (4.34c);
(ii) the free energy is calculated in the SPM.

In view of the high dimension results, point (i) is natural, since one expects first a θ collapse transition, followed at low temperature by a freezing transition driven by the off-diagonal (in replica space) terms. The procedure exactly parallels the one of the randomly hydrophilic-hydrophobic (LB) chain of the preceding section, except for the variational wave function that enters the Rayleigh-Ritz principle. This wave function should present RSB. A simple form was proposed by Shakhnovich and Gutin,[80,81] (see also Edwards and Muthukumar[54]), and reads:

$$\Psi(\mathbf{r}_1, \ldots, \mathbf{r}_n) = \frac{(\det K)^{d/4}}{(2\pi)^{nd/4}} \exp\left(-\frac{1}{4}\sum_{a,b} \mathbf{r}_a K_{ab} \mathbf{r}_b\right) \tag{4.35}$$

where K is a $n \times n$ Parisi-like hierarchical matrix,[13] and d is the dimension of space. The variational free energy is extremized with respect to K. The results suggests that the freezing transition in low enough dimensions is not of the REM type, and that a full (infinite-step) replica symmetry-breaking scheme is then needed. It is unclear to us whether this analysis applies for $d = 3$, but a variational wave function of the form (4.35) has been used in other disordered condensed matter situations: vortex lattice with impurities in superconductors,[82–84] interface in a random potential,[85–88] amorphous solidification of vulcanized macromolecules,[89] among others.

At this point, a word of caution is in order. A variational approach is by definition not exact. Specific difficulties arise in disordered systems, as illustrated by the problem of directed polymers in a random medium in two dimensions: a variational approach without replica symmetry breaking[56] yields a roughening exponent $\zeta = 3/4$, while a variational approach with full replica symmetry breaking[86] yields $\zeta = 3/5$. The exact result[90] is $\zeta = 2/3$.

4.5. *The Random Sequence Polymer and Coding*

We now turn to the last case, namely the random sequence of monomers. This model illustrates the coding problem presented in Appendix 5 (for a different coding procedure, linked to a random field, see de Gennes[91]). Each monomer is of a certain type $p = 1, \dots, Q$, and the short-range interaction between monomer i of type p_i and j of type p_j is taken as:

$$v_{ij}(\mathbf{r}_i, \mathbf{r}_j) = -v_0\delta(\mathbf{r}_i - \mathbf{r}_j) + \beta V_{p_i p_j}\delta(\mathbf{r}_i - \mathbf{r}_j) \tag{4.36}$$

where V_{pq} is a $Q \times Q$ matrix. The δ interaction has been chosen to simplify the calculations. As in the previous section, v_0 represents the overall effect of the solvent, as well as the direct non-random pair interactions (we again follow the notations of Eq (4.2)). As in Sec. 4.4, we assume that the chain is collapsed, and we shall take v_0 to be positive. In addition, we shall assume the presence of a repulsive three-body term to stabilize the system.

We first specialize to the case of a separable matrix. Namely, we shall take:

$$v_{ij}(\mathbf{r}_i, \mathbf{r}_j) = -(v_0 + \beta q_i q_j)\delta(\mathbf{r}_i - \mathbf{r}_j) \tag{4.37}$$

where $\{q_i\}$ is an ensemble of generalized charges, which are taken as independent random variables, distributed with the same probability $P(q)$. We assume that the average of q_i is zero, since a non-zero average can be included in the v_0 term.

As in the previous section, we shall discuss first the high dimensional case, then reintroduce the chain constraint.

4.5.1. *High Dimension Approach*

Let us consider the Hamiltonian:

$$\beta H = -\sum_{i<j}(v_0 + \beta q_i q_j)\ \delta(\mathbf{r}_i - \mathbf{r}_j) + \frac{w_3}{6}\sum_{i \neq j \neq k} \delta(\mathbf{r}_i - \mathbf{r}_j)\ \delta(\mathbf{r}_j - \mathbf{r})\,. \tag{4.38}$$

Using the algebra of the previous section, we can write:

$$Z(\{q_i\}) = \int \mathcal{D}\mathbf{r}_i \prod_i g(\mathbf{r}_i, \mathbf{r}_{i+1}) e^{-\beta H} \tag{4.39}$$

where the function g enforces the chain constraint.

As we saw before, in the high dimension limit, we may neglect the chain constraint, and write:

$$Z(\{q_i\}) = \int \mathcal{D}\rho(\mathbf{r})\mathcal{D}\phi(\mathbf{r})\mathcal{D}m(\mathbf{r})$$
$$\times \exp\left(-\frac{1}{2}\sum_{\mathbf{r}} m^2(\mathbf{r}) + \frac{v_0}{2}\sum_{\mathbf{r}} \rho^2(\mathbf{r}) - \frac{w_3}{6}\sum_{\mathbf{r}} \rho^3(\mathbf{r}) + i\sum_{\mathbf{r}} \phi(\vec{r})\rho(\vec{r}) + \sum_{i=1}^{N} \log\zeta(q_i)\right) \tag{4.40a}$$

where

$$\zeta(q_i) = \sum_r e^{-i\phi(\mathbf{r}) - \sqrt{\beta} q_i m(\mathbf{r})}. \tag{4.40b}$$

In addition to the usual density ρ and conjugate field ϕ, there appears a new order parameter $m(\mathbf{r}) = \sum_{i=1}^{N} q_i\delta(\mathbf{r} - \mathbf{r}_i)$, which links the spatial concentration with the generalized charge sequence.[27] In the large N limit, we can use the central limit theorem and write:

$$Z = \int \mathcal{D}\rho(\mathbf{r})\mathcal{D}\phi(\mathbf{r})\mathcal{D}m(\mathbf{r})$$
$$\times \exp\left(-\frac{1}{2}\sum_{\mathbf{r}} m^2(\mathbf{r}) + \frac{v_0}{2}\sum_{\mathbf{r}} \rho^2(\mathbf{r}) - \frac{w_3}{6}\sum_{\mathbf{r}} \rho^3(\mathbf{r}) + i\sum_{\mathbf{r}} \phi(\mathbf{r})\rho(\mathbf{r}) + N\int dq\ P(q)\ \log\zeta(q)\right). \tag{4.40c}$$

This last form is evaluated by the SPM.

The saddle-point equations are easily solved, and we only quote the results. Since we are in the collapsed phase, we have a finite monomer density $\rho = \frac{2v_0}{w_3}$. There is a phase transition in the collapsed phase, between a high temperature phase with $m = 0$, (there is no correlation between the charge "pattern" and the spatial distribution), and a low temperature phase, with $m \neq 0$, where this correlation is non-zero.

Defining the second moment of the "generalized" charge as $\Delta = \int dq \cdot P(q)q^2$, the critical temperature is given by $T_c = \Delta\ \rho$. The existence of this order parameter m can also be viewed as a microphase separation, since $m(\mathbf{r}) = \rho_+(\mathbf{r}) - \rho_-(\mathbf{r})$, where ρ_+ and ρ_- are the concentrations of positively and negatively charged monomers respectively: a non-vanishing m denotes a spatial segregation of oppositely charged monomers.

The above model can easily be generalized to the case where the interaction matrix V_{pq} is a finite sum of M separable interactions, and reads:

$$v_{ij}(\mathbf{r}_i, \mathbf{r}_j) = -\left(v_0\delta(\mathbf{r}_i - \mathbf{r}_j) + \beta w_{ij}\delta(\mathbf{r}_i - \mathbf{r}_j)\right) \tag{4.41a}$$

where

$$w_{ij} = \sum_{p=1}^{M} q_i^{(p)} q_j^{(p)} . \tag{4.41b}$$

The parameters $\{q_i^{(p)}\}$, $p = 1, \ldots, M$ can be viewed as a set of M generalized charges characterizing each amino acid i. We again assume that the average of all the $q_i^{(p)}$ is zero, and denote $\Delta_p = \int dq\, P_p(q)\, q^2$. In this case, it can be easily shown that there are M order parameters $m_p(\mathbf{r})$ defined as above; the system undergoes a phase transition in the collapsed phase, at a temperature $T_c = \max_p\{\Delta_p\, \rho\}$. The physics is very similar to the "one pattern" case, with a progressive condensation of the order parameters m_p as the temperature decreases.

When the number of charges M goes to infinity, it is easy to see that we recover the random-bond model of Sec. 4.4. This situation is similar to the Hopfield model of Appendix 4, where for $M/N > 0.14$, there is a qualitative change of the type of ordering, with no correlation with any of the patterns. Here, the analytic calculation is much harder, and this result was seen numerically.[78] Physically, it can be understood by showing that, as M increases, the correlation between the interaction terms w_{ij} goes to zero. Correlations between w_{ij} and w_{kl} occur when they share common indices. For instance:

$$\begin{aligned} \overline{w_{ij}} &= 0 \\ \overline{w_{ij}^2} &= \sum_{p=1}^{M} \Delta_p^2 \sim M \\ \overline{w_{ij}^2 w_{jk}^2} &= \left(\sum_{p=1}^{M} \Delta_p^2\right)^2 + 2\sum_{p=1}^{N} \Delta_p^4 \end{aligned} \tag{4.41c}$$

so that

$$\frac{\overline{w_{ij}^2 w_{jk}^2} - \overline{w_{ij}^2}.\overline{w_{jk}^2}}{\overline{w_{ij}^2}.\overline{w_{jk}^2}} = O\left(\frac{1}{M}\right) . \tag{4.41d}$$

This shows that for large M, the $\{w_{ij}\}$ behave as independent variables.

To be complete, we mention that Dotsenko *et al.*[92] have studied Hopfield-like models of a chain on a fully connected lattice, where each pattern corresponds to a permutation of the nodes. (In other words, each pattern corresponds to a "shape"of the heteropolymer). This model has been further considered by Dorotheyev *et al.*[93]

4.5.2. *Finite Dimension Approach*

In the finite dimensional approach, one has to resort again to replica calculations. Assuming a simple separable interaction of the form (4.37), with Gaussian distributed charges, the quenched average of Z^n can be computed as in Sec. 4.4. One needs to introduce now three order parameters, we get:

$$
\begin{aligned}
\rho_a(\mathbf{r}) &= \int_0^N ds\ \delta(\mathbf{r}_a(s) - \mathbf{r})\,,\\
m_a(\mathbf{r}) &= \int_0^N ds\ q(s)\ \delta(\mathbf{r}_a(s) - \mathbf{r})\,,\\
q_{ab}(\mathbf{r}, \mathbf{r}') &= \int_0^N ds\ \delta(\mathbf{r}_a(s) - \mathbf{r})\ \delta(\mathbf{r}_b(s) - \mathbf{r}')\,.
\end{aligned}
\tag{4.42}
$$

The free energy can again be expressed in terms of these order parameters. The analysis by Shakhnovich and Gutin[28,94] and Sfatos *et al.*[74] suggests that a freezing transition (of the REM type) induced by the chain stiffness, may prevent the occurrence of the high dimension scenario, at least for rigid enough chains. For flexible chains, the physics is very similar to that of the high dimensional case.

4.6. *Conclusions*

We have studied various models of random polymers, undergoing collapse and/or freezing transitions. The solution of these models requires severe approximations, which render the comparison with realistic polymers (or proteins) somehow delicate. In particular, the link between the randomness along the chemical sequence (or primary structure) and the three-dimensional spatial structure in the low temperature phases is not clearly apparent, except perhaps in the random sequence model of Sec. 4.5.

This model shares many properties of Hopfield's neural network model. In particular, the low temperature phase can be characterized (in the absence of freezing) by a globular state with strong correlation with the "charge pattern" $\{q\}$. One of the problems of this approach is that this is a ferromagnetic-like transition, with finite conformational entropy in the ordered phase, so that it cannot be a very good candidate to describe the native state of a protein. On the contrary, the frozen state (if it exists) is a natural candidate to the native state, with entropy $\simeq 0$, although it has weak correlations with the charge patterns.

This neural network approach may be useful to design simplified folding potentials, which have the known protein X-ray structures as local minima. These potentials could then be used to predict low-energy structures.[75,95–98]

5. Summary

We have presented, in the framework of equilibrium statistical mechanics, a polymer-oriented view of the folding problem. Our approach was twofold.

First of all, we have studied in Sec. 3, the influence of secondary structures (or local order) on a globally compact chain. This study yields an almost unique crystalline ground state for the chain, where the secondary structures invade the whole system. The defects of this crystalline order are rejected on the surface of the crystal. In marked contrast with usual crystals, the chain constraint implies that the bulk of this ground state is (almost) frozen, up to a phase transition towards a kind of "molten globule", compact but with only short and mobile secondary structures. At still higher temperatures, this compact globule will undergo a transition towards a swollen coil phase. A subtle question concerns the role of the surface in our approach, and its influence on the transitions. This point clearly deserves more work. Furthermore, we have not considered dynamical processes, which should be particularly slow in any compact phase, and it may well be that the ground state is not reachable on any reasonable timescale, and that one ends up with a kind of glass at low temperatures. It is interesting to note that the properties of the real glass transition[99–102] depends on the average number of neighbors.[103–106] The average number of neighbors was the crucial parameter in the approach of Sec. 3.

We have then considered the collapse and/or freezing transitions of various random heteropolymer models (Sec. 4). These models are concerned with the tertiary structure. As in the first case, the role of the non-extensive terms of the free energy may be extremely important. Take as an example a REM-like freezing transition. These terms determine the dominant ground states, that is the dominant folds. Our present understanding, as well as our analytical techniques, are clearly insufficient in this context. Finally, we have sketched how some random sequence models may exhibit a coding *à la* Hopfield (i.e a non-zero overlap between the chain ground state configuration(s) and pattern(s) characterizing the primary sequence).

Our approach does suggest a resemblance between proteins and glassy states. The main question of the uniqueness and stability of the folding for a given sequence remains largely unanswered. The influence of the surface in the above models, and in real proteins, seems to be a promising track. Despite all these limitations, it appears to us that the study of various minimal models of proteins can be useful in clarifying global features of protein folding.

Appendix 1

In the text, we shall frequently use some mathematical formulas, and particularly Gaussian integrals. In this appendix, we review, without any demonstration, some of these integrals which are used thoroughly throughout the text. In the next section, we present Wick's theorem, which is a simple and powerful method to compute averages of polynomials with Gaussian weights.

A1.1. *Useful Formulae*

The simple one-dimensional Gaussian integral is given by:

$$\int_{-\infty}^{+\infty} dx\, e^{-ax^2+bx} = \left(\frac{\pi}{a}\right)^{1/2} e^{\frac{b^2}{4a}} \tag{A1.1}$$

where $a > 0$. In the following, we shall omit the bounds on the integral, and the variables will run over the whole real axis.

For multiple integrals, this formula can easily be generalized. Let A be a positive definite $N \times N$ symmetric matrix (i.e. all its eigenvalues are strictly positive) and $\{u_i\}$ an N-dimensional vector. We have then:

$$\int dx_1 \cdots dx_N\, e^{-\frac{1}{2}\sum_{i,j} x_i A_{ij} x_j + \sum_i u_i x_i} = \frac{(2\pi)^{\frac{N}{2}}}{\sqrt{\det A_{ij}}}\, e^{\frac{1}{2}\sum_{i,j} u_i A^{-1}_{ij} u_j} \tag{A1.2}$$

where $\det A$ denotes the determinant of A and A^{-1} its inverse.

Finally, we shall frequently use the Fourier representation of the δ-function. We have the identity:

$$\delta^d(\mathbf{r}) = \int \frac{d\mathbf{k}}{(2\pi)^d}\, e^{i\mathbf{k}\mathbf{r}}\,. \tag{A1.3}$$

A2.2. *Wick's Theorem*

We shall often have to calculate averages of polynomials, with Gaussian weights. Typical quantities to be calculated are:

$$\langle x_{i_1} x_{i_2} \cdots x_{i_p} \rangle = \frac{1}{Z_0} \int dx_1 \cdots dx_N \; x_{i_1} x_{i_2} \cdots x_{i_p} \; e^{-\frac{1}{2} \sum_{i,j} x_i A_{ij} x_j} \tag{A1.4a}$$

where

$$Z_0 = \int dx_1 \cdots dx_N \; e^{-\frac{1}{2} \sum_{i,j} x_i A_{ij} x_j} \,. \tag{A1.4b}$$

G. C. Wick (see e.g. Negele and Orland[107]) has devised the following prescription to evaluate this integral: the expectation value (A1.4a) is given by the sum of all possible complete sets of contractions of the variables, where the contraction $\overline{x_i.x_j}$ is given by:

$$\begin{aligned} \overline{x_i.x_j} &= \langle x_i x_j \rangle \\ &= A^{-1}_{ij} \,. \end{aligned} \tag{A1.4c}$$

We see that averages of an odd number of x is zero. As an example, using this method, we get easily that:

$$\langle x_1 x_2 x_3^2 \rangle = \overline{x_1.x_2}\;\overline{x_3.x_3} + \overline{x_1.x_3}\;\overline{x_2.x_3} + \overline{x_1.x_3}\;\overline{x_2.x_3} \,. \tag{A1.5}$$

This method has a simple graphical interpretation: represent each x_i by a line leaving point i. A contraction $\overline{x_i.x_j} = A^{-1}_{ij}$ can be represented as a line joining x_i and x_j. It denotes a propagator. Wick's theorem states that the average (A1.4a) is the sum over all graphs where all pairs of points are connected pairwise by propagators. This theorem (which is the starting point for the introduction of Feynman diagrams) plays a crucial role in many-body physics.

Appendix 2

In this appendix, we display the explicit form of the classical CHARMM energy function, currently used in Molecular Dynamics or Monte Carlo calculations.[108] Other widely used functions such as AMBER, OPLS, GROMOS, MOIL, among others, have similar forms.

The total energy is of the form

$$E = \sum_{\text{bonds}} k_b(b-b_0)^2 + \sum_{\text{angles}} k_\theta(\theta-\theta_0)^2 + \sum_{\text{dihedrals}} k_\phi(1+\cos(n\phi-\delta))$$

$$+ \sum_{\text{impropers}} k_\nu(\nu-\nu_0)^2 + \sum_{i<j}{}' 4\varepsilon_{ij}\left\{\left(\frac{\sigma_{ij}}{r_{ij}}\right)^{12} - \left(\frac{\sigma_{ij}}{r_{ij}}\right)^{6}\right\} + \sum_{i<j}{}' \frac{332}{\varepsilon_r}\frac{q_i q_j}{r_{ij}} \tag{A2.1}$$

where distances are in Angströms, angles in radians, and E in kcal/mol (at room temperature, $k_B T$ is of order 0.6 kcal/mol).

The first four terms represent bonded terms (i.e. the energy of respectively valence bonds, valence angles, dihedrals and improper dihedrals).

The last two terms represent respectively the Van der Waals energy (through a 6–12 Lennard–Jones potential) and the Coulomb energy (with partial charges). The $\sum'_{i<j}$ means a sum over non-bonded terms only. The relative dielectric constant is ε_r. The various parameters entering Eq. (A2.1) have been discussed by Brooks *et al.*,[4] and Smith and Karplus.[5] They are obtained semi-empirically, and adjusted to fit the structure of small organic molecules (alkanes, etc.), and their vibrational and rotational frequencies. There is no guarantee however that the native protein shape is an absolute, or even a local minimum of the energy (A2.1). Long Molecular Dynamics runs have indeed shown that starting from a protein in its native state, the molecule will eventually escape to unphysical structures (Elber, private communication).

Appendix 3

In this appendix, we review the properties of free Brownian chains, and show their relation to diffusion and Schrödinger equations. For interacting chains, we present the self-consistent field approximation (SCF), and show how it can be further simplified when there is a gap in the energy spectrum, by using the ground state dominance approximation.

A3.1. *Markovian and Brownian Chains*

Let us consider a discrete chain, made of N monomers. The first atom of the chain, numbered 0, is fixed at point 0. By definition of a Markov chain, the position $\mathbf{r}_n$ of atom n depends only on the position $\mathbf{r}_{n-1}$ of the previous atom $n-1$ along the chain. Namely, a Markov chain is characterized by the conditional probability distribution $g(\mathbf{r}, \mathbf{r}')$ that atom n is at point $\mathbf{r}$, given that atom $n-1$ is at $\mathbf{r}'$. If we further assume translational invariance, this function will only depend on the difference, and reads $g(\mathbf{r} - \mathbf{r}')$.

The probability of finding the end of the chain (atom number N) at point $\mathbf{R}$, given that the origin is fixed at $\mathbf{0}$, is thus given by:

$$G(\mathbf{R}, N|\mathbf{0}, 0) = \int d\mathbf{r}_1 \cdots d\mathbf{r}_{N-1}\, g(\mathbf{R} - \mathbf{r}_{N-1})\, g(\mathbf{r}_{N-1} - \mathbf{r}_{N-2}) \cdots g(\mathbf{r}_1 - \mathbf{0})\,. \tag{A3.1}$$

The Fourier transform Γ of G is thus given by:

$$\begin{aligned} \Gamma(\mathbf{k}, N) &= \int d\mathbf{R}\, e^{i\mathbf{k}\mathbf{R}}\, G(\mathbf{R}, N|\mathbf{0}, 0) \\ &= \gamma^N(\mathbf{k}) \end{aligned} \tag{A3.2a}$$

where

$$\gamma(\mathbf{k}) = \int d\mathbf{r}\, e^{i\mathbf{k}\mathbf{r}}\, g(\mathbf{r})\,. \tag{A3.2b}$$

Performing an inverse Fourier transform, we obtain:

$$G(\mathbf{R}, N|\mathbf{0}, 0) = \int \frac{d\mathbf{k}}{(2\pi)^d} e^{-i\mathbf{k}\mathbf{R}}\, \gamma^N(\mathbf{k})\,. \tag{A3.2c}$$

In the limit of long chains, $N \to \infty$, we apply the stationary-phase method to (A3.2c). Since $g(\mathbf{r})$ is a probability distribution, its integral is one, and thus it is easy to see that:

$$|\gamma(\mathbf{k})| \leq \gamma(\mathbf{0}) = 1$$

so that we may expand (A3.2c) around $\mathbf{k} = \mathbf{0}$. We can write, to second order in $\mathbf{k}$:

$$\gamma(\mathbf{k}) = 1 - \frac{a^2}{2d}\mathbf{k}^2 \tag{A3.3a}$$

where a, which has the dimension of a length, is called the persistence length, and can be interpreted as the effective monomer length. We have:

$$\begin{aligned} G(\mathbf{R}, N|\mathbf{0}, 0) &= \int \frac{d\mathbf{k}}{(2\pi)^d} \exp(-i\mathbf{k}\mathbf{R} - N\frac{a^2}{2d}\mathbf{k}^2) \\ &= \left(\frac{d}{2\pi N a^2}\right)^{d/2} e^{-\frac{dR^2}{2Na^2}}\,. \end{aligned} \tag{A3.3b}$$

We see that, for long enough chains, regardless of the function $g(\mathbf{r})$, the asymptotic probability distribution for the end of the chain is Gaussian. As an outcome, we see that the entropy (more precisely the entropy reduction) of a chain, with origin constrained at $\mathbf{0}$ and extremity constrained at $\mathbf{R}$ is given by:

$$S = -\frac{dR^2}{2Na^2}, \tag{A3.3c}$$

a result that we shall frequently use.

The expression (A3.3b) of $G(\mathbf{R}, N|\mathbf{0}, 0)$ shows that it satisfies a diffusion-like equation:

$$\left(\frac{\partial}{\partial N} - \frac{a^2}{2d}\nabla^2\right) G(\mathbf{R}, N|\mathbf{0}, 0) = \delta(\mathbf{R})\delta(N) \tag{A3.4a}$$

which is nothing but an imaginary-time Schrödinger equation. Therefore, using the standard notations of quantum mechanics, we see that we can write:

$$G(\mathbf{R}, N|\mathbf{0}, 0) = \langle \mathbf{R}|e^{-N\mathcal{H}}|\mathbf{0}\rangle \tag{A3.4b}$$

where $\mathcal{H}$ is a quantum-like Hamiltonian equal to:

$$\mathcal{H} = -\frac{a^2}{2d}\nabla^2 . \tag{A3.4c}$$

It is well-known in quantum mechanics (see Negele and Orland[107]) that matrix elements of the form (A3.4b) have a path-integral representation, called the Feynman integral, which is exactly the continuous representation of (A3.1). We may thus write:

$$G(\mathbf{R}, N|\mathbf{0}, 0) = \int_{\mathbf{r}(0)=\mathbf{0}}^{\mathbf{r}(N)=\mathbf{R}} \mathcal{D}\mathbf{r}(s) \exp\left(-\frac{d}{2a^2}\int_0^N \mathrm{d}s\, \dot{\mathbf{r}}^2(s)\right) . \tag{A3.4d}$$

This continous chain is called a Brownian chain. It is the prototype of all long chains defined by Markov processes.

These expressions can easily be generalized to the case of a chain in an external potential. Assuming that each atom $\mathbf{r}_n$ feels the potential $U(\mathbf{r}_n)$, the distribution function of the chain at inverse temperature β reads:

$$G(\mathbf{R}, N|\mathbf{0}, 0) = \int_{\mathbf{r}(0)=\mathbf{0}}^{\mathbf{r}(N)=\mathbf{R}} \mathcal{D}\mathbf{r}(f) \times \exp\left(-\frac{d}{2a^2}\int_0^N \mathrm{d}s\, \dot{\mathbf{r}}^2(s) - \beta\int_0^N \mathrm{d}s\, U(\mathbf{r}(s))\right) . \tag{A3.5a}$$

It has a form similar to (A3.4b) with

$$\mathcal{H} = -\frac{a^2}{2d}\nabla^2 + \beta\, U(\mathbf{r}) \tag{A3.5b}$$

and satisfies the Schrödinger equation:

$$\left(\frac{\partial}{\partial N} - \frac{a^2}{2d}\nabla^2 + \beta\, U(\mathbf{r})\right) G(\mathbf{R}, N|\mathbf{0}, 0) = \delta(\mathbf{R})\delta(N)\,. \tag{A3.5c}$$

The solution of (A3.5c) can easily be expressed in terms of the eigenstates and eigenvalues of the quantum Hamiltonian $\mathcal{H}$. Denoting by $\Psi_n(\mathbf{r})$ the eigenfunction of $\mathcal{H}$ with eigenvalue E_n, which satisfy the equation

$$\mathcal{H}\Psi_n(\mathbf{r}) = E_n\Psi_n(\mathbf{r})\,, \tag{A3.6a}$$

we can rewrite:

$$G(\mathbf{R}, N|\mathbf{0}, 0) = \sum_{\{n\}} e^{-NE_n}\Psi_n(\mathbf{R})\Psi_n(\mathbf{0})\,. \tag{A3.6b}$$

A3.2. *Ground State Dominance*

An important simplification occurs when

(i) the length N approaches infinity; and
(ii) there is a gap in the energy spectrum of $\mathcal{H}$.

Indeed, in that case, the sum (A3.6b) is dominated by the ground state Ψ_0 with energy E_0, and the distribution function reduces to:

$$G(\mathbf{R}, N|\mathbf{0}, 0) \sim e^{-NE_0}\Psi_0(\mathbf{R})\Psi_0(\mathbf{0})\,. \tag{A3.7}$$

The precise condition of validity of (A3.7) is $N\Delta \gg 1$, where Δ is the gap between the ground state and the first excited state. Such is the case when the Hamiltonian $\mathcal{H}$ has bound states, and the chain is long enough.

As an example, consider the problem of the the adsorption of a polymer chain on a substrate. The substrate is modeled as an impenetrable wall in the $x < 0$ direction. The monomer substrate interaction is taken to be a square well of depth $-U_0$ and of width w. We are thus led to study the ground state of a quantum particle in a square potential $-\beta U_0$, subject to the boundary condition $\Psi_0(0) = 0$, which enforces impenetrability of the wall. It is easily seen that for large enough β (low temperature), the system will have (at least) one bound state, with a concentration profile localized around the wall. In that case, for long enough chains, the ground state dominance approximation is justified. For small β (high temperature), the system has no more bound state, and it corresponds to a delocalised case. The spectrum is continuous, with no finite gap, and in order to do calculations, one has to solve the full time-dependent Schrödinger equation.

To conclude this section, let us recall that the ground state of a Hamiltonian can be deduced from a variational principle, namely the Rayleigh–Ritz principle. This principle states that the ground state energy E_0 of a Hamiltonian $\mathcal{H}$ is given by:

$$E_0 = \min_{\{\Psi(\mathbf{r})\}} \frac{\langle\Psi|\mathcal{H}|\Psi\rangle}{\langle\Psi|\Psi\rangle} \tag{A3.8a}$$

where the minimization runs over the space of square-integrable functions. Using the energy as a Lagrange multiplier to enforce normalization, it is equivalent to minimize the functional $\mathcal{E}$:

$$\mathcal{E} = \int \mathrm{d}\mathbf{r}\ \Psi(\mathbf{r}) \left(-\frac{a^2}{2d}\nabla^2 + \beta\, U(\mathbf{r}) - E_0\right) \Psi(\mathbf{r})\,. \tag{A3.8b}$$

A3.3. *Self-Consistent Field Approximation (SCF)*

We have seen in the previous sections how one can generate path integral representations for polymer problems. In the case of a self-interacting chain, however, there is no equivalent to the Schrödinger equation, and one has to resort to approximations to solve the problem. One such powerful approximation has been devised by Edwards,[19] and is called the self-consistent field approximation (SCF). To illustrate the method, let us consider a chain in a bad solvent. With reference to Sec. 2.2d, the partition function of this model reads:

$$\begin{aligned} Z = \int \mathcal{D}\mathbf{r}(s)\ \exp\left(-\frac{d}{2a^2}\int_0^N \mathrm{d}s\ \dot{\mathbf{r}}^2\right) \\ \times \exp\left(\frac{v}{2}\int_0^N \mathrm{d}s\mathrm{d}s'\delta(\mathbf{r}(s)-\mathbf{r}(s')) - \frac{w}{6}\right. \\ \left.\times \int_0^N \mathrm{d}s\mathrm{d}s'\mathrm{d}s''\delta(\mathbf{r}(s)-\mathbf{r}(s'))\delta(\mathbf{r}(s)-\mathbf{r}(s''))\right)\,. \end{aligned} \tag{A3.9a}$$

It is very useful to make a change of variable on (A3.8a), so that the partition function of the probem is expressed as a functional integral over all possible monomer concentration $\{\rho(\mathbf{r})\}$. To do so, we enforce the variables $\{\rho(\mathbf{r})\}$ by inserting the identity:

$$1 = \int \mathcal{D}\phi(\mathbf{r})\mathcal{D}\rho(\mathbf{r}) \exp\left(i\int \mathrm{d}\mathbf{r}\phi(\mathbf{r})\rho(\mathbf{r}) - i\int_0^N \mathrm{d}s\ \phi\left(\mathbf{r}(s)\right)\right) \quad \text{(A3.9b)}$$

which expresses that:

$$\rho(\mathbf{r}) = \int_0^N \mathrm{d}s\ \delta(\mathbf{r} - \mathbf{r}(s))\,. \quad \text{(A3.9c)}$$

Inserting these identities in (A3.8a) yields:

$$Z = \int \mathcal{D}\phi(\mathbf{r})\mathcal{D}\rho(\mathbf{r}) \exp\left(i\int \mathrm{d}\mathbf{r}\phi(\mathbf{r})\rho(\mathbf{r}) + \frac{v}{2}\int \mathrm{d}\mathbf{r}\rho^2(\mathbf{r}) - \frac{w}{6}\int \mathrm{d}\mathbf{r}\rho^3(\mathbf{r})\right)\zeta(\phi) \quad \text{(A3.10a)}$$

where

$$\zeta(\phi) = \int \mathcal{D}\mathbf{r}(s) \exp\left(-\frac{d}{2a^2}\int_0^N \mathrm{d}s\ \dot{\mathbf{r}}^2 - i\int_0^N \mathrm{d}s\ \phi\left(\mathbf{r}(s)\right)\right)\,. \quad \text{(A3.10b)}$$

According to (A3.5), ζ can be expressed as:

$$\zeta(\phi) = \int \mathrm{d}\mathbf{r}\langle\mathbf{r}|e^{-N\mathcal{H}}|\mathbf{0}\rangle \quad \text{(A3.10c)}$$

where

$$\mathcal{H} = -\frac{a^2}{2d}\nabla^2 + i\phi(\mathbf{r})\,. \quad \text{(A3.10d)}$$

In a bad solvent, we expect the chain to collapse at some temperature, and therefore, there should be some bound state in the system. We thus assume *ground state dominance*, so that:

$$\zeta(\phi) = \Psi_0(\mathbf{0})\left(\int \mathrm{d}\mathbf{r}\Psi_0(\mathbf{r})\right)\exp(-NE_0) \quad \text{(A3.11a)}$$

where Ψ_0 is the ground state of $\mathcal{H}$, with energy E_0. Using Eqs. (A3.8), we have seen that the extensive part of ζ can be written as:

$$\zeta = \exp\left(-N \min_{\{\Psi(\mathbf{r})\}} \frac{\langle\Psi|\mathcal{H}|\Psi\rangle}{\langle\Psi|\Psi\rangle}\right)$$

or equivalently

$$\zeta = \exp\left(-N \min_{\{\Psi(\mathbf{r})\}} \left[\int \mathrm{d}\mathbf{r}\ \Psi(\mathbf{r}) \left(-\frac{a^2}{2d}\nabla^2 + i\phi(\mathbf{r})\right) \Psi(\mathbf{r}) \right.\right.$$
$$\left.\left. -E_0 \left(\int \mathrm{d}\mathbf{r}\ \Psi^2(\mathbf{r}) - 1\right)\right]\right) \tag{A3.11b}$$

where E_0 appears as a Lagrange multiplier which constrains the norm of Ψ^2 to 1.

Replacing (A3.11b) in (A3.10a) yields:

$$Z = \int \mathcal{D}\phi(\mathbf{r})\mathcal{D}\rho(\mathbf{r}) \exp\left(-N\mathcal{F}(\rho,\phi)\right) \tag{A3.12a}$$

where

$$N\mathcal{F}(\rho,\phi) = -i\int \mathrm{d}\mathbf{r}\phi(\mathbf{r})\rho(\mathbf{r}) - \frac{v}{2}\int \mathrm{d}\mathbf{r}\rho^2(\mathbf{r}) + \frac{w}{6}\int \mathrm{d}\mathbf{r}\rho^3(\mathbf{r})$$
$$+ N \min_{\{\Psi_0(\mathbf{r})\}} \left[\int \mathrm{d}\mathbf{r}\ \Psi(\mathbf{r}) \left(-\frac{a^2}{2d}\nabla^2 + i\phi(\mathbf{r})\right) \Psi(\mathbf{r}) \right.$$
$$\left. -E_0 \left(\int \mathrm{d}\mathbf{r}\ \Psi^2(\mathbf{r}) - 1\right)\right]. \tag{A3.12b}$$

Since (A3.12a) cannot be evaluated exactly, we must use some approximation. A natural approximation is the saddle-point method (SPM), which consists in expanding $\mathcal{F}(\rho,\phi)$ around its minimum.

The minimization equation with respect to $\phi(\mathbf{r})$, $\rho(\mathbf{r})$ and $\Psi(\mathbf{r})$ reads:

$$\begin{aligned} \rho(\mathbf{r}) &= N\Psi^2(\mathbf{r}), \\ i\phi(\mathbf{r}) &= -v\rho(\mathbf{r}) + \frac{w}{2}\rho^2(\mathbf{r}), \\ E_0\Psi(\mathbf{r}) &= \left(-\frac{a^2}{2d}\nabla^2 + i\phi(\mathbf{r})\right)\Psi(\mathbf{r}). \end{aligned} \tag{A3.13}$$

The first equation shows that the monomer concentration is N times the square of the normalized wave-function. The second equation expresses the mean-field potential seen by each monomer, and finally the last equation can be recast in the form:

$$\left(-\frac{a^2}{2d}\nabla^2 - vN\Psi^2(\mathbf{r}) + \frac{w}{2}N^2\Psi^4(\mathbf{r})\right)\Psi(\mathbf{r}) = E_0\Psi(\mathbf{r}) \tag{A3.14}$$

where the energy E_0 is chosen so that the square wave function is normalized. This result in turn leads to Eqs. (2.21) of Sec. 2. The above equation is a nonlinear Schrödinger equation, which cannot in general be solved analytically. Let us note however that the two-body term v plays the role of an attractive self-consistent field, whereas w plays the role of a repulsive one, which forbids the collapse onto a finite region. In dimensions larger than two, this equation will have a bound state if v is large enough, and the associated energy will be finite, even in the limit $N \to \infty$.

To get more analytic information, we can restrict the space of normalized wave functions $\{\Psi(\mathbf{r})\}$ to the set of Gaussian wave-functions, depending on a single parameter R, which measures the spatial extent of the polymer globule:

$$\Psi(\mathbf{r}) = \left(\frac{1}{2\pi R^2}\right)^{d/4} \exp(-\frac{\mathbf{r}^2}{4R^2})\,. \tag{A3.15a}$$

Replacing (A3.15a) in (A3.12b) yields the simple equation for R:

$$0 = \frac{a_0}{R^3} - va_2\frac{N}{R^{d+1}} + wa_3\frac{N^2}{R^{2d+1}} \tag{A3.15b}$$

where a_0, a_2 and a_3 are simple numerical constants. This equation shows that for large N and any attractive two-body interaction, the system is collapsed in a globular state, with finite concentration, with exponent $\nu = 1/d$.

To be complete, let us mention how the SCF method can be generalized when the ground state dominance approximation cannot be applied. To simplify, let us consider the case of a two-body repulsive monomer–monomer interaction. In that case, one cannot use (A3.11a), and one must consider the full expression for the free-energy:

$$N\mathcal{F}(\rho,\phi) = -i\int d\mathbf{r}\phi(\mathbf{r})\rho(\mathbf{r}) + \frac{v}{2}\int d\mathbf{r}\rho^2(\mathbf{r}) - \log(\zeta(\phi))\,. \tag{A3.16a}$$

We evaluate again (A3.12a) using the SPM. We skip the calculations, which are slightly more involved than above, referring the reader to Edwards[19] for more details. The monomer concentration can be expressed in terms of two wave-functions $\psi_+(\mathbf{r},s)$ and $\psi_-(\mathbf{r},s)$ as:

$$\rho(\mathbf{r}) = \int_0^N ds \frac{\psi_+(\mathbf{r},s)\psi_-(\mathbf{r},s)}{\int d^d r'\psi_+(\mathbf{r}',s)\psi_-(\mathbf{r}',s)}\,. \tag{A3.17a}$$

These wave functions satisfy the equation:

$$\left(-\frac{\partial}{\partial s}+\mathcal{H}\right)\psi_+(\mathbf{r},s)=0$$

$$\left(\frac{\partial}{\partial s}+\mathcal{H}\right)\psi_-(\mathbf{r},s)=0 \tag{A3.17b}$$

with

$$\mathcal{H}=-\frac{a^2}{2d}\nabla^2+v\rho(\mathbf{r}) \tag{A3.17c}$$

and with the boundary conditions:

$$\psi_+(\mathbf{r},N)=1\,,$$

$$\psi_-(\mathbf{r},0)=\delta(\mathbf{r})\,. \tag{A3.17d}$$

Within the SCF approximation, it is possible (but quite involved), to show that one recovers Flory's results, namely a swelling exponent $\nu = 3/5$ in three dimensions, and also to calculate other quantities, such as the scattering function, the free energy, and the osmotic pressure.

Appendix 4. Spin Glasses

A4.1. *Introduction*

The spin glass (SG) problem originally arose in the study of disordered magnetic alloys, where magnetic impurities (e.g. Mn), quenched at random positions in the host lattice (e.g. Cu), interact with one another through competing (oscillating) interactions. This model has widely been extended,[12,13] and the interactions between the spins are accordingly of various sorts (exchange, dipolar, Dzialoshinsky-Morya, etc.). There is however a general consensus that the very essence of the SG problem (i.e. frustration + quenched disorder) is captured by the Edwards–Anderson Hamiltonian:

$$\mathcal{H}=-\sum_{i<j}J_{ij}\;\mathbf{S}_i\mathbf{S}_j \tag{A4.1}$$

where $\mathbf{S}_i$ is the spin of impurity i at position $\mathbf{r}_i$ and $J_{ij} = J(\mathbf{r}_i, \mathbf{r}_j)$ is the random coupling between impurities i and j, distributed with a law $P(\{J_{ij}\})$. The frustration stems from the fact that the couplings $P(\{J_{ij}\})$ may take positive (ferromagnetic) or negative (antiferromagnetic) values. Furthermore, the position $\mathbf{r}_i$ of the impurities (and therefore the J_{ij} themselves) are quenched variables, which means that they evolve with a time scale infinitely longer than the time scales of the thermodynamic variable $\mathbf{S}_i$. This implies that for a given set of $\{J_{ij}\}$, the free energy $F(\{J_{ij}\})$ at temperature T is obtained as:

$$F(\{J_{ij}\}) = -T \log Z(\{J_{ij}\}) \tag{A4.2a}$$

with

$$Z(\{J_{ij}\}) = \mathrm{Tr}_{\{\mathbf{S}_i\}} \exp\left(-\mathcal{H}/T\right) . \tag{A4.2b}$$

Note that $F(\{J_{ij}\})$ is also a random variable. For short range interactions, we now show that, when the number N of spins approaches infinity, $F(\{J_{ij}\})$ is sharply peaked around its average value $\overline{F}$, where

$$\overline{F} = \int \prod P(\{J_{ij}\}) F(\{J_{ij}\}) d(\{J_{ij}\}) . \tag{A4.3}$$

This property is called self-averageness of the free energy. The argument goes as follows: Suppose we divide the system of N spins $\mathbf{S}_i$ into subsystems containing each m spins with $1 \ll m \ll N$. The system's total free energy is the sum of two contributions:

(i) a contribution from each subsystem, of order m per "domain";
(ii) a contribution from each "domain wall" between neighboring subsystems, of order $m^{\frac{d-1}{d}}$ for short-range interactions.

For large m, the second contribution is negligible compared with the first. Equation (A4.3) results by considering each subsystem to represent a realization of the couplings $\{J_{ij}\}$.

It can also be shown that the self-averageness of the free-energy in the thermodynamic limit also holds for the infinite range (mean-field) version of (A4.1).

It is not our purpose to review here all the details and subtleties of the spin-glass paradigm. It suffices to say that spin-glasses have become the prototypical model with a complicated phase space, slow dynamical processes and memory effects etc., and a new kind of phase transition. It has invaded or pervaded many fields (including protein folding) such as combinatorial optimization, neural networks, Darwinian Evolution, among others, and its influence is growing. For our purposes, we shall stick to the mean-field models, where the random couplings $\{J_{ij}\}$ have an infinite range (i.e. each spin i interacts with each spin j). Already at this level, there is a rich variety of models, among which we quote:

* the Sherrington–Kirkpatrick model, where the spins $\{\mathbf{S}_i\}$ are classical m-component spins, and the $\{J_{ij}\}, i < j$ are independent (bond) random variables.
* the Mattis model, with similar $\{\mathbf{S}_i\}$ variables, but where the $\{J_{ij}\}$ have the form: $J_{ij} = J\epsilon_i\epsilon_j$, where the $\{\epsilon_i\}$ are independent (site) random variables. Note that here the sign of the constant J is quite relevant.
* the Hopfield model for neural networks, which generalizes the Mattis model by parametrizing the couplings as: $J_{ij} = J\sum_{p=1}^{M}\epsilon_i^p\epsilon_j^p$, and the $\{\epsilon_i^p\}$ are independent (site) random variables. This model, as well as the preceding one, will be considered in Appendix 5.
* Potts spin glasses, where the variables $\{\mathbf{S}_i\}$ are replaced by q-state Potts variables σ_i with Hamiltonian:

$$\mathcal{H}_q = -\sum_{i<j} J_{ij}\ \delta_{\sigma_i,\sigma_j} \tag{A4.4}$$

* p-spin glasses ($p > 2$) where the S_i variables are Ising spins $\{S_i = \pm 1\}$ which interact through the Hamiltonian:

$$\mathcal{H}_p = -\sum_{1\le i_1<i_2<\cdots<i_p\le N} J_{i_1\cdots i_p}\ S_{i_1}\cdots S_{i_p} \tag{A4.5}$$

with independent random couplings $J_{i_1\cdots i_p}$.

The mean field solutions of these models have been worked out, with great difficulties due to the fact that one has to average the free energy over the random coupling distributions (in the following, $\overline{A}$ denotes an average of the physical quantity A over the random couplings), which leads to the use of replicas through:

$$\log Z = \lim_{n\to 0}\frac{Z^n - 1}{n} \tag{A4.6}$$

and

$$\overline{F} = -T\lim_{n\to 0}\frac{\overline{Z^n} - 1}{n}\,. \tag{A4.7}$$

One must therefore consider n copies ($\alpha = 1,\ldots,n$) of the system, average over the random coupling distribution, and take the limit $n \to 0$ (not the same as in Sec. 3!). In order to illustrate the use of these ideas in the context of protein folding, we shall present the simplest spin glass calculations, which correspond to the p-spin glasses (A4.5), with $p \to \infty$ (taken after the thermodynamic limit $N \to \infty$). This limit also corresponds to the case of a q-state Potts glass, in the limit $q \to \infty$ (again after $N \to \infty$). Both can be mapped onto the Random Energy Model (REM) of B. Derrida, which appears to be central in the spin glass approach of proteins (in high enough dimensions). More generally, p-spin glasses ($p > 2$) and q-Potts glasses ($q > 4$) have a lot in common and may well bridge the gap between glasses and spin glasses.[109–113]

A4.2. *Quick Solution of the Simplest Spin Glass*

In this section, we closely follow the paper by Gross and Mézard.[31] Let us consider the Hamiltonian (A4.5)

$$\mathcal{H}_p = - \sum_{1 \le i_1 < i_2 < \cdots < i_p \le N} J_{i_1 \cdots i_p} \, S_{i_1} \cdots S_{i_p}$$

where the N variables S_i are Ising spins, that is, $S_i = \pm 1$, and the random independent coupling $\{J\}$ are distributed according to a Gaussian law:

$$P(\{J_{i_1 \cdots i_p}\}) = \sqrt{\frac{N^{p-1}}{\pi \, p!}} \exp\left(- \frac{N^{p-1}}{p!} \frac{J^2_{i_1 \cdots i_p}}{J^2} \right) . \tag{A4.8}$$

In (A4.8), the N-dependent normalization ensures that the free energy (A4.2) is extensive. The simplest spin-glass is defined by the limit $p \to \infty$ (after one has taken the thermodynamic limit $N \to \infty$), and is exactly soluble as we now sketch.

Denoting by $\alpha, \beta, \ldots$ the replica indices, we have:

$$Z^n = \mathrm{Tr}_{\{S_i^\alpha = \pm 1\}} e^{-\frac{1}{T} \sum_{\alpha=1}^n \mathcal{H}_p^\alpha} \tag{A4.9}$$

and thus

$$\overline{Z^n} = \mathrm{Tr}_{\{S_i^\alpha = \pm 1\}} \exp\left(\frac{J^2}{4 N^{p-1} T^2} \left\{ n N^p + \sum_{\alpha \ne \beta} \left(\sum_i S_i^\alpha S_i^\beta \right)^p \right\} \right) . \tag{A4.10}$$

The Edwards–Anderson order parameter is defined as ($\alpha < \beta$):

$$q_{\alpha\beta} = \frac{1}{N} \sum_i S_i^\alpha S_i^\beta . \tag{A4.11}$$

Using a Lagrange multiplier $r_{\alpha\beta}$ to enforce Eq. (A4.11), we have:

$$\overline{Z^n} = e^{\frac{nNJ^2}{4T^2}} \int_{-\infty}^{+\infty} \prod_{\alpha<\beta} dq_{\alpha\beta} \int_{-i\infty}^{+i\infty} \frac{dr_{\alpha\beta}}{2i\pi} \exp(-NG(q_{\alpha\beta}, r_{\alpha\beta})) \quad \text{(A4.12a)}$$

where

$$G(q_{\alpha\beta}, r_{\alpha\beta}) = -\frac{J^2}{2T^2} \sum_{\alpha<\beta} q_{\alpha\beta}^p + \sum_{\alpha<\beta} q_{\alpha\beta} r_{\alpha\beta} - \log\left(\mathrm{Tr}_{\{S_\alpha\}} \exp(\sum_{\alpha<\beta} r_{\alpha\beta} S_\alpha S_\beta)\right). \quad \text{(A4.12b)}$$

As N approaches infinity, the free energy, given by the dominant saddle point of G reads:

$$\frac{\overline{F}}{NT} = \lim_{n\to 0} \max_{\{q,r\}} \left(\frac{G}{n}\right) - \frac{J^2}{4T^2}. \quad \text{(A4.13)}$$

As usual in spin glasses, this dominant saddle point corresponds to a maximum of the free energy.

A4.2.1 *Replica Symmetric Solution*

We try a solution of the form:

$$\begin{aligned} q_{\alpha\beta} &= q \\ r_{\alpha\beta} &= r \end{aligned} \quad \text{(A4.14)}$$

and get:

$$q = \int_{-\infty}^{+\infty} \frac{dz}{\sqrt{2\pi}} \tanh^2(z\sqrt{r}) \exp\left(-\frac{z^2}{2}\right) \quad \text{(A4.15a)}$$

and

$$r = \frac{J^2}{2T^2}\, p\, q^{p-1}. \quad \text{(A4.15b)}$$

In the limit $p \to \infty$, this gives $q = r = 0$, yielding a free energy:

$$\frac{\overline{F}}{N} = -\frac{J^2}{4T} - T\log 2. \quad \text{(A4.16)}$$

This solution suffers from a negative entropy below a temperature $T_c = \frac{J}{2\sqrt{\log 2}}$. Since we are dealing with discrete spin variables, a negative entropy is not physically acceptable, and the replica symmetric solution must be rejected below T_c.

A4.2.2. *The Simplest Replica Symmetry-Breaking Solution*

Thus, the low temperature phase of the simplest spin-glass must have a replica symmetry-breaking (RSB) solution. It is not our aim to go into the technicalities of RSB, and we shall only mention a few points of interest in the context of protein-oriented models. The simplest RSB solution is obtained as follows: one divides the n replicas into m clusters of $\frac{n}{m}$ replicas.

* If replicas α and β belong to the same cluster, we take $q_{\alpha\beta} = q_1$ and $r_{\alpha\beta} = r_1$.
* Otherwise, we take $q_{\alpha\beta} = q_0$ and $r_{\alpha\beta} = r_0$.

With this parametrization of the matrices q and r, we find

$$\frac{\overline{F}}{N} = -\frac{J^2}{4T} - T\log 2$$

$$+ T \max_{\{q_0,q_1,r_0,r_1,m\}} \left[\frac{J^2}{4T^2}[mq_0^p + (1-m)q_1^p]\right.$$

$$\left. -\frac{1}{2}(mq_0r_0 + (1-m)q_1r_1) + \frac{1}{2}r_1 - K\right] \tag{A4.17a}$$

where

$$K = \frac{1}{m}\int_{-\infty}^{+\infty} \frac{dz_0}{\sqrt{2\pi}} \exp\left(-\frac{z_0^2}{2}\right)$$

$$\times \log\left\{\int_{-\infty}^{+\infty} \frac{dz_1}{\sqrt{2\pi}} \exp\left(-\frac{z_1^2}{2}\right) \cosh^m[z_0\sqrt{r_0} + z_1\sqrt{r_1 - r_0}]\right\}. \tag{A4.17b}$$

Extremizing the above expression with respect to $\{q_0, q_1, r_0, r_1, m\}$, we find that there is a *freezing transition* at the temperature T_c defined above.

* For $T > T_c$, the only solution is $q_0 = q_1 = r_0 = r_1 = 0$, yielding the free energy (A4.16).

* For $T \leq T_c$, the only physical solution (with $q_0 < q_1$) is:

$$\begin{aligned} q_0 &= 0\,, \\ q_1 &= 1\,, \\ r_0 &= 0\,, \\ r_1 &= \infty\,, \\ m &= \frac{T}{T_c}\,, \\ \frac{\overline{F}}{N} &= -\frac{J}{2T_c}\,. \end{aligned} \tag{A4.18}$$

The low temperature phase of the system is completely frozen, with constant free energy and order parameters. It may seem from (A4.18) that only two states are necessary to describe the simplest spin glass.

In fact, using the so-called TAP (Thouless, Anderson and Palmer[114]) mean field equations, it can be shown[12,13] that the spin glass phase of this model has an exponentially large number of metastable states. Among these states, only those of lowest free energy contribute to the low temperature thermodynamics. Their number M is not exponentially large, but still diverges in the thermodynamic limit ($N \to \infty$). Denoting by $\{\mu\}$ this set of states and by F_μ their corresponding free energies, it can be shown that the extensive part Nf_0 of the free energy of these states is the same,

$$F_\mu = Nf_0 + f_\mu \tag{A4.19a}$$

and that they differ by subtle non-extensive terms f_μ. The Boltzmann weight associated with state μ is thus:

$$p_\mu = \frac{e^{-f_\mu/T}}{\sum_{\{\mu'\}} e^{-f_{\mu'}/T}} \tag{A4.19b}$$

showing that the probability of the different low free energy states is linked to the non-extensive part f_μ of their free energies. Moreover, involved replica calculations lead to the following relation:

$$\overline{\sum_\mu p_\mu^2} = \left(1 - \frac{T}{T_c}\right)\,. \tag{A4.19c}$$

Together with the normalization $\sum_\mu p_\mu = 1$, Eq. (A4.19c) proves that only a small number of states exhaust the sum rule. Indeed, if they were all quasi-degenerate, the p_μ would be of order $\frac{1}{M}$, and (A4.19c) could not be satisfied. This shows that only a few number of states $\{\mu_0\}$ have finite Boltzmann weights p_{μ_0}. Once again, these states, called dominant states, are determined by non-extensive terms of the free energy. This point will be of relevance in the study of proteins.

Consequently, many physical quantities in this model lack self-averageness in the limit $N \to \infty$. We shall not elaborate on this point, and refer the interested reader to the previously mentioned references. Neither will we consider the important issue of ultrametricity,[115] which is central in spin glasses, and perhaps relevant to some Darwinian Evolution models.[116,117]

A4.3. *Simplest Spin Glass and Random Energy Model*

Another aspect is crucial for linking spin glasses and protein folding models. We again consider Hamiltonian (A4.5).

Let us consider a spin configuration $\{S_i^1\}$, with Hamiltonian $\mathcal{H}_p(\{S_i^1\})$. This configuration has energy E, with probability $P(E)$, where:

$$P(E) = \overline{\delta(E - \mathcal{H}_p(\{S_i^1\}))}\,. \tag{A4.20a}$$

Using Eq. (A4.8), we obtain:

$$P(E) = \frac{1}{\sqrt{N\pi J^2}} \exp(-\frac{E^2}{NJ^2}) \tag{A4.20b}$$

independently of p and of the spin configuration $\{S_i^1\}$. Considering now two spin configurations $\{S_i^1\}$ and $\{S_i^2\}$, we can calculate the probability $P(E_1, E_2)$ that their respective energies are E_1 and E_2 by:

$$P(E_1, E_2) = \overline{\delta(E_1 - \mathcal{H}_p(\{S_i^1\}))\, \delta(E_2 - \mathcal{H}_p(\{S_i^2\}))}\,. \tag{A4.21a}$$

Introducing the overlap q_{12} between the two configurations by:

$$q_{12} = \frac{1}{N} \sum_i S_i^1 S_i^2 \tag{A4.21b}$$

one finds:

$$\begin{aligned} P(E_1, E_2) &= \left(N\pi J^2(1 + q_{12}^p) N\pi J^2(1 - q_{12}^p)\right)^{-1/2} \\ &\quad \times \exp\left(-\frac{(E_1 + E_2)^2}{2NJ^2(1 + q_{12}^p)} - \frac{(E_1 - E_2)^2}{2NJ^2(1 - q_{12}^p)}\right)\,. \end{aligned} \tag{A4.21c}$$

If the two spin configurations are macroscopically different, then $|q_{12}| < 1$, and, as p approaches infinity, the probability distribution $P(E_1, E_2) \rightarrow P(E_1)\, P(E_2)$. The energies of different spin configurations are independent (uncorrelated) random variables.

The physics of random energy levels with $P(E)$ given by (A4.20b) was introduced by Derrida,[29,30] who also demonstrated the freezing transition of the preceding section. In the context of proteins, this model was first introduced phenomenologically by Bryngelson and Wolynes.[26] We mention for completeness that correlations between energy levels can be introduced, yielding richer statistical mechanical behavior.[(118,119)]

A4.4. *Conclusion*

In conclusion, we mention that p-spin glass models, with $p > 2$ and q-Potts glasses, with $q > 4$, have similar phase transitions and properties. Noteworthily, q_1 jumps at T_c from 1 to 0, as shown by (A4.15) and (A4.18), in a way similar to a first order transition (without latent heat though). This discontinuity suggests that there may exist a second temperature T_d in the model, above which no metastable states exist. Since all barrier heights diverge in mean-field theory, this temperature will have dynamical significance. In particular, it has been pointed out,[111,112] that the $p = 3$ case bears a striking resemblance to the (dynamic) glass transition (see the review on the glass transition by Götze[102]).

The special case of the Sherrington–Kirkpatrick model ($p = 2$ spin glass) requires a more complicated RSB scheme.[13] The one step RSB scheme with q_0 and q_1, depicted above, must be iterated an infinite number of times, and the order parameter $q_{\alpha\beta}$ becomes a function $q(x)$ defined on the interval $0 \leq x \leq 1$. This function is related to the probability distribution $P(q)$ of the overlap $q = \frac{1}{N} \sum_i S_i^1 S_i^2$ of two copies of the system. This function $P(q)$ is defined by:

$$P(q) = \sum_{\{S_i^1, S_i^2\}} \frac{e^{-\beta\mathcal{H}(S_i^1)}}{Z} \frac{e^{-\beta\mathcal{H}(S_i^2)}}{Z}\, \delta\left(q - \frac{1}{N}\sum_i S_i^1 S_i^2\right) \qquad \text{(A4.22)}$$

where Z is the partition function of the system. This function plays a central role in the Sherrington–Kirkpatrick model. More generally the introduction[120] of two copies of the same system, and the subsequent calculation of the function $P(q)$[121] is crucial in the study of all models with a complex phase space.

Appendix 5. Neural Networks: The Hopfield Model

For our purposes, the problem of neural networks can be simply stated as follows. Consider a set of M patterns, labeled by an index p. These patterns can be viewed as a set of N Ising spins (or bits of information), and can thus be represented by the set of M vectors $\{\xi_i^p\}$, $i = 1, \ldots, N$, with $\xi_i^p = \pm 1$. Can one find a Hamiltonian $\mathcal{H}(S_i)$ which has a minimum for each one of the M patterns? In other words, can one code the patterns $\{\xi_i^p\}$ in the interactions present in $\mathcal{H}$? If this is the case, a standard relaxational dynamics would be able, starting from an initial S_i configuration which is a perturbed pattern (pattern + noise), to retrieve the original patterns.

For the sake of simplicity, we shall consider only two-body interactions, and parametrize the Hamiltonian as:

$$\mathcal{H} = -\sum_{i<j} J_{ij} S_i S_j \,. \tag{A5.1}$$

Let us first consider the case $M = 1$, in a mean-field geometry, where the $\{J_{ij}\}$ are infinitely-ranged. This model has been solved by Mattis[122]: in order for the pattern $\{\xi_i\}$ to be the ground state of $\mathcal{H}$, one can use:

$$J_{ij} = \frac{\xi_i \xi_j}{N} \,. \tag{A5.2}$$

Note that the sign in (A5.2) is crucial. With this choice, the Hamiltonian can be rewritten (up to non-extensive terms) as:

$$\mathcal{H} = -\frac{1}{2N} \left(\sum_i \xi_i S_i \right)^2 \tag{A5.3}$$

so that the ground state is indeed the given pattern $S_i = \xi_i$.

Since the quenched variables ξ_i are equal to ± 1, we can define new Ising spin variables $\sigma_i = \xi_i \; S_i$. In terms of these new variables, the partition function of the model reads:

$$Z = \sum_{\{\sigma_i = \pm 1\}} \exp \left(\frac{\beta}{2N} (\sum_i \sigma_i)^2 \right) \,. \tag{A5.4}$$

It is of a usual ferromagnetic kind and thus has a ferromagnetic phase transition at $T_c = 1$ in our units. Above T_c, the magnetization $\langle \frac{1}{N} \sum_i \sigma_i \rangle =$

$\langle \frac{1}{N} \sum_i \xi_i S_i \rangle$ vanishes, so that there is no macroscopic overlap between the average state $\{S_i\}$ and the pattern ξ_i. Below T_c, a finite magnetization builds up, and therefore a macroscopic overlap between the system's magnetization and the pattern develops. At zero temperature, this overlap is one, and the spins are aligned (or anti-aligned, since the ground-state is twice degenerate) on the pattern. One can say that the pattern is coded in the interactions $\{J_{ij}\}$.

As mentioned above, any initial state of the spins S_i will evolve towards the stored pattern (or to its opposite) under relaxational dynamics: the pattern has a finite basin of attraction.

The Mattis model has been generalized by J. J. Hopfield,[32,33] in the case of a large number M of uncorrelated patterns $\{\xi_i^p\}$. By uncorrelated, we mean:

$$\frac{1}{N} \sum_{i=1}^{N} \xi_i^p \xi_i^q = \delta_{pq} \,. \qquad \text{(A5.5)}$$

Hopfield has studied the associated Hamitonian (A5.1), with couplings given by:

$$J_{ij} = \frac{1}{N} \sum_{p=1}^{M} \xi_i^p \, \xi_j^p \,. \qquad \text{(A5.6)}$$

This kind of coupling is referred to as the Hebb rule. Detailed analytic calculations can be found in Amit *et al.*[123,124] We only quote the main results.

Depending on the ratio $\alpha = \frac{M}{N}$, there are two regimes:

(i) $\alpha < 0.14$: there is a phase transition at temperature $T_c < 1$. Note that for $\alpha = 0$, there are finite M, $T_c = 1$. Above T_c, the ground state of the system is paramagnetic; its average magnetization (or spin configuration) has no finite correlation with any of the patterns. Below T_c, the system can be found in $2M$ ground states (if one includes the reversal of magnetization), each of these states having a finite overlap with one of the patterns. Each of the patterns has a finite basin of attraction: they are coded in the interactions. In addition to these lowest energy states, there exists, at low temperature, an exponential number of metastable states. These metastable states are mixtures of patterns: they have finite overlaps with several patterns. As a consequence, in the low temperature phase, starting from an initial spin configuration, the system may evolve either towards one of the stored patterns, or towards one of these metastable spurious states.

(ii) $\alpha > 0.14$: the low temperature phase is a spin-glass phase, with no average magnetization: there is no finite correlation between the spin configuration and any of the patterns; the basin of attraction of each pattern vanishes. The spin-glass phase does not code for the patterns. As usual in spin-glasses, the relaxational dynamics is extremely slow, due to the ruggedness of phase space.

The Hopfield model can be generalized in several ways:

(a) One may use other type of variables (Potts spins, vectorial spins, etc.).
(b) One may generalize the couplings to incorporate three- and higher-body terms.
(c) Instead of using the Hebb rule (A5.6), one may treat the $\{J_{ij}\}$ as dynamical variables, and device algorithms to "learn" the patterns. The main goal of neural networks is to construct such learning algorithm, so as to store the maximum number of patterns, without storing too many spurious states.
(d) Finally, to mimick more closely the synaptic interactions in the brain, one may introduce non-Hamiltonian dynamics, that is non-symmetric couplings $J_{ij} \neq J_{ji}$.

In the context of proteins, neural networks have been used to predict secondary structures,[125] and side-chain conformations.[126] For example, the idea is to learn some secondary structures associated with known primary sequences (these are the patterns), and generate the Hamiltonian $\mathcal{H}$ which stores this information. Then, a primary sequence with unknown secondary structures is taken as an initial condition for a relaxational dynamics with Hamiltonian $\mathcal{H}$. The success rate of these methods is roughly 65 %, and represents a slight improvement over the often used empirical Chou-Fasman rules.[127]

References

1. T. E. Creighton, *Proteins* (W. H. Freeman, New York, 1984).
2. T. E. Creighton, *Protein Folding* (W. H. Freeman, New York, 1992).
3. K. A. Dill, *Biochemistry* **29**, 7133 (1990).
4. B. R. Brooks, R. E. Bruccoleri, B. D. Olafson, B. States, S. Swaminathan and M. Karplus, *J. Comput. Chem.* **4**, 187 (1983).

5. J. C. Smith and M. Karplus, *J. Am. Chem. Soc.* **114**, 801 (1992).
6. L. Pauling and R. B. Corey, *Proc. Natl. Acad. Sci. USA* **37**, 235, 251, 272, 729 (1951).
7. J. S. Richardson, *Adv. in Prot. Chem.* **34**, 167 (1981).
8. P. J. Flory, *Statistical Mechanics of Chain Molecules* (Interscience-Wiley, New York, 1969).
9. P. G. de Gennes, *Scaling Concepts in Polymer Physics* (Cornell Univ. Press, Ithaca, 1979).
10. M. Levitt, *Current Opinion in Structural Biology* **1**, 224 (1991).
11. H. S. Chan and K. A. Dill, *Physics Today*, February issue, 24 (1993).
12. K. Binder and A. P. Young, *Rev. Mod. Phys.* **58**, 801 (1986).
13. M. Mézard, G. Parisi and M. A. Virasoro, *Spin Glass Theory and Beyond* (World Scientific, Singapore, 1987).
14. P. J. Flory, *Proc. Roy. Soc.* **A234**, 60, 73 (1956).
15. J. H. Gibbs and E. A. Di Marzio, *J. Chem. Phys.* **28**, 373 (1958).
16. M. Karplus and E. I. Shakhnovich 1992, in Creighton T. E. (1992), p. 127.
17. J. P. Hendricks and F. U. Hartl, *Ann. Rev. Biochem.* **62**, 349 (1993).
18. D. Thirumalai, in *Statistical Mechanics, Protein Structure and Protein-Substrate Interaction*, ed. S. Doniach (Plenum Press, New York, 1994).
19. S. F. Edwards, *Proc. Phys. Soc. London* **85**, 613 (1965).
20. P. G. de Gennes, *Phys. Lett.* **A38**, 339 (1972).
21. J. Bascle, T. Garel and H. Orland, *J. Phys.* **A25**, L1323 (1992).
22. J. Bascle, T. Garel and H. Orland, *J. Phys.* (France) **I**, 3, 259 (1993).
23. J. Bascle, T. Garel and H. Orland, *J. Phys.* (France) **II**, 3, 245 (1993).
24. O. B. Ptitsyn, *J. Protein Chem.* **6**, 273 (1987).
25. O. B. Ptitsyn, in Ref. 2, p. 243.
26. J. Bryngelson and P. G. Wolynes, *Proc. Natl. Acad. Sci. USA* **84**, 7524 (1987).
27. T. Garel and H. Orland, *Europhys. Lett.* **6**, 307 (1988); T. Garel and H. H. Orland, *Europys. Lett.* **6**, 597 (1988).
28. E. I. Shakhnovich and A. M. Gutin, *Europhys. Lett.* **8**, 327 (1989).
29. B. Derrida, *Phys. Rev. Lett.* **45**, 79 (1980).
30. B. Derrida, *Phys. Rev.* **B24**, 2613 (1981).
31. D. J. Gross and M. Mézard, *Nucl. Phys.* **B240**, FS12, 431 (1984).
32. J. J. Hopfield, *Proc. Natl. Acad. Sci. USA* **79**, 2554 (1982).
33. J. J. Hopfield, *Proc. Natl. Acad. Sci. USA* **81**, 3088 (1984).
34. P. J. Flory, *Principles of Polymer Chemistry* (Cornell Univ. Press, Ithaca, 1971).
35. M. Doi and S. F. Edwards, *The Theory of Polymer Dynamics* (Oxford Univ. Press, Oxford, 1986).
36. F. W. Wiegel, *Introduction to Path Integral Methods in Physics and Polymer Science* (World Scientific, Singapore, 1986).
37. J. des Cloizeaux and G. Jannink, in *Les Polymères en Solution*, ed. de Physique (Les Ulis, 1987).

38. K. Freed, *Renormalization Group Theory of Macromolecules* (Wiley, New York, 1987).
39. M. E. Fisher, *Physics* (Long Island City, N.Y.) **3**, 255 (1967).
40. J. P. Cotton, *J. Phys.* (France) **41**, L231 (1980).
41. H. Orland, *J. Phys.* (France) **I**, 4, 101 (1994).
42. I. M. Lifchitz, *JETP* **28**, 1280 (1969).
43. M. A. Moore, *J. Phys.* **A10**, 305 (1977).
44. A. Y. Grosberg and A. R. Khokhlov, *Sov. Sci. Rev.* **A8**, 147 (1987).
45. E. Raphael and J. F. Joanny, *Europhys. Lett.* **13**, 623 (1990).
46. P. G. Higgs and J. F. Joanny, *J. Chem. Phys.* **94**, 1543 (1991).
47. J. Wittmer, A. Johner and J. F. Joanny, *Europhys. Lett.* **24**, 263 (1993).
48. Y. Kantor, H. Li and M. Kardar, *Phys. Rev. Lett.* **69**, 61 (1992).
49. Y. Kantor, M. Kardar and H. Li, *Phys. Rev.* **E49**, 1383 (1994).
50. B. H. Zimm and S. D. Levene, *Quart. Rev. Biophysics* **25**, 2 171 (1992).
51. T. Garel, D. A. Huse, S. Leibler and H. Orland, *Europhys. Lett.* **8**, 9 (1989).
52. T. Garel, H. Orland and E. I. Shakhnovich, *J. Chem. Phys.* **93**, 2043 (1990).
53. A. Grossberg, S. Izrailev and S. Nechaev, *Phys. Rev.* **E50**, 1912 (1994).
54. S. F. Edwards. and M. Muthukumar, *J. Chem. Phys.* **89**, 2435 (1988).
55. D. Thirumalai, *Phys. Rev.* **A37**, 269 (1988).
56. J. D. Honeycutt and D. Thirumalai, *J. Chem. Phys.* **90**, 4542 (1989).
57. P. Le Doussal and J. Machta, *J. Stat. Phys.* **64**, 541 (1991).
58. T. Hwa and D. S. Fisher, *Phys. Rev.* **E49**, 3136 (1994).
59. M. Kléman and J. F. Sadoc, *J. Phys.* (France) **40**, L569 (1979).
60. M. Kléman, *J. Phys.* (France) **46**, L723 (1985).
61. D. R. Nelson and F. S. Spaepen, *Solid State Phys.* **42**, 1 (1988).
62. A. G. Murzin and A. V. Finkelstein, *J. Mol. Biol.* **204**, 749 (1988).
63. J. I. Lauritzen and R. Zwanzig, *J. Chem. Phys.* **52**, 3740 (1970).
64. H. Orland, C. Itzykson and De Dominicis, *J. Phys.* (France) **46**, L353 (1985).
65. C. J. Camacho and D. Thirumalai, *Phys. Rev. Lett.* **71**, 2505 (1993).
66. M. Gordon, P. Kapadia and A. Malakis, *J. Phys.* **A9**, 751 (1976).
67. T. R. Kirkpatrick and D. Thirumalai, *J. Phys.* **A22**, L149 (1989).
68. J. F. Nagle, P. D. Gujrati and M. Goldstein, *J. Phys. Chem.* **88**, 4599 (1984).
69. S. Doniach, J. Bascle, T. Garel and H. Orland, *J. Mol. Biol.* **254**, 960 (1995).
70. D. Poland and H. A. Scheraga, *Theory of Helix Coil Transitions in Biopolymers*, (Academic Press, New York, 1970).
71. S. P. Obukhov, *J. Phys.* **A19**, 3655 (1986).
72. T. Garel, L. Leiber and H. Orland, 1994, *J. Phys.* (France) **II**, 4, 2139 (1994).
73. J. Bryngelson and P. G. Wolynes, *J. Phys. Chem.* **93**, 6902 (1989).
74. C. D. Sfatos, A. M. Gutin and E. I. Shakhnovich, *Phys. Rev.* **E48**, 465 (1993).
75. M. S. Friedrichs and P. G. Wolynes, *Science* **246**, 371 (1989).
76. M. Sasai and P. G. Wolynes, *Phys. Rev. Lett.* **65**, 2740 (1990).
77. M. Sasai and P. G. Wolynes, *Phys. Rev.* **A46**, 7979 (1992).

78. J. R. Garel, T. Garel and H. Orland, *J. Phys.* (France) **50**, 3067 (1989).
79. D. J. Gross, I. Kanter and H. Sompolinsky, *Phys. Rev. Lett.* **55**, 304 (1985).
80. E. I. Shakhnovich and A. M. Gutin, *J. Phys.* **A22**, 1647 (1989).
81. E. I. Shakhnovich and A. M. Gutin, *Biophys. Chem.* **34**, 187 (1989).
82. J. P. Bouchaud, M. Mézard and J. S. Yedidia, *Phys. Rev. Lett.* **67**, 3840 (1991).
83. J. P. Bouchaud, M. Mézard and J. S. Yedidia, *Phys. Rev.* **B46**, 14686 (1992).
84. T. Giamarchi and P. Le Doussal, *Phys. Rev. Lett.* **72**, 1530 (1994).
85. M. Mézard and G. Parisi, *J. Phys.* **A23**, L1229 (1990).
86. M. Mézard and G. Parisi, *J. Phys.* (France) **I**, 1, 809 (1991).
87. M. Mézard and G. Parisi, *J. Phys.* (France) **I**, 2, 2231 (1992).
88. M. Mézard and G. Parisi, *J. Phys.* **A25**, 4521 (1992).
89. P. M. Goldbart and A. Zippelius, *Phys. Rev. Lett.* **71**, 2256 (1992).
90. M. Kardar, *Phys. Rev. Lett.* **55**, 2235 (1985).
91. P. G. de Gennes, in *Directions in Condensed Matter Physics* (World Scientific, Singapore, 1986), p. 83.
92. V. S. Dotsenko, S. Franz and M. Mézard, *J. Phys.* **A25**, 6631 (1992).
93. E. A. Dorotheyev, V. S. Dotsenko and B. Tirozzi, *Int. J. Mod. Phys.* **B7**, 2509 (1993).
94. E. I. Shakhnovich and A. M. Gutin, *J. Phys.* (France) **50**, 1843 (1989).
95. M. S. Friedrichs, R. A. Goldstein and P. G. Wolynes, *J. Mol. Biol.* **222**, 1013 (1991).
96. R. A. Goldstein, Z. A. Luthey-Schulten and P. G. Wolynes, *Proc. Natl. Acad. Sci. USA* **89**, 4918 (1992).
97. R. A. Goldstein, Z. A. Luthey-Schulten and P. G. Wolynes, *Proc. Natl. Acad. Sci. USA* **89**, 9029 (1992).
98. S. J. Wodak and M. J. Rooman, *Current Opinion in Structural Biology* **3**, 247 (1993).
99. C. A. Angell, *Nucl. Phys.* **B5A**, 69 (1988).
100. J. Souletie, *J. Phys.* (France) **51**, 883, 1990.
101. J. Souletie and D. Bertrand, *J. Phys.* (France) **I**, 1, 1627 (1991).
102. W. Götze 1991, in *Liquids, Freezing and Glass Transition*, eds. J. P. Hansen, D. Levesque and J. Zinn-Justin (Elsevier, 1989).
103. J. C. Philips, *J. Non. Cryst. Sol.* **34**, 153 (1979).
104. M. F. Thorpe, *J. Non. Cryst. Sol.* **57**, 355 (1983).
105. M. Tatsumisago, B. L. Halfpap, J. L. Green, S. M. Lindsay and C. A. Angell, *Phys. Rev. Lett.* **64**, 1549 (1990).
106. D. S. Franzblau and J. Tersoff, *Phys. Rev. Lett.* **68**, 2172 (1992).
107. J. W. Negele and H. Orland, *Quantum Many Particle Systems* (Addison-Wesley, Menlo Park, 1987).
108. M. Karplus and G. A. Petsko, *Nature* **347**, 631 (1990).
109. T. R. Kirkpatrick and P. G. Wolynes, *Phys. Rev.* **B36**, 8552 (1987).
110. T. R. Kirkpatrick and P. G. Wolynes, *Phys. Rev.* **A35**, 3072 (1987).
111. T. R. Kirkpatrick and D. Thirumalai, *Phys. Rev. Lett.* **58**, 2091 (1987).

112. T. R. Kirkpatrick and D. Thirumalai, *Phys. Rev.* **B36**, 5388 (1987).
113. D. Thirumalai and T. R. Kirkpatrick, *Phys. Rev.* **B38**, 4881 (1988); T. R. Kirkpartrick, D. Thirumalai and P. G. Wolynes, *Physics Rev.* **A40**, 1045 (1989).
114. D. J. Thouless, P. W. Anderson and R. G. Palmer, *Phil. Mag.* **35**, 593 (1977).
115. R. Rammal, G. Toulouse and M. A. Virasoro, *Rev. Mod. Phys.* **58**, 765 (1986).
116. B. Derrida and L. Peliti, *Bull. Math. Biol.* **53**, 355 (1991).
117. P. G. Higgs and B. Derrida, *J. Mol. Evol.* **35**, 454 (1992).
118. B. Derrida and E. Gardner, *J. Phys.* **C19**, 2253 (1986).
119. B. Derrida and E. Gardner, *J. Phys.* **C19**, 5783 (1986).
120. A. Blandin, M. Gabay and T. Garel, *J. Phys.* **C13**, 403 (1980).
121. G. Parisi, *Phys. Rev. Lett.* **50**, 1946 (1983).
122. D. C. Mattis, *Phys. Lett.* **56A**, 421 (1976).
123. D. J. Amit, H. Gutfreund and H. Sompolinsky, *Phys. Rev.* **A32**, 1007 (1985).
124. D. J. Amit, H. Gutfreund and H. Sompolinsky, *Ann. Phys.* **173**, 30 (1987).
125. N. Qian and T. J. Sejnowski, in *Evolution, Learning and Cognition*, ed. Y. C. Lee, (World Scientific, Singapore, 1988).
126. J. Irwin, H. Bohr, K. Mochizuki and P. G. Wolynes, *Int. J. Neur. Syst.* **3**, 177 (1992).
127. P. Y. Chou and G. D. Fasman, *Ann. Rev. Biochem.* **47**, 251 (1978).

CHAPTER 5

ATOMIC BIOLOGY, ELECTROSTATICS, AND IONIC CHANNELS

R. S. EISENBERG

Department of Molecular Biophysics and Physiology
Rush Medical College
1750 West Harrison Street
Chicago IL 60612, USA

Contents

1. Molecular Biology

The identification and manipulation of the molecules of life is one of the triumphs of man, putting biological systems under his control, embedding them into his technology, allowing the control of many of life's functions. In the next few years some of the dreams of centuries will become realities, as we learn to control disease and our biological life cycle in ways that only poets and novelists have imagined, by controlling proteins.

The triumph of this molecular biology is based on the identification and manipulation of proteins, and of their blueprints, DNA and some kinds of RNA. The technology of molecular biology manufactures these blueprints and thereby allows manipulation and bulk production of the encoded protein. This impressive technology is possible because so many biological functions can be controlled, or influenced, simply by the presence or absence of specific proteins in sufficient amounts. Evolution has used this simple method to control many systems and we can use it too, probably because it is easier to make adaptations this way, depending on the mutation and selection of single genes, rather than in other ways that depend on changes in many genes, and therefore occur with much lower probability, if the mutation of each of the genes is independent, as traditionally assumed.

If biological systems have such simple built-in controls, we can take advantage of this simplicity and govern those systems ourselves, even if we do not understand them. A car can easily be controlled by a driver who knows little of how it works. Similarly, a metabolic pathway governed by a single enzyme can be controlled technologically long before the pathway is known or understood. Molecular biology, as presently practiced, allows the control of biological functions that depend on the concentration of one protein. It will do less in systems that depend on the presence and properties of many proteins. Let us hope these are few and far between (for the sake of our own health). The quip is that it will be possible to grow a nervous system long before we understand it!

Molecular biology is also a useful tool for analyzing the function of a protein. Molecular genetics allows the rational manipulation of a protein's primary (and, perhaps someday, secondary and tertiary) structure, an experimental technique of the greatest importance. However, the simple replacement of one amino acid by another (e.g. in site-directed mutagenesis and its congener techniques) cannot tell us how the protein works and, how that amino acid influences its function, any more than the simple replacement of the parts of a machine, (eg. an automobile) will tell us much about how it works. Replacement of amino acids may determine which one controls a function of an enzyme, just as replacement of the spark plugs will show that they (along with many other parts of the engine) control ignition. But the identity of the other parts will remain unknown, and so will their interactions with the spark plug. And how that amino acid (or spark plug) works will often remain a mystery, as well.

Changing one amino acid is much more likely to have a specific effect on function than on the physical properties of proteins. Functions may be controlled by a single amino acid, even if that amino acid has a subtle, distant, or wide-ranging effect on the physical properties of the protein. There are too many amino acids in a protein; any one amino acid is expected to have subtle, diffuse and unimportant overall effects on physical properties. A particular amino acid residue may, nonetheless, be able to control function of an entire protein for essentially evolutionary, not physical reasons, even if most individual amino acids in the protein have a small effect on the properties or function of the whole protein. After all, the fundamental mechanism of evolutionary change is the substitution of one amino acid by another. One must expect many properties of proteins to be controlled by specific amino acids. Proteins evolve by randomly changing amino acids until one produces a function with selective advantage. Changes in each amino acid are (more or less) independent random events. Thus, the probability of multiple useful changes is very much less than the probability of a single useful change. For this reason, many functionally significant properties of proteins are likely to be controlled by particular amino acids; indeed, those are the amino acids found by evolution to produce the functions needed!

Yet, saying that an amino acid controls a function says nothing about how it controls that function. A light switch controls the light, and knowing where the switch is and how to use it is of immense practical importance, but locating the light switch tells nothing about how the light or switch works, or how they are coupled together. The light might be incandescent, fluorescent, or whatever; the switch might be mechanical, electronic, or something else; the coupling might be electrical, electromagnetic (e.g. an infrared remote control), or even mechanical. And the switch might be inches, feet, or yards (indeed, miles, in special cases) away from the light.

Controlling protein function, to improve our technology, is one thing, of great practical importance; but understanding the function, to extend our knowledge, is another. To understand, we need a specific physical model with atomic resolution, I believe. The techniques of molecular biology have too coarse resolution to resolve many of the roles of atoms in protein function.

2. Atomic Biology

I believe atomic biology is needed to supplement present day molecular biology, if we are to design and understand proteins, as well as define, make, and use them.

The challenge of understanding proteins in general (the grand challenge of atomic biology) is a formidable one, one I fear that is beyond us (at least, me) for now. Most significant proteins are enzymes that catalyze chemical reactions. Such reactions occur in solutions, in condensed phases with little space between atoms, and they almost always involve changes in covalent bonds. The underlying physics of enzymes is the physics of chemical reactions in solutions, of quantum chemistry in condensed phases. Here, experimental and theoretical problems are too formidable to solve, even for unstructured chemical reactions in homogeneous bulk solutions. The chemical reactions of enzymes would be just as difficult to understand, even if the reactions were spatially homogeneous like chemical reactions in bulk solutions. But most enzymes have a specific three-dimensional structure which is anything but homogeneous. Indeed, enzymes direct structured chemical reactions, achieving many of their remarkable properties this way. It is unlikely that the structured chemical reactions typical of enzymes can be understood until chemists better understand chemical reactions in the condensed phase in general, until experimental techniques allow the visualization of the directed structural changes of enzymes and their substrates, and until experimental techniques resolve the time course of those structures and their transformations. This science is certainly beyond our grasp, and will even stay out of our reach for quite some time, I fear.

The promise of atomic biology then seems a distant trumpet compared to the clarion call of molecular biology. Why even try to answer such a feeble summons? The reason is that a feeble sound becomes loud when focused on an appropriate object, on ionic channels,[1] a group of proteins much simpler than enzymes. Trumpet's sound becomes loud when focused on an appropriate object, even loud enough to bring walls tumbling down, so they say. The promise of atomic resolution may be reached if we focus on a class of proteins simpler than enzymes, on ionic channels.[1]

Ionic channels are proteins that catalyze[2] the movement of solutes from one place to another, from outside a cell to inside a cell, or outside an organelle to inside an organelle. Channels do not catalyze changes in covalent bonds and so understanding their function does not require understanding the (quantum) theory of chemical reactions in solutions. Yet the biological importance of these channel proteins is hard to overstate. They are nearly as important as enzymes. They sit in membranes in series with the metabolic machinery of cells. Just as security in an airport is best provided by controlling access, so is control of cellular functions easily provided by specialized gates, selectively filtering entering ions and metabolites, thereby restricting access to the interior of cells and organelles. Channels control access to the cell interior the way doors control access to buildings. Channels provide control by regulating the diffusion and drift (the electrodiffusion) of ions across membranes. In this way they produce the electrical function of cells, for example, signaling in the nervous system. They coordinate muscle contraction, allow the heart to pump blood, and are involved in the transport and secretion of most substances in cells and organs, including substances that are uncharged. Channels govern much transport across membranes and are in series with the enzymatic machinery of the cell.

They are ideally placed to control the life of the cell, and the large number of drugs that act on channels usefully provide such control to physicians and scientists—they control a substantial fraction of the functions of life.

Ionic channels are important, yet they have a simple structure that controls function in a simple way. Channels have an aqueous pore down their middle, a pore we might call a channel within a channel if it were not so confusing.[a] In this pore ions move following laws of electrodiffusion much as they do in solution. In classical work (before 1981, say) on channels and membranes, currents through individual molecules were not resolved. The currents recorded from macroscopic amounts of membrane were usually interpreted as flows through a fixed area of homogeneous membrane. We now know that the area through which macroscopic currents flow is anything but fixed. The number of open channels varies with a wide range of conditions and thus so does the area for current flow. Channels open and close in response to stimuli, and when they are open, molecules move by electrodiffusion.

The classical interpretation of macroscopic currents has been discarded in the face of these new highly resolved measurements. Many of the phenomena attributed to "the membrane" in classical theories arise from changes in the number of open channels, and in the area available for current flow. Such gating phenomena are not the subject of this paper (except peripherally); rather we concentrate on the properties of the channel's pore protein, and of the channel, pore, once it is open. Classical theories are still relevant, however, even if classical descriptions of data are not. The *classical theories now describe the flow of ions through one open channel*, not through a membrane. Flowing through a channel, each ion follows a random stochastic trajectory starting in the surrounding baths. Collectively, the drift component of the random trajectories of individual ions produce a diffusive flux driven by the (bath) concentrations of permeating ions as in the baths that is described (in the mean) by classical theories.

The opening and closing of channel — called "gating" in the literature of channology — is not a simple process. Gating determines much of the biological function of channels and is thought to be a conformation

[a]"Channel" in dictionaries usually means "a passage through which something can be directed ... a tubular passage ... a conduit".[3] Biologists, however, sensibly identify a channel as **the protein**, not the pore, perhaps because biologists are more interested in the protein than its contents, just as a steelworker is more interested in the metal of his channel than the air it encloses.

change, a substantial change in the position of many of the atoms of the channel protein. If so, it is not immediately amenable to the careful, high resolution analysis of atomic biology. Conformation changes are hard to observe directly (by structural methods); they can rarely be measured on the physiological time scale, and those conformation changes that can be observed directly are seldom directly relevant to biological function—usually observed only indirectly if at all. When direct structural methods do resolve movements in proteins, those movements are often of (artificially or naturally) "tagged" parts of the protein that may not participate in biological function in a central way and so may not follow a simple, reasonably robust set of phenomenological laws. Thus, it is not surprising that the physics of (functionally relevant) conformation changes is not known. Indeed, we do not know if there are any general principles that govern conformation change. Conformation changes are described and named in molecular biology, but they are not often analyzed or measured directly or analyzed. The physical forces and atomic interactions that produce these changes in protein shape are hardly ever known. Conformation changes of channels cannot yet be analyzed with a physical theory at the present time at atomic resolution.

For this reason, if we insist on the agenda of atomic biology (the agenda of physical understanding of biological function at atomic resolution) we must restrict analysis of channels to their pores, to the already open channel, at least to begin with.[b] We lose much interesting biology, including the analysis of traditional mechanisms of gating. In return we gain simplicity, hoping that permeation through the open channel is a humble enough process occurring in a well-enough defined structure that we can reduce it to a physical problem with atomic resolution.

Perhaps the open channel will show what atomic biology can do.

3. Electrodiffusion in the Open Channel

Movement of ions through channels is driven by diffusion in a concentration gradient and drift in an electric field. Both diffusion and drift are significant in every channel, in nearly every situation. Neither can be approximated, because both are important in determining biological function. The driving

[b]However, permanently open channels of one conformation can gate (at least in principle) by changing the shape of their electric field (see below, and Chen and Eisenberg.[4,5])

force for both must be understood and included in any consistent model of a channel relevant to biological function or experimental data.

3.1. *Drift in the Electric Field*

The electric field is the driving force for this motion and the electric field is determined by the types and distribution of charges; all are important, and must be included in an atomic theory.

3.1.1. *Charge in the Channel's Pore*

The contents of the channel are one kind of charge, namely, the ions in the channel's pore. These have much the same concentration as the permanent protein charge on the wall of the channel (because they more or less neutralize the protein charge). They serve to shield or screen the permanent charge of the protein and so cannot be neglected. Indeed, they will change if the flux through the channel is changed by experimental manipulations of (trans-membrane) potential or concentrations (in the bath). The charged contents of the channel have often been neglected in treatments of ion movement through channels, although they have not been neglected in the analogous problem of hole and electron conduction in semiconductors.[6,7] There, the contents of the semiconductor "channel" can change the electric field (e.g, in a thyristor[8,9]) and make the semiconductor device turn ON or OFF.

Almost any molecule in a channel's pore has a significant charge. Water is a highly polar molecule: each of its atoms has significant charge. Even nonelectrolytes soluble in water have significant distribution of electric charge. That is why they dissolve in a polar solvent! In particular, biological nonelectrolytes like glucose have nonuniform distributions of electric charge in their chemical bonds. Most of those bonds are quite polar, and partial electric charge on most of the atoms of glucose is substantial, of the same order of magnitude as the charge on an ion. The partial charges on glucose sum to neutrality, to be sure—that is why glucose is a nonelectrolyte—but they are separated by angstrom distances, roughly the same as the distance between an ion and the channel wall. The partial charges within nonelectrolytes cannot be neglected. Even a nonelectrolyte like glucose must be viewed as a collection of charge. Also, it is better to describe that charge directly as a charge density than to use a few terms (typically, monopole and dipole, sometimes quadrupole) in a multipole expansion, which, all too often, takes hundreds of terms to converge.

Close to a (bounded uniform) distribution of charge, where the electric field is strong and the charge interacts most significantly with ions, proteins, etc., the multipole expansion [Refs. 10 (pp. 82–87) and 11] has an nth term that behaves (in magnitude) like $(R/r)^n(1/n)$ where r is the distance from the center of charge to the observation point and R is the distance to the charge itself $(R/r) \leq 1$. Thus, the rate of convergence is proportional to R/r. This is indeed weak convergence, particularly just where the field is largest and most important, near the charge distribution, when $r \cong R$. Close to the charge distribution, hundreds of terms are needed to reach even 10 % accuracy.

3.1.2. *Protein's Permanent Charge*

The channel wall, the surface of the channel protein, is also a collection of charges. The channel wall is a distribution of permanent protein charge determined by the structure of the protein, a charge that does not change unless the structure of the protein changes.

This charge resides in nearly all the bonds of the protein, not just in those with formal charges. Most of the bonds of a protein are polar and the partial electric charge on most of the atoms of a protein is substantial, of the same order of magnitude as the charge on an ion or an atom of the protein with a full formal charge[12–14]. The partial charges are separated by angstrom distances, the same distance between an ion and the channel wall, and so the partial charges are of great significance and cannot be neglected. A protein is a collection of charges.

Because a protein is a collection of permanent charges implies—collection of permanent charges, its interaction with substrates is quite different from the interactions described by usual enzyme kinetics (e.g. Michaelis–Menten kinetics, e.g., Fersht[15], Walsh[19]). Indeed, the interactions are rather different from those implied when a chemical reaction, or ion permeation, is described as the movement of a particle in a "potential of mean force". The usual treatment of ion permeation over potential barriers is one example of the latter.

If the surface of a protein is described as a potential of mean force (one of the mean-force potentials described by Hill[17]) that potential of mean force is usually taken as independent of the concentration of particles crossing it. Indeed, the potential structure (i.e. electric field) is usually treated as a constant, independent of concentration of solutes in the bath, independent of trans-membrane potential (if the protein is a channel), or electrochemical potential (if the protein is an enzyme), independent of the permeating particle. However, if the protein is modeled as a surface of permanent charge, the potential of mean force cannot be a constant independent of these variables, because anything that changes the distribution of charge in the system will change the shape of the potential of mean force, even if the conformation of the protein remains strictly fixed, as we will discuss in some detail later in this paper.

Under one set of experimental conditions, the surface of a protein can be represented as a distribution of charge, or the *corresponding* distribution of potential; the two representations are strictly equivalent,[c] for that set of conditions, but no other. If concentrations are changed, or if the trans-membrane potential is changed, the distribution of net charge is changed, even if the conformation of the protein, and thus the distribution of permanent charge on the protein remain strictly fixed. Also the potential of mean force cannot remain the same, unless charge is supplied or withdrawn from the protein. This cannot happen (since the protein is an isolated object, not connected to a battery in these systems). The permanent charge stays the same as experimental conditions change. But the net charge does not change and so neither does the potential. Thus, on fundamental grounds, the *potential of mean force is a function of experimental conditions* even if the structure of the protein does not change.

The field equations have a unique solution if (for example) charge is specified on the boundary, that is to say if Neumann boundary conditions are imposed on the field in the first place. The solution to the Neumann problem then describes all the properties of the field—the distribution of potential and charge (etc.) everywhere, including at the boundary. (That is what "unique solution" means!) The potentials (etc.) that are derived at the boundary from the unique solution to the Neumann problem are the only values compatible with the charges specified on the boundary in the first place. Anything that changes conditions, even if it leaves the boundary

[c]That is to say, the field equations have a unique solution if either boundary condition is imposed. That solution then implies the other properties of the field, distribution of charge, etc. at the boundary. Those implied properties are the **only** boundary "conditions" (really, boundary properties) compatible with the imposed boundary condition.

conditions (i.e., charges in the Neumann problem) untouched, will change the potential everywhere, including at the boundary. The relation between charge and potential on the boundary, and everywhere else, depends on the entire problem. An analogous argument applies to any set of boundary conditions that imply a unique solution of the field equations, e.g., Dirichlet boundary conditions that specify the potential on the boundary or the rather complex boundary conditions necessary to describe a biological channel.[129,130]

This dependence of the electric field and potential of mean force on experimental conditions has many implications for both channels and enzymes. Most models of both kinds of proteins seek to explain data taken under a range of experimental conditions, for example different concentrations of substrate or permeant ions, different potentials, and sometimes different ionic strengths. These models typically describe the protein as moving from state to state, usually following a simple rate law given by Eyring rate theory as a function of the potential of mean force on the surface of the protein (e.g., Hill,[17–19] Hille,[1]). The rate constants in these rate laws are, of course, taken as independent of concentration of substrate (that is why they are called constants) and the dependence on potential is taken as exponential. But the potential of mean force on the surface of a protein or channel must in general depend on the (average concentration of) substrate in the channel's pore or on the protein's surface. That substrate shields the permanent charge to a variable extent, depending in a nonlinear way on its concentration and the overall electric field. That shielding (or screening as it is sometimes called) changes the net charge (i.e. substrate charge plus protein charge) and so it changes the potential of mean force. Assuming that the potential of mean force is independent of substrate concentration is equivalent to assuming that the potential of mean force is independent of charge density; it cannot be true.[d]

Thus, self-consistent models of channels that fit properties measured under different conditions need to use different potentials of mean force under different conditions. Similarly, self-consistent models of enzyme reactions need to use different potentials of mean force under different conditions (if they are to be consistent with the properties of the electric field). Rate constants of such models are likely to vary with substrate concentration

---

[d]It may be a decent approximation under special conditions. Of course, the validity of the approximation must be actually shown, not just assumed, in each special case.

(and other experimental conditions), because the potential barriers that produce them vary.

Of course, these different potentials of mean force under different conditions cannot just be assumed; they must be calculated from an electrostatic model of the protein, one that describes the protein's surface as a distribution of permanent charge, as (for example) does the PNP theory described later in this paper. The variation of the potential of mean force with experimental conditions has profound implications for traditional biochemical analysis of enzyme rates. In particular, the rate constants determined from a Michaelis–Menten analysis of enzyme kinetics will depend on substrate concentration, even if the conformation of the enzyme is independent of concentration. Perhaps some of the allosteric effects and complex kinetics found so often in traditional analysis of enzyme kinetics (Walsh, 1979) represent a failure in Michaelis–Menten theory, not a complex set of conformational changes of the enzyme.

3.1.3. *Distant Charge and the Trans-Membrane Potential*

Not only do channel proteins and permeating ions have electric charge, but so **does the bathing solution they are in**, which surrounds the cell membrane.

A trans-membrane potential is found across the outer membrane of nearly all cells, and across the membranes of many organelles, and that potential is used for biological functions. For example, the cell membrane potential controls the opening and closing of many types of channels (so-called voltage gated channels). The trans-membrane potential is also one of the determinants of ionic motion in the channel's pore, once a channel is open.

The electrical environment of channels is controlled by the entire cell: the lipid membrane in which channels sit supports an electrical potential produced by surface charge and by a variety of mechanisms in the cell and its membrane. In particular, the membrane potential is determined by the ionic conductance of the membranes (what we now recognize as the ensemble properties of single channels) and by the concentration gradients of ions.

Concentration gradients between cell inside and outside are maintained by specialized active systems that use energy ultimately derived from ATP. Some concentration gradients are maintained by the energy stored in other concentration gradients. For example, the substantial gradient of Ca^{++} ion—some 10^{-3} M (*outside*) $/10^{-7}$ M (*inside*) with typical membrane potential of -90mV, inside negative—is maintained in part by a transporter protein (residing in the cell membrane) that exchanges Na^+ for Ca^{++}. The gradient for Ca^{++} is also maintained by an active transport protein in the membrane (the Ca^{++} pump) that hydrolyzes ATP and uses that energy to pump Ca^{++} out of the cell, against its concentration and electrical gradient. The ultimate energy source for most concentration gradients (in most cells) is the ATP hydrolyzed by the Na/K ATPase as it pumps Na^+ and K^+ out of and into the cell.

The potential across a membrane is determined by processes far away, as is seen whenever a voltage-dependent channel is studied in its natural state, for example, when a Na^+ channel is studied in a nerve. There, the potential and the electric charge on a membrane *millimeters* away is the dominant determinant of the conformation of a protein! Because the membrane is such a good insulator, even without its myelin wrapper, the charge producing the membrane potential is spread over macroscopic distances. These distant charges certainly cannot be ignored: they help control the electric field within the channel protein and are responsible for its natural function; the conduction of the nerve impulse. *Consideration of nearby charges is not sufficient if a protein is sensitive to the electric field.*

3.1.4. *Interfacial Charge*

The charge producing the membrane potential resides chiefly at the "interfaces" of the membrane. Each of the several interfaces in the system has a charge, both permanent (i.e. independent of the local electric field) and induced (i.e. dependent on the local electric field, absent when the field is zero). The induced charge in turn can depend linearly on the local electric field, and thus be traditionally described by a dielectric constant; or it can reside in the ionic atmosphere adjacent to the interface. In the latter case it depends in a complex nonlinear way on electric field (and many other variables) as described by Debye–Hückel theory in the equilibrium case or the PNP model in a continuum description of the non-equilibrium situation.[4,5]

Consider the interface between the lipid and the bulk solution. Permanent charge is present there because the carbonyl bonds of the aqueous head groups have a substantial permanent polar charge and the nearby carboxylic acid moieties carry a permanent formal charge (at biological pH).

Induced charge is present because of the ion atmosphere (the Debye–Hückel or Gouy–Chapman effect, depending on the choice of idealized geometry), the dielectric effects of the water and the lipid, and the trans-membrane potential. Charge near the lipid/solution interface can create a significant offset in the potential within the channel protein and the channel's pore (even though the interface can be distant from the channel) because the "far-field" potential contributes an important term to the near field potential.[20]

Modification of this far-field is likely to produce a dramatic change in channel properties simply by changing the electric field within the channel protein and its pore. Indeed, Chen and Eisenberg[4] have proposed this field coupling as the mechanism linking accessory proteins like receptors to ionic channels. They suggest that binding of an agonist to a receptor, or a subsequent conformation change of the receptor, or a chemical reaction involving a receptor (e.g., phosphorylation), is very likely to change the charge distribution of that receptor protein, thereby changing the electric field far from the channel itself. The far-field contributes a term to the potential within the channel even in the absence of specific chemical or allosteric interaction. The far-field would thus have a dramatic effect on gating, opening (or closing) a channel, and on permeation, modifying the rates of permeation. In this way *second messenger systems could be coupled to nearby channels by the electrostatic field.* Allosteric conformation changes of vague physical mechanism do not have to be invoked to explain the coupling of receptor to channel. The electric field can be the link, just as it is the link between the potential on the (distant) gate of a field-effect transistor and the potential in its conducting "channel".

3.2. *Ends of the Channel: Dehydration, Solvation, and Charge*

Charge can also exist at the end of the channel, that is, at the interface between the pore and the bulk solution. Surface potentials exist in that region that have direct effects on channel gating and permeation; the surface charge producing those potentials cannot be ignored.

Other large effects certainly occur at this interface between pore and solution because the solvation shell of an ion changes dramatically as it enters the channel. Such desolvation and resolvation effects may be dominant determinants of selectivity and flux coupling.

Ions in bulk solution are solvated by water. Ions in a channel's pore are solvated by the channel protein (mostly). They also interact with the two adjacent water molecules within the pore. As an ion enters a channel, the ion and the surrounding atoms undergo a large rearrangement, akin to a change in phase. The energies involved are huge (say 60 kT) compared to the energies important in permeation (say 6 kT). A few kT has a large effect on permeation because rates of permeation tend to be exponential functions of energy barriers, with exponents given by the barrier height, in units of kT. Yet tiny mismatches between the energy of hydration (in the bulk material) and the energy of solvation (inside the channel) would have a huge effect on ion transport.

A phase boundary of this sort between channel and solution is likely to support complex phenomena, as it does in electrodes or gas/solid interfaces. Electrochemists describe the nonequilibrium properties of such interfaces by the Butler/Volmer equation,[21] while physical chemists describe the transport properties of the gas/solid interface by the closely related expressions of statistical rate theory (e.g. Ward,[22] Ward and Elmoselhi[23]). We should not be surprised if this interface in channels supports many complex phenomena and needs a description akin to other interfaces. The entry interface of channels may be responsible for the selectivity, and perhaps flux coupling[4,5] of channels and transporters.

The last interface we consider is between the pore and the channel protein, where permanent charge and induced charge are present. The permanent charge has already been identified and analyzed as part of the contents of the channel, the permanent protein charge. Induced charge is found here and elsewhere in the channel.

The potential across a membrane is determined by processes far away, as is seen whenever a voltage dependent channel is studied in its natural state, for example, when a Na^+ channel is studied in a nerve. There, the potential and the electric charge on a membrane *millimeters* away is the dominant determinant of the conformation of a protein! Because the membrane is such a good insulator, the charge producing the membrane potential is spread over macroscopic distances. These distant charges certainly cannot be ignored: they help control the electric field within the channel protein and are responsible for the natural function: the conduction of the nerve impulse. *Consideration of nearby charges is not sufficient if a protein is sensitive to the electric field.*

3.2.1. *Induced Charge*

Induced charge is present at the interfaces, in the protein of the channel, the lipid of the membrane, and the water of the bulk solution. The local electric field displaces electrons in atoms, and other charge, from their "rest" position (i.e., their position when the local electric field is zero). This induced charge is the dipolar polarization charge of dielectrics, described in traditional macroscopic theories by a polarization vector, an ideal dipole proportional to the local electric field.[10,24,25] Polarization charge in reality usually has complex time and field dependence.[26–34] However, it is always induced and exists only when the electric field is non-zero. For some reason, the word "polar" is also used to describe a chemical bond with a permanent charge distribution, independent of the electric field, causing confusion; conventional usage might force one to say that "a polar bond contains no polarization charge".

Each type of charge must appear in a theory of channel permeation because each makes a significant contribution to the electric field that controls channel gating and open channel permeation. Otherwise, the theory will be inconsistent with the properties of matter or the electric field. Conventional macroscopic theory, in which electricity is a continuous fluid, allows easy mathematical description of all these types of charge and their interactions with each other and the electric field they produce. That is just the point of the *PNP* model including boundary conditions,[4,5] as we shall discuss later in this paper.

Electricity is not a fluid, however. It consists of discrete charges making stochastic movements, and the discreteness of charge is clearly relevant in systems as small as ionic channels. The description of each type of charge, and their interactions with each other and the electric field they produce, is much harder in discrete consistent models. Indeed, it has not been done satisfactorily in bulk ionic solutions (e.g. mixtures of Na^+Cl^- and K^+Cl^-) and H_2O let alone in protein solutions.

3.3. *Diffusion Through Channels*

Having described the electrical driving force, namely, the types of charge, we turn to the other driving force for flux through channels, concentration, or really activity (i.e., free energy per mol) for which the word "concentration" is a widely used surrogate, an approximate stand-in.

A condensed phase like a protein in a liquid contains little empty space. Indeed, "...in a liquid a typical molecule is usually within the force fields of, say, 10 nearest-neighbor molecules and is never completely free of the influence of neighbors." (Berry, Rice, and Ross, 1980, p. 845). The thermal velocity of matter at room temperature is some 6 Å per picosecond, more or less the speed of sound. Enormous thermal velocity in a system with little empty space implies a very high number of "collisions" (using the word loosely) some 10^{13} times per sec if one dares to use a kinetic model of hard spheres to represent a liquid. The pores of channels are tiny, with diameters of (say) 4 Å and lengths of some 25 Å. Ions probably take some 100 nsec to cross that length, certainly they take more than 10 nsec (illustrated in the figures of Barcilon, Chen, Eisenberg, and Ratner[35]).* Thermal motion leads to a great deal of thrashing about of a particle as it permeates the channel. It moves some 60 μm in 100 nsec, or (in other words) the particle moves randomly some 24,000 channel lengths before it crosses the channel. All this motion (60,000 nm total path length of random motion) produces little net displacement (2.5 nm net drift) because the ion reverses direction so many times (colliding) with neighboring atoms. On the microscopic[e] scale,[10] friction and collisions dominate — as shown by elegant scaling arguments[36,37] – producing a Brownian motion approximated by the random walk of a drunken man, the (limit of the) sum of frequent tiny displacements made by the ion with each atomic collision.[38] The path of motion is a continuous function of unbounded variation, with an infinite number of maxima and minima, or zero crossings, in ***any*** finite interval, ***no matter how small*** the interval!

The random motion of atoms is very different from the smooth motion we see on the macroscopic time scale. It is even more different from the straight-line motion of the molecules of ideal gases (in which molecules hardly ever interact with each other and follow straight line ballistic paths (much larger than their diameter) between well defined collisions, collisions that themselves last negligible time).

This random motion is described by a Langevin equation[39] on the microscopic scale of individual atoms. On a macroscopic scale, it is described by

*See footnote t.

[e]Universally called "microscopic" although microscopes do not resolve atoms! I am told, however, by Tom DeCoursey that the usage fits with the Greek roots of the word "microscopic", viz. $\mu\iota\kappa\rho\acute{o}\zeta$ = small and $\sigma\kappa o\pi\varepsilon\hat{\iota}\nu$ = to view.

the diffusion equation.[40–43] The diffusive motion of ions through channels is at least as important as the drift motion in the electric field. Diffusive flux is driven by the concentration gradients of ions across channels; the corresponding description of individual random walks driven by concentration gradients is needed to link the (stochastic) motion of individual ions to (deterministic) macroscopic fluxes. As simple and important as this problem sounds, it has caused some difficulty because the relevant mathematical theory usually describes the motion of just one random walker, and does not in fact even use concentration as a variable! Only recently has concentration been included as a driving force in a stochastic theory of diffusion for the general situation of interest to biologists, namely for motion over a potential barrier of any shape (Eisenberg, Klosek, and Schuss,[44] described later in this paper).

4. Electrodiffusion in Mixed Electrolytes

The use of concentrations to approximate activities (i.e. free energy per mole) is nearly universal in biology and widespread in chemistry. It is often felt that the activity coefficient (the ratio of activity to concentration) is a fairly trivial correction, implying a different, albeit non-linear scale of concentrations but not having a qualitative effect on the equilibrium or nonequilibrium properties of channels (or solutions, for that matter). Ions are usually treated as independent entities with activities that depend on their own concentration, corrected by an activity coefficient that depends on total ionic strength of all species. The activity (in solutions of strong electrolytes, i.e., fully dissociated electrolytes) is nearly always assumed independent of the chemical nature of the other species in solution (i.e., independent of the concentration of a particular type of ion). This is a reasonable approach when dealing with solutions of one type of strong electrolyte (e.g. NaCl), but not when dealing with the mixed strong electrolyte solutions nearly always found in biology as Duan Pin Chen has recently taught me. The usual assumption fails when

(1) ionic strength is varied;
(2) the composition of mixed solutions is varied (even at fixed ionic strength) as in "anomalous mole fraction" experiments[1];
(3) one constituent of a mixed solution far exceeds another, as is usually the case in biological solutions (e.g., NaCl $\cong$ 150 mM far

exceeds KCl $\cong$ 5 mM in physiological solutions; Ca^{++} is almost always much less concentrated than monovalent cations in physiological solutions. It is typically $2 \times 10^{-3} M$ outside cells and some $10^{-7} M$ inside cells).

Significant deviations from independence occur in all mixtures (even of "ideal" gases: Mayer and Mayer,[46] p. 203; Rowlinson and Swinton[154]), not only electrolyte solutions, although there the effects are much larger. Mixtures of gases show strong departures from ideal behavior (e.g., Chapman and Cowling[47]). Liquid mixtures cannot be understood "... solely from a knowledge of the pure components. Such attempts rest upon the fallacy that the forces $(\alpha - \beta)$ between two molecules of species α and β are always determinable from the strengths of the forces $(\alpha - \alpha)$ and $(\beta - \beta)$. If it were true that the $(\alpha - \beta)$ forces were always some "average" of the $(\alpha - \alpha)$ and $(\beta - \beta)$ forces, then the properties of a binary mixture would be predictable in principle solely from a knowledge of those of the two pure components. However, such averaging is not universally valid. It is true that for very simple substances and for the prediction of relatively crude properties there are suitable averages .... However, such averaging is unsatisfactory for many classes of substances and inadequate for the detailed interpretation even of the simplest mixtures. One should rather take the observed properties of a binary mixture ..." (Rowlinson and Swinton,[154] p. 86).

Ionic solutions show strong departures from ideal behavior in both their equilibrium and non-equilibrium properties. Mixed electrolyte solutions, such as the Ringer solutions that mimic the natural environment of cells have large excess free energy — a free energy of mixing beyond the ideal. In fact, a substantial fraction of the experimental work in classical electrochemistry was devoted to measuring and trying to understand this excess free energy and its variation with the species, concentration, and charge of ion.[f] Non-ideal behavior is even more apparent in the non-equilibrium properties of mixed electrolyte solutions, e.g., in their conductance. "It

[f] I found the review of Anderson and Wood[48] (887 references!) and book of Tyrrell and Harris[45] a great help in dealing with the vast literature (cf., Franks[49]). Substantial part of Harned and Owen[50] and Robinson and Stokes[51] are devoted to mixed ion effects and their early analysis. Friedman,[52] and Conway, Bockris, and Yaeger[53] discuss more modern theories.

has long been known (Bray and Hunt[54]), that the conductances of ***even very dilute solutions*** of mixed electrolytes are ***not additive***, so that Kohlrausch's law of independent ionic mobilities is ***not valid for mixtures***." (Kortüm,[55] p. 211).[g] "Kohlrausch's rule of the independent migration of ions, valid as a limiting law for simple binary electrolytes, does not apply to mixtures, nor are the conductances additive."[155]

Part of the non-ideality comes from a "relaxation effect" arising from the ever present disturbance of the ionic atmosphere as an ion moves in an electric field. The resulting asymmetry in the ionic atmosphere means that the charge density in front of the ion will be slightly lower, and behind slightly higher, than at equilibrium. Because ionic atmosphere has the opposite sign from the central ion, relaxation effects decrease the drift velocity.

Movement is also retarded by a "electrophoretic effect" caused by the action of the electric field on the ionic atmosphere of a central ion. The ionic atmosphere itself is dragged in the opposite direction from the central ion, increasing their relative velocity, and so increasing the frictional retarding force on the ion. The properties of the ionic atmosphere depend on the species of ion and so both the relaxation effect and the electrophoretic effect are different for different species, even if they have the same charge.

These are just two of the many interaction effects in mixed ionic solutions. Others (appearing in higher order terms of the expansions of cluster theory: Anderson and Wood,[48] Friedman,[52] Mayer and Mayer[46]) may be at least as important and yet not have been named or recognized as such. Indeed, present day theories do not fit the dependence of excess free energy on concentration, charge, or ionic species very well, so clearly something is missing! In my opinion, the essential fact is that the activity and diffusion coefficient of one electrolyte is strongly affected by the concentration of another, even at fixed ionic strength. If the relative concentrations of ions were fixed, the effect of one ion on the activity of another might remain hidden from experimental view because it would be lumped into an activity coefficient assumed (incorrectly) to depend only on the ionic strength, i.e., assumed (incorrectly) to be independent of the types of ions present in the

[g]Emphases added. Note that 'very dilute' means less than 1 mM (!), Kortüm, 1965, p. 189.

mixed solution. But relative concentrations are not fixed in most biological experiments. Rather, they are varied and the effects of such variation are usually interpreted as properties of channels, not of interactions in the bulk solution.

The fact that the excess free energy, and conductivity, of mixed ionic solutions vary significantly with concentration and the resulting changes in driving force and conductance (of the bulk phase) certainly complicate traditional interpretations of physiological experiments. Indeed, some of the so-called anomalous mole fraction effect in channels[1] could arise from the excess free energy in the bulk solution, and not in channels at all (Chen, Ratner, and Eisenberg, personal communication).

The excess free energy $\Delta_m G$ of a solute depends on many variables. It depends on the amount of solute A, its mole fraction y_A in a weight of solvent W (in kilograms); and it depends on the second solute B, present in mole fraction y_B. The dependence is given by an empirical formula,[48] also derivable from cluster theory[52]

$$\frac{\Delta_m G}{W} = RT \cdot I_s \cdot y_A y_B \left[g_0 + (y_A - y_B)\, g_1\right] \tag{1}$$

where fixed ionic strength is I_s, g_0 and g_1 are constants, characteristic of the type of ion and solution, independent of the concentrations (i.e., mole fractions) of solutes y_A or y_B, but dependent on ionic strength. RT is the thermal energy per mole, with R the gas constant and T the absolute temperature. In simple solutions, like mixtures of KCl and NaCl, the g_1 term is negligible and the ionic strength dependence of it and g_0 is not significant.

The significance of Eq. (1) is clear; some of the driving force for the diffusion of one species is provided by the concentration of another and that excess driving force depends on the concentration of both species. The free energy of one species is coupled to the free energy of the other. *Coupling is a phenomena of bulk solutions*, not just channels. In particular, if one solute in a solution is present in considerable excess (like NaCl in normal Ringer) or is a great deal more mobile than another, quite unexpected phenomena occur: "... [coupling] effects ... are spectacular when the component is a strong acid .... The presence of the acid causes the salt to move against its own concentration gradient." (Tyrrell and Harris,[45] p. 377).

The full implications of this excess free energy of bulk solutions of mixed electrolytes for the interpretation of channel experiments is not yet known; Duan Pin Chen uncovered this literature and its relevance to coupling only recently.[h] Even if the effects are relatively small in bulk solution, they may be large at the entrance of channels, where they influence offset potentials (i.e., the Donnan or built-in potentials, see Chen and Eisenberg[4,5]) and concentrations, or the effects may be large within the channel itself. After all, the physical processes producing excess free energy in bulk solution are likely to occur in channels and at their entrances as well, albeit in different forms, because in channels the ionic atmosphere cannot translate spatially,[156] as it does in solution, although it certainly can deform.

5. Models of Permeation

Having identified the driving forces for ionic movement, present in any model of permeation, we turn to models of the permeation process itself. Models of permeation are used whenever experiments are planned, data taken, or results analyzed. Whether the model is an explicit set of differential equations and boundary conditions, a computer program, a verbal analysis of states, or a vague image, the model exists in the mind of the experimenter, motivating (and thus governing) his choice of experiments and data to collect, and his analysis.

This role of theory is so embedded in the physical sciences that it often goes unstated. Physicists and astronomers since the time of Galileo and Newton have designed their experiments and chosen their observations with a theory in mind, usually an explicit theory at that. Chemists in the middle of the last century classified the atoms into a periodic table, a theoretical structure, and then designed experiments to test that structure. Since about the same time, chemists interested in molecules have designed their experiments to study their component atoms and their bonding. Indeed, "molecules" are themselves theoretical constructs of atoms and the bonds between them.

One of triumphs of science was the discovery that the periodic table was an output of the physics of the atom (quantum mechanics) and that chemical bonds arose from the same theory. One cannot imagine experi-

[h] January, 1994.

mental chemistry today without the idea of a chemical bond. What once was a controversial theory has become the foundation of an experimental science, used (often unconsciously) by every scientist, even those exclusively interested in the cut and thrust of laboratory work.

Indeed, the role of scientific theory is just as pervasive in some parts of our daily life. The computers and communication equipment we all depend on are the most striking examples, perhaps, because they are so clearly the reasonably direct result of rather abstract mathematics. But another example is less well known, although much older, and more supportive of our comfort and existence. The buildings we live in are designed on paper (and nowadays on computer screen) using simple formulae to deal with gravity, and (in the utilities of the building), electricity, water, and air flow. No one thinks of making a model of a building; the theory suffices. Indeed, it would be difficult to build a building if a realistic model were needed first. The model would cost nearly as much as the building itself and yet might differ from the real structure in significant ways. Of course, the theory of the building is not perfect. We all know of buildings where important functions have been left out (regrettably a common error of the more artistic and thus theoretical architects); and occasionally even the laws of gravity are ignored (for a while, e.g., until the next earthquake). But all-in-all, construction is an applied science which has fully integrated its theoretical foundation.

Biology is not like that. For a variety of reasons, both historical and logical, many biologists pride themselves on the absence of biological theory, even those who deal with ideas much more than with laboratory results. Why biologists have so separated themselves from a main scientific tradition is an interesting question to be pursued in a historical or social context. Suffice it to say here, that theory flourishes in science after experiments are specific, quantitative and reproducible; after the underlying structures are known and understood, after major questions are agreed on. Before then, knowledge is vague and theories should be vague too. Vague ideas are usually worse than explicit theories, but sometimes it is the other way around. Vague ideas are better than specific theories when little is known, and when much must simply be described. Vague ideas are better than specific theories if the theories are wrong, or even worse, tautological, misleading, irrelevant or incomprehensible. However, enough is now known about the open channel that vague theories are unnecessary, indeed they are an impediment, at least in my view: experiments are specific, quantitative and reproducible; underlying structures are almost known although not understood; and the major questions are agreed on. The stage is almost set for a useful theory, that will embed itself as deeply into experimental practice as theory is embedded in the experiments of physics, chemistry, and engineering.

5.1. *State Models of Permeation are Inconsistent with the Electric Field*

Until recently, channology was appropriately more devoted to the acquisition of data than its analysis. Indeed, such is still the case in most of molecular biology where the simple identification of the proteins of life is of the greatest importance and will certainly preoccupy most biologists for many years to come.

If the same approach is used to analyze the mechanism of proteins, however, it will not succeed. Analysis of mechanism requires an explicit physical model, one built on *known* properties and interactions of atoms.

The most popular models of permeation — indeed of any property of proteins — are state models (e.g. Hill,[17,18]; Hille[1]). In these, an experimental observation is identified with a state of a protein, transitions from state to state are described by rate theory adapted from the Eyring theory of gas phase chemical reactions, and predictions are made and compared to experiments.

The difficulty with this class of theories is both fundamental and operational. Fundamentally, the states and rate constants of these theories are vaguely defined, and they are rarely related to the fundamental physical processes that determine the properties of condensed phases. The functional dependence of states is usually not specified. How a state or rate constant depends on the concentration, charge (permanent and mobile); on the trans-membrane potential and potential distribution; on the shape, size, structure, dielectric properties, flexibility (etc.) of a protein are the essential issues. Ignoring them is not helpful, I believe, if the goal is a physical or atomic understanding of biological function, yet few of these variables appear in state models. Indeed, state models never (to my knowledge) compute the electric field. They assume (tacitly but all the more frighteningly for that) that fluxes can vary as concentrations and trans-membrane potential change without a change in the shape of the electric field. I believe such models are fundamentally flawed; they are inconsistent[i] within themselves and with first-order properties of the electric field.

State theories are hard to develop, to extend, and to make more precise. Their vagueness means they can easily fit a range of data, but cannot be easily verified. They tend to confuse description with understanding, and form a closed universe quite disparate from the physical world known and described by chemists and physicists working in condensed phases. Indeed, it is often not clear how a state could be defined if the trajectory of every atom in a protein and pore were known.

Without a physical basis, rate constant models are both too flexible and too rigid. They are too flexible because additional states and rate constants can be added to fit awkward data. They are too rigid because the states and rate constants are treated as constants even if physics says they are variables. They are inconsistent with the first-order properties of the electric field. These are not desirable characteristics in a theory, let alone a paradigm motivating (and thus governing) most experimental work. In my view, a Kuhnian (scientific) revolution (Kuhn, 1970) is needed to replace the vague ideas of state models with the well defined, falsifiable, and thus experimentally useful ideas of the stochastic theory of electrodiffusion.

None of this implies that proteins do not have states, nor that their flux may not follow exponential laws in some cases. But those states should arise from a theory, with known functional dependence, or (much better yet) the states should be identified experimentally by *direct measurement.*[j] Otherwise, it is impossible to guess the states, or the laws which govern the rates by which the states change, and investigators will eventually be frustrated by the non-uniqueness of their theories and move on to other problems.

---

[i]In this paper, "consistent theory" is one that predicts an electric field consistent with the set of concentrations and charges in the theory, under all experimental conditions. More specifically, a theory is "consistent" if, under any set of experimental conditions, the concentrations of ions in the system (i.e. channel, boundaries, and solution), when substituted into Poisson's equation (continuum models) or Coulomb's law (discrete models) predict the electric field actually used in that theory under those conditions.

[j]The open and closed conductances of single channels are good examples of states well-defined by direct measurement. The multiple closed states used to fit the kinetics of Na^+ conductance in nerve (the delay after a step in depolarizing, i.e., positive going potential) are examples of unobserved, less well-defined states.

To be useful, a theory in biology (or most other reductionist sciences) should have a vocabulary and syntax provided by another underlying science that has studied related (usually simpler, better defined, and better understood) systems in more detail. In that way, the vocabulary, syntax, and underlying ideas will have survived theoretical scrutiny and experimental test in the underlying field. Thus, physics draws its rules from mathematics (although it can choose which type of mathematics, and occasionally invent some, e.g., the Dirac delta function). Chemistry uses the laws of the electric field determined by physics. And molecular and atomic biology use the rules for molecular and atomic interactions in condensed phases. In particular, the grammar of molecular and atomic biology must explicitly include the electric field because that is the dominant force in most molecular and atomic interactions. Indeed, the electric field determines all the interactions of atoms and molecules, (according to the Hellmann-Feynmann theorem[157] as unlikely as that sounds) if the electric field is computed from the distribution of charge (i.e., electron density) that solves the *Schrödinger* equation[158,159] and thereby minimizes the appropriate energy functional.[160,161]

In the case of proteins, the appropriate grammar is physical chemistry of the condensed phase,[56–59] electrochemistry,[55,60,61] and the dynamics of proteins.[62,63] That grammar uses physical variables as nouns, like potential, location, charge, concentration, and flux. It uses as verbs the differential equations of condensed phase physics, basically a combination of Newton's laws and Maxwell's equations with a smattering of quantum mechanics added. Indeed, the verbs are simply the laws of electrodiffusion (on one length scale or another) if we consider only the open channel, as we do here.

The need to consult an underlying science, to use its vocabulary and syntax, is not just a moral rubric. The lack of knowledge of an underlying field can have direct effects on every day experimentation. Physiologists like me have spent a good fraction of their careers changing the mixture of components in electrolyte solutions (blissfully) unaware of the work of electrochemists on such mixtures. Indeed, biologists in general, including physicians, and some chemists, are taught to treat ionic solutions ideally, as mixtures of independent components, although ionic solutions of biological interest do not behave ideally in the laboratories of electrochemists. Not knowing of these non-ideal properties of bulk solutions, we could not evaluate their significance to biological phenomena. We do not know what part of the effects of solution changes used routinely in channology occur in the bath! Knowledge of the underlying science is needed to find out.

In a similar spirit, the theory of condensed phases can sometimes provide justification for state theories. Such state descriptions are extremely helpful if they are correct, and if they can be derived from the grammar of physical theories described with explicit functional dependence on the parameters of the molecules, etc. Such is sometimes possible, even for quite general situations (see Eisenberg, Klosek, and Schuss,[44] discussed below). But such derivations cannot be generally expected, nor can they ever be guessed with any certainty. States should be the mathematically derived output, not the assumed input of analysis.

If states are used as inputs to a theory, it is easy to assume properties that proteins cannot have. For example, if states are assigned *a priori* without physical basis, it is natural to make the corresponding rate constants independent of the concentration of permeating ions. But, as we have seen, such independence would imply a potential of mean force independent of substrate concentration and that is impossible under any range of conditions. The substrate shields the permanent charge of the protein; the shielding is a function of substrate concentration and the potential of mean force and the corresponding rate constant will be a function of substrate concentration.

6. Making Models in Atomic Biology

Evidently kinetic models with rates and states are not to my taste. What then is? How do we go about making a consistent model of a channel? What are the objects of the model, its nouns? What are the dynamics of the model, its verbs? Those are the questions we now turn to.

Ever since atoms were imagined and temperature was understood to be their random motion,[65] physicists have struggled to relate macroscopic observables to microscopic quantities. The difficulties are formidable in the gas phase, for spherical atoms.[56,66] In the liquid phase, they are much worse.[57] For a solution of charged particles in a highly polar and asymmetrical solvent,[58,67] the problems are insurmountable in my opinion: the system deviates from ideal in too many different ways and each deviation is very hard to deal with individually — in fact, few of the deviations can be approximated to more than a leading term, yet higher order terms may be expected to (collectively) dominate, as they do for multipole expansions of the electric field. Further, as the *coup de grâce*, each non-ideality interacts with the others in ways that are likely to be large, yet very hard to formulate, analyze, or solve *a priori*.[k]

Understanding these problems, many despaired of a (general, accurate, reasonably simple) theory of ionic solutions. However, such a theory may not be necessary. Computers may allow us to reconstruct the actual atomic motions of a solution, and thus simulate the properties of the solution, without a general theory. That is the goal of molecular dynamics.

7. Molecular Dynamics

Molecular dynamics simulates a solution by describing each of its atoms and their interactions.[l] The nouns of the simulation are the atoms and chemical bonds of the molecules. The verbs are Newton's laws of motion. The forces between atoms (either nouns or verbs depending on your choice) are described by a force-field, given by the derivative of so-called potentials (not to be confused with electrical or thermodynamic potentials) which summarize the properties of chemical bonds.[69–71] These potentials can be determined reasonably well when the distribution of charge in each molecule is quite constant, when "polarization" of an atom or molecule (corresponding to what we have called induced charge earlier in this paper; in this case, the induced charge is within an atom or molecule and is caused by movement of its cloud of electron orbitals) is small and the range of temperatures and phenomena are limited (e.g., to exclude quantum effects like changes in covalent bonds). In that case, molecular dynamics is able to predict the properties of pure liquids quite well.[68,72,73] If the molecules are asymmetrical, difficulties arise, because long range effects begin to occur: raisins are found at the top of boxes of cereal because shaking of a random homogeneous mixture of inhomogeneous objects does **not** necessarily produce a homogeneous result. If long range electric fields are present, these probably

---

[k]Indeed, how can we be sure that all significant interactions have been included?

[l]Ref. 68 contains a most informative series of review articles on molecular dynamics of many different systems.

need to be separately incorporated into a simulation[m] as we shall see. If solvent and solute are present, asymmetrical, and charged, the calculations become difficult to manage, because of the needed size of the system and duration of the integrations. However, workers are hopeful that the exponential development of computers will permit the simulation of solutions, and thus the solution of solutions, in the near, if not immediate future.

Extending these techniques to proteins[62,63] is irresistible, inevitable, and valuable, albeit approximate, and so approximate in the view of some that it has attracted serious criticism. In my view, the critics may have a point, but they should stimulate, not discourage work: the molecular dynamics of proteins provides a more realistic and detailed view of atomic motions than many of us ever thought possible, and that view will be of enormous value in understanding the function of proteins, as well as in constraining lower resolution self-consistent models to decently approximate atomic reality.

The question for channels is then obvious: why not just calculate the motion of an ion as it permeates a channel? Indeed, why not just simulate a system with one concentration of ions on one side of the membrane, another concentration on the other, with an electrical potential as well? Why not calculate enough trajectories to determine the flux and be done with the problem once and for all?

The answers are all the same: "I wish we could, but we cannot."

One of the reasons we cannot is that oversimplified "potentials" are used to describe atom–atom interactions. The choice of potentials determines the force-field around atoms and so significantly modifies the electrostatic barriers for permeation. It also determines the difficulty of the calculation and so simplifications are usually made, which neglect induced charge, and which lump hydrogens together with a more massive neighbor into a composite pseudo-atom, whose force-field implicitly includes the effects of the hydrogens. However, *ab initio* quantum mechanical analysis using density functional theory (D. P. Chen, C. Lee, and H. Chen, personal communication) shows that polarization of hydrogens by the electric field dramatically reduces energy barriers for permeation. Computations of gramicidin that include all hydrogens explicitly (e.g., the hydrogen in the -CH_2 groups of the protein) predict an order of magnitude smaller barrier than traditional

[m]Probably by linking to another level in our hierarchy of models (see below).

calculations in which hydrogens are not explicitly present and do not polarize (for the most part). Kumpf and Dougherty,[162] have shown that "cation-π" interactions between permeating ion and aromatic residues of channel proteins are large and specific, suggesting that they may be involved in ionic selectivity. It is obvious that quantum mechanical calculations of channels, being only recently conceived, are just reaching infancy. Rapid growth and a productive childhood are expected, along with a stormy, hopefully brief adolescence, and a maturity that cannot yet be imagined.

The other problems of molecular dynamics discussed here arise directly from limitations in the size and duration of the calculation. Molecular dynamics is extremely useful—indeed, indispensable for the understanding of short range interactions, like those that probably dominate ion entry into channels, selectivity, and perhaps flux coupling. Molecular dynamics is also needed to determine the meaning, limitations, and parameters of other more approximate treatments. But we shall see that molecular dynamics *per se* is restricted, like any other technique. I spend disproportionate time highlighting these limitations (see below) trying to curb over-enthusiasm, trying to focus molecular dynamics on the questions it can answer, using other techniques to address other questions. But highlighting limitations is also a step towards understanding and removing the limitations: I hope some readers will be stimulated enough to remove these limitations that seem unavoidable to me!

7.1. *Temporal Limitations*

It seems likely that an ion takes some 10 to 100 nsec to permeate a channel, that is to say, to reach the opposite side for the first time (see figures in Barcilon, Chen, Eisenberg, and Ratner,[35]). A direct simulation of permeation would then need to be at least this long to observe one elementary event. In practice, it might have to be much longer, if ion movement in the bath must be included, if the approach and retreat of the ion from either side of the channel are also of interest.

Simulation of permeation requires computation of an ensemble of permeation events under each condition of interest. An ensemble might mean 2,500 events if an accuracy of $1/\sqrt{2500} = 1/50 = 2\ \%$ were acceptable. A typical current-voltage "curve" would contain at least one hundred such points and so would require simulations of 12.5 msec altogether, calculated as shown below:

$$50\frac{\text{nsec simulation}}{\text{trajectory}} \times 2,500\frac{\text{trajectories}}{\text{predicted points}} \times 100 \text{ points} = 12.5 \text{ msec}\,.$$

A present-day high speed workstation calculates about 10 psec of simulation per hour of computer time, for a channel like gramicidin with some 200 atoms. One current-voltage curve would thus take about a billion hours. Computing would have to speed up about a billion times to make this calculation practical.

Of course, prolonged calculations are not just limited by computer time; they are often irreproducible as well because of the discrete nature of computer arithmetic. Round-off error is a significant limitation.[72,73] In our (Chen and Eisenberg) experience, individual trajectories of the channel protein gramicidin longer than (say) 10 psec are likely to be sensitive to details of machine arithmetic. Long calculations also show behavior reminiscent of chaotic systems.[73,163] Individual trajectories are exponentially sensitive to the assumed initial positions of atoms after about 10 psec. The initial positions are of course not known with much accuracy, and the initial velocities are more or less completely unknown: even a fine X-ray structure might have an RMS error in position of 1 Å, and X-ray measurements barely constrain (and do not determine) velocities at all. These uncertainties are enough to ensure striking divergence in individual trajectories after some psec of dynamics.

Individual trajectories should not be the focus of attention. Rather, it is the ensemble of trajectories that determines the prediction of experimental events and data. The assumption widely made (e.g., Hoover,[163]) is that averaging over an ensemble of trajectories removes the errors of discrete arithmetic and the sensitivity to initial position, etc. This assumption is inevitable and underlies any large scale calculation of mechanical systems (like molecular dynamics). It seems certainly reasonable and likely to be true in many cases. Nevertheless, it cannot always be true.

Systems with different qualitative behavior in different regions of phase space and exquisitely sensitive to initial conditions, are called "chaotic" nowadays and are common in large scale nonlinear mechanical[74,75] and chemical systems,[76,77] some quite like channels.[163] The differential equations of molecular dynamics form such a system, exceedingly non-linear by the standards of mechanical systems. Sometimes chaotic systems have periodic behavior, yet have trajectories that seem a random function of time.[n] Sometimes one region of phase space cannot be accessed from another at all, or not in finite time.[73] It is not clear how a numerical simulation would average over such systems. It is not clear what the average value of a quantity would mean, and estimators of variance and other sample statistics (like power spectra, correlation functions, or coherence functions) would be even more problematic. Of course, such a system would have bizarre behavior in the experimental world as well as in the world of simulation. Thermodynamic experiments, *performed in finite time*, might well not produce estimates of the thermodynamic parameters of such a system. It might take infinite time either to compute or observe those properties. So it is not clear how to study this kind of system, given the time available to most of us. What is clear is that theory, simulations and experiments must describe the same state space and that state space should be sampled in the same way in each.

To summarize, perhaps in vivid words: the computational ensemble (of round-off error) is not always the thermodynamic ensemble (of everything), and neither may coincide with the ensemble measured in experiments.

7.2. *Spatial Limitations*

The size of the system computed is also a serious limitation on molecular dynamics.[72,73] Calculations today are limited to something like 10,000 atoms. That certainly seems, initially, to be sufficient to calculate the short ranged interactions in a system. However, systems of interest are three-dimensional and so even a 27,000 atom system would have just 30 atoms in each direction. Obviously, at second blush, one would wish a bigger system. Just as obviously, exponential growth in computer capability will give us much bigger systems very soon.

What is not often realized, is just how big the system needs to be[o] to

[n]E.g., pseudorandom noise generators made from shift registers are a practical example.[78] They produce an apparently random signal, with a (nearly) flat power spectrum, yet the signal is *strictly* periodic.

[o]Periodic boundary conditions TC are often used to increase the spatial size of simulations of equilibrium (but not non-equilibrium) systems.[72,73,163] Periodic boundary conditions assume that all the independent variables of the calculation (e.g., the positions of the atoms and their velocities) are the same on each corresponding face of an array of space filling solids. That is to say, the system is decomposed into cubes and the independent variables are assumed to have the same value at a particular location on all corresponding faces of the cubes. Periodic boundary conditions cannot be

simulate a system like ionic channels involving concentrations and potentials. It is clear that the system needed to simulate channels has to be big enough to define an average concentration and average electrical potential. That is unfortunately very big. Certainly, one would wish 2,500 solute atoms in the system — that means some 600,000 water atoms on both sides of the channel if the solute is Na^+ in typical Ringer solution, much more if the solute is K^+ or Ca^{++}. That is incredibly large, but perhaps computable.

The volume needed to define a concentration can be frighteningly large, but the volume needed to define an electric field can be still much larger. The electric field needs to be computed in the volume of space containing relevant charge—whatever volume that is. A cubic volume some 20 Debye lengths on a side might seem sufficient under biological conditions, say a volume 20×3 Å $= 60$ Å on a side, containing somewhat less than the number of atoms necessary to define a concentration. That may be a reasonable estimate of the relevant volume in free solution. But that estimate ignores the effect of boundaries and the charge along them. The volume enclosed by these boundaries can be very large indeed (cm^3) in cases where the system is surrounded by an insulator, like a cell membrane. And neither that charge nor the potential it creates at the boundary is always fixed; rather they vary and interact nonlinearly with (potentially) all the other charges in the system. Simplifications separating boundary from system may be possible, but are not obvious and need analysis for justification.

Consider for example, a real biological application, the simulation of a single Na^+ channel in a nerve cell, a nerve membrane wrapped into an indefinitely long cylinder, a cable 10 μm in diameter if it is a vertebrate

applied to the nonequilibrium case (at least as usually implemented) because they do not permit a gradient of concentration, electrical potential, or electrochemical potential between corresponding boundaries. Periodic boundary conditions are dangerous even in the equilibrium case (at least in my view) for two reasons. First, they replace actual physical boundary conditions with artificial periodic conditions. The physical boundary conditions often have large, even dominant effects on the properties of the system (as we see repeatedly in this paper) and so replacing them is likely to change the system qualitatively. Second, periodic boundary conditions force the system to be artificially rigid. They ensure that the system is (spatially) periodic in the velocity and position of every one of its atoms at every time. That is to say more precisely, **all the coordinates of phase space**, at all times are exactly the same (at corresponding positions on corresponding boundaries), thus imposing an impressive rigidity, a rigidity that real matter does not have, not even a crystal at a temperature of absolute zero, in which only the average positions of the atoms are periodic.

nerve, or 500 μm diameter if it is a squid nerve. The opening and closing of that channel, and the current through the channel once it is open, are found experimentally to depend on the potential across the membrane, if that is controlled by the standard electronics of a patch clamp on one side of the membrane,[79] and a voltage clamp of one sort or another,[1,80] on the other. But if the nerve is left in its natural state, without the voltage clamp, then the current through the single channel, and the gating of the channel, will depend on the potential within (roughly speaking) one length constant of the channel. In the mammalian nerve, the length constant is several millimeters, while for the squid nerve, it is on the centimeter scale. A direct simulation of either case is out of the question — it would involve perhaps 10^{21} ions, and 100 times more particles, for the simulation of the squid! This explosion in size occurs because a nerve fiber is designed to act as a transmission line, an imperfect capacitor, which stores charge along a macroscopic length λ, with little decrement in potential. In other words, the charge that determines the potential in a particular channel molecule is spread over a millimeter to centimeter distance of the nerve fiber.

Transmission lines like nerve fibers have been studied for a long time (even before Maxwell wrote his equations of the electric field) ever since Kelvin[81,82] analyzed the spread of charge and potential over thousands of miles to help design the transatlantic (telegraph) cable.[164,165] And similar equations (without inductance, however) have been used by physiologists to model the (electrically) linear properties of nerve and muscle fibers for nearly that long. Hodgkin and Huxley,[83] showed how to use the differential equations[p] of cable theory to describe the long distance active conduction of charge along a nerve fiber (the nerve impulse or action potential (nearly a soliton) that propagates and carries information) while using a localized description of (active nonlinear) current flow through the nerve membrane, or channels. They realized that two different theories were needed with two different resolutions. One theory (e.g., their theory of ionic conductances) described the channels, while the other described the cable of the nerve. The output of the high resolution theory (of ionic conductances) was a trans-membrane (i.e., radial) current that served as the input of the low resolution theory of conduction itself.

[p]The same equations describe the transmission lines of electrical engineering,[84,85] and are called the telegrapher's equations in applied mathematics (e.g., Zauderer[41]).

I feel that the same approach, using models with multiple resolutions, is needed to construct an atomic theory of ionic channels and connect it to biological reality. It seems obvious that the flux through a single channel will not depend on the microscopic properties of all the ions needed to simulate its electric field. Surely, the flux will depend in some average way on the ions far from the channel. The problem is how to make such a low resolution theory of the charges far from the channel, how to link it to the model of the channel with its atomic resolution, how to be sure the models are mutually consistent.

8. Hierarchy of Models of the Open Channel

A hierarchy of models is needed—each on a different scale, each consistent with the electric field on that scale—to cope with the immense range of sizes and times between a channel (angstroms) and a nerve fiber (centimeters), between the vibrations of a proteins bonds (femtoseconds) and the average time course of gating (milliseconds to seconds). Only with a hierarchy of models can the atomic structure of a channel be linked to its macroscopic functions.

Molecular dynamics is indispensable for the analysis of short range forces and phenomena. These are of great importance in understanding channels, including their critical properties of selectivity, flux coupling, and ionic interactions. But molecular dynamics is only one of a hierarchy of models needed to deal with experimental data, to link structure and biological function. Molecular dynamics is the foundation of the analysis of channels and proteins and that foundation needs to be firm if the superstructure (the house of the atomic biology of channels) is to stand tall; everything depends on the foundation, but channels (and people, if I may continue to torture the metaphor) live and function in the house, the structure itself, not in the basement, and so we ought not neglect the higher levels. Each level of the hierarchy rests on its foundations, namely the lower levels. The parameters of each higher level (an analysis with less resolution) will be the results of a lower level.

In particular, the outputs of a simulation of molecular dynamics will be the inputs of a theory of stochastic dynamics, as we shall now see.

8.1. *Stochastic Motion in the Channel: Langevin Dynamics*

Our next landing on the staircase of the hierarchy of models is stochastic dynamics, or Langevin dynamics. Here, the dynamics of a permeating ion are described by 1) the Langevin equation,[39,87–91] an equation giving the position $X(t)$ of the particle at time t or 2) the intimately related Fokker–Planck equation giving the probability density function of finding a particle near position x with velocity $\dot{x}$. The Langevin equation is a form of Newton's second law (that the sum of the forces equals the [acceleration] ×[mass]) with an added random term. The forces in the Langevin equation are (a) the (deterministic) electric field that produces the drift of charged particles; (b) the (deterministic) retarding force of friction; (c) the random term describing the fluctuations in force produced by the microscopic collisions that generate the deterministic friction.

$$\overbrace{-m\frac{d^2X}{dt^2} = m\beta(x)\frac{dX}{dt} + ze\Phi'(x) - \sqrt{2mkT\beta(x)\dot{w}}}^{\text{Langevin Equation}}\ . \tag{2}$$

I write this expression in the customary terse form found in the literature with the amusing mixture of notations we have inherited for derivatives: (Boyer,[166] e.g., p. 252; Cajori,[167] e.g., p. 257; Grabiner[168]), writing some derivatives with Leibniz differential quotient notation $\frac{d\Phi(x)}{dt} = \Phi'(x)$, others with Newton's fluxion notation (superscript dot) $\dot{w} \equiv \frac{dw}{dt}$, and still others with Lagrange's prime notation $\Phi'(x) \equiv \frac{d\Phi(x)}{dx}$. The effective mass of the moving particle is m; $m\beta(x)\frac{dX}{dt}$ is the friction term, with $\beta(x)$ the friction coefficient (per unit mass), $ze\Phi'(x)$ the electrical force, z the valence of the particle (a positive or negative number), e the charge on a proton (1.6×10^{-19} coulomb), and $\Phi(x)$ the potential of mean force (also a functional of many other variables besides the location x). $\sqrt{2mkT\beta(x)\dot{w}}$ is the random force, where kT is the thermal energy (typically some 4×10^{-21} joules $\cong$ 0.96 kcal $\cong$ 0.025 e volts), k the Boltzmann constant, T the absolute temperature, and $\dot{w}$ the derivative of standard Brownian motion as defined in, e.g., Refs. 39, 90, 99 and 111.

Condensed phases like biological solutions are dominated by collisions and friction because they contain little empty space to accommodate the large thermal velocity of their atoms and molecules. An enormous number of atomic collisions occur in the shortest time of interest and so it is natural to consider large friction expansions of the relevant dynamical equations. The usual high friction expansion of the full Langevin equation drops the second derivative term from equation (2). The resulting reduced Langevin equation is the starting point for analysis of stochastic diffusion phenomena and is thought to describe the trajectories that account for the macroscopic diffusion described by Fick's law. It is a great deal easier to handle both numerically and analytically than the full Langevin equation (2).

$$\overbrace{m\beta(x)\frac{dX(t)}{dt} = -ze\Phi'(x) + \sqrt{2mkT\beta(x)}\dot{w}}^{\text{Reduced Langevin Equation}} \,. \tag{3}$$

The equivalent high friction expansion of the Fokker–Planck equation yields the Smoluchowski equation (of the same form as the Nernst–Planck equations given later in this paper) and predicts a Maxwellian distribution of velocities.

The variables of the Langevin Eqs. (2) and (3) are written here in the conventional, if misleading manner. The variables are not as independent as the notation, naively applied, implies. In particular, the electric field $-ze\Phi'(x)$ is a functional (e.g., a solution of Poissons equation with boundary conditions) that depends on nearly everything else in the problem, just as several of the types of charge described previously in this paper. The structure of the protein determines the permanent charge that helps determine $\Phi(x)$: note it does not determine the potential directly since a protein is a distribution of charge not potential. The protein is not connected to a source of energy and charge like a battery so it cannot maintain a potential in the face of changing charge, whether that is caused by changes in experimental conditions, boundary conditions, or properties of the protein itself.

Dynamics of the protein itself would enter chiefly in better approximations than equation (3), e.g., the dynamics of the protein would determine the memory kernel (describing the impulse response of the friction function) in a generalized Langevin equation.[39,97] The friction and mass terms may also have complex dependencies, although these are more obscure since their description is not as clear as the electric field. The electric field is produced by charge, and only charge (in the present problem), but the effective mass and friction reflect a multitude of interactions and dynamics in the system, evident in simulations although not well understood in analytical theories of liquids and solutions.

8.2. *Simulations of the Reaction Path: The Permion*

The Langevin equation determines a stochastic path that describes one ion flowing in a channel, like a particle in a chemical reaction. The ensemble of those random paths is the distribution described by the Fokker–Planck equations. The lowest energy path of that distribution is called a reaction path and can be computed by methods developed (for proteins) by Ron Elber[169,170] that estimate the path of steepest descent in the potential surface describing a chemical reaction.

The reaction path through a channel is often assumed to be a straight line, but calculations of the reaction path within gramicidin show rich behavior. The path found in the calculations[q] is quite different from a straight line, with deviations of the order of 1 Å, enough to have energetic effects of many kT. The simulations show that the straight-line Cartesian co-ordinate $X(t)$ of the Langevin equations (2) and (3) needs to be replaced with a curvilinear reaction path.

Along the reaction path, there is significant motion of all the waters within the channel's pore and most of the atoms of the channel protein, as the permeating ion moves through the channel. It is useful to consider all these motions together, defining a quasi-particle — a *permion* — much like the quasi-particles holes, phonons, and electrons (for that matter) defined in solid state physics[7,8,95,96] to represent correlated motions of many atoms of a semiconductor.

Many properties of the permion can be evaluated along its reaction path. The effective mass (as customarily normalized and defined, e.g., Czerminski and Elber,[169]) is much less than the mass of the corresponding permeating ion. The friction is quite complex, strongly dependent on location within the channel, and with interesting time dependence as well. Finally, the calculated free energy is an estimate of short-ranged interactions in this region of the channel. More realistic descriptions of the free energy are problem-

[q]Excluding the entrance and exit regions which we have not yet simulated.

atic since it is not clear how to simulate the entry process or the different types of electric charge described previously in this paper. Direct evaluations of the noise term — its frequency spectrum and position dependence — would also be interesting.

Calculations of dielectric effects are needed. Perhaps some insight could be developed by calculating the free energy of (hypothetical) ions of different (fractional) charge, but otherwise similar properties. Such ions can be created in the simulations of molecular dynamics by progressively "turning off" the charge on the atom, leaving all its other parameters the same. The dependence of free energy on charge is the physical property that defines dielectrics and allows the definition of a dielectric constant, when the dependence is linear. Analysis of that dependence in complex nonideal systems like proteins may allow generalization of the idea of dielectric "constant".[97] An operational definition of dielectric constant for complex systems might be a useful description of the energy stored in the distortion of its electric field, that is, in its induced charge. We can seek such a definition by studying the variation of free energy as the charge of an atom is changed in a simulation of molecular dynamics, all other properties of the atom being kept the same. If the free energy is a linear function of the assumed charge, and robust—i.e., reasonably insensitive to the details of the calculation—the slope of the function should allow definition of an effective dielectric constant. If the slope is not linear, or is overly sensitive to details, such a definition is not likely to be very useful; indeed, it might be misleading.

In this way, molecular dynamics can supply paths, parameters and functions (like the potentials, frictions, and masses that depend on position) for use in a Langevin equation. Of course, molecular dynamics also provides physical insight (e.g., the nature of interactions, the existence of a curvilinear path) and an atomic perspective on protein motions (the nature of dehydration and resolvation on entry; the existence of a permion) that cannot be acquired from other techniques. But *molecular dynamics needs to be supplemented with solutions of the Langevin equation if macroscopic variables like concentration and membrane potential are to be introduced* into a theory of the open channel, so we can try to describe experimental phenomena (like current flow) that so clearly depend on those driving forces.

8.3. *Langevin Dynamics*

Langevin dynamics[39,87–89] — usually in the high friction limit, Eq. (3) — is the standard starting place for most stochastic theories of chemical reactions.[59,98,99] It has the considerable advantage of simplicity, compared with molecular dynamics — it appears to be an analytical theory, although the stochastic nature of w, and the bewildering properties[90,100] of $\dot{w}$ (it is a continuous function of unbounded variation, with an infinite number of maxima and minima, or zero crossings, in any finite interval, no matter how small the interval!) makes exploitation of the analytical theory difficult, for example, boundary conditions are difficult to prescribe or interpret for a function of unbounded variation, even if it is continuous.

The direct application of the reduced Langevin Eq. (3) to channel problems has been impeded by a simple but frustrating problem: current through channels is driven by gradients of concentration (as we have often said), but concentration does not appear in either Langevin Eqs. (2) or (3); rather they describe the trajectories of the particles one at a time and the corresponding Fokker–Planck equations describe the probability that *one particle* is found in some region at some time. This problem is particularly frustrating in view of the central role of concentration in diffusion and the enormous literature[92] of stochastic motion over potential barriers. So many difficult issues have been confronted and solved in that literature that it seems as if the problem of introducing concentration into a stochastic model of diffusion should have been solved long ago.

The problem has, however, plagued simulations of stochastic dynamics of channels, ever since they were started by Kim Cooper working in Erik Jakobsson's laboratory.[35,101–107] It is not clear how to introduce particles into the simulation in a way that mimics ion entry from a solution of fixed concentration, or how to remove particles from the simulation in a way that mimics ion exit from a channel into a solution of fixed concentration.

When we first looked at this problem, we found that most papers in the literature of stochastic dynamics assumed large barriers for diffusion[39,59,92,93,98,99] to reduce the problem to a study of diffusion at just the top of a barrier. Rates over the top of the barrier were then scaled to the appropriate concentrations in a way we did not fully understand but no doubt can be justified (e.g., see Barcilon, Chen, Eisenberg, and Ratner[35]: Section VIII, Eq. 8.6). Our analysis could not use that approximation because flux through channels does not depend just on properties at the top of potential barriers within the channel. It is known experimentally to also depend strongly on the potential (i.e., charge) and concentrations far away from the channel interior, for example, at the boundaries of the system, as we have discussed at length.

Furthermore, I believe on evolutionary grounds that barriers in channels are probably not large, and certainly should not uncritically be assumed large early in the development of a theory: many channels are designed to conduct as much current as possible for their biological function, and those are the ones we can observe with presently available patch clamp amplifiers. It seems likely that the channels we study today will have small barriers, not large ones, particularly those involved in the nerve action potential.[r] We were also reluctant to follow this approach because of our previous experience with other boundary value problems. In those, boundary conditions proved to be of the greatest importance, often determining the main properties of the system actually observed experimentally. In the present problem, of stochastic motion over a potential barrier, we knew if the usual assumptions were made about the potential, the direct effect of boundary conditions would disappear; submerged beneath the high barrier (Barcilon, Chen, Eisenberg, and Ratner[35]: Section VIII, Eq. 8.6). We feared the consequential loss of physical meaning. It seemed clear that we would have to derive our own theory, starting with the existing literature of the stochastic motion of a single particle and extending it to concentrations of particles moving over barriers of any shape. Surely, we thought, it would be easy to generalize the theory of one particle to a theory of N particles, at least if they were assumed not to interact. All we thought necessary was to multiply the probabilities of the usual theory by N, the number of particles.

The trouble was the more we understood of the problem, the more were the possible choices of "the probability" and the "number of particles". Which probability? The probability of which ion, where, conditional on what? Which number of particles? The number of which ion, on which

[r]Evolution has gone through considerable challenges to maximize the conduction velocity of the propagating nerve impulse by (for example) creating giant nerve fibers in invertebrates or wrapping nerve fibers in insulating myelin in vertebrates. It seems likely that evolution would go to at least as much challenge to maximize the flow of Na^+ ions (which is the energy and current source for propagation), if it possibly could.

side of the channel, conditional on what? We could not guess. Indeed, as we learned more, we found other ways to compute flux in the literature, for example, by determining mean first passage times (MFPT) and channel contents.[108] We hastened to determine these times and contents using some apparent identities and analytic tricks[35] but found apparent paradoxes. The MFPT could be defined in different ways, each with different mathematically imposed boundary conditions (none of the boundary conditions were determined by the physics of the problem) and each definition of MFPT gave conflicting expressions for the fluxes, some nonsensical. Probability theory is well-known to be subtle and difficult[109,110] but it certainly should give unique results in a well-defined physical set-up like this, which is, after all, the experimental system originally analyzed by Fick, 150 years ago!

The problem was particularly frustrating because the history of diffusion theory suggested the solution should be straightforward. After all, the original problem posed by Fick involved diffusion from one concentration to another (actually through a biological system, it is amusing to note in the present context[s]). The description of the stochastic trajectories underlying this classic diffusion should be straightforward, we thought, if we simply followed the classical work of probabilists, starting with Einstein, on diffusion and Brownian motion.

For some six years, Victor Barcilon and I struggled with this problem, quite unsuccessfully. Finally, we found a way to avoid it, to solve part of it without understanding the stochastic basis of the solution, by computing the average flux using an analytical trick (the construction of a formal adjoint) which was clearly true, depending only on the accuracy of our algebra, but which was empty of stochastic meaning. We (now joined by Mark Ratner and Duan Pin Chen) compared the analysis with simulations with satisfactory results. Duan Pin Chen showed how to simplify the simulations, deriving statistical formulae of tantalizing simplicity, but whose probabilistic meaning we could only guess, not derive.[35]

Understanding required a stochastic theory, later constructed with Zeev Schuss and Malgorzata Klosek.[44] We tried for nearly two years just to formulate the problem in stochastic language, seeking to describe a concentration boundary condition at the ends of the channel by invoking the special

[s]Fick was a physiologist interested in membranes, as well as diffusion.[1,111]

properties of that location (e.g., Chen and Eisenberg,[4,5]) in our channel proteins. The problem was that trajectories do two different things at a boundary with a defined concentration. Some trajectories start there and others end (i.e., are absorbed) there. The problem was how to describe these behaviors in the language of stochastic trajectories. Only a few boundary behaviors of stochastic trajectories have been analyzed or are thought to be possible.[112–114,171–173] The problem was to map **both** the starting and ending behavior of trajectories (at one location of known concentration) into those boundary behaviors. We tried for a long time to separate starting and ending trajectories by invoking special properties of the channel protein, i.e. barriers or "selectivity filters" at the ends of the channel's pore. But we did not succeed because (as it turned out) we were always working entirely in the high friction limit.

The key to solving the problem was realizing that it had nothing to do with channels! In fact, traditional stochastic theories of diffusion could not describe diffusion from a region of one concentration to a region of another, whatever the link between them, whether it was a channel or free solution. In other words, the classical stochastic theory of diffusion did not solve the original problem of diffusion posed by Fick.

The difficulty arose from the classical (and our previous) description of diffusion in the high friction limit. We (along with many others studying Brownian motion in condensed phases since Einstein) had started with the reduced Langevin Eq. (3) as a description of diffusion. We had assumed that because diffusion in a solution is so entirely dominated by friction (collisions occurring every 100 femtoseconds, with observations being made at intervals of one microsecond) that the distribution of velocity was in its steady state equilibrium configuration, namely Maxwellian, implying that the second derivative could be dropped from the full Langevin Eq. (2) leaving the reduced Eq. (3), describing trajectories with a distribution of position, but not velocity. If we described the particle by a reduced Langevin equation, with only one spatial derivative, we were entitled to only one boundary condition. Yet at a boundary of fixed conentration, trajectories have two behaviors: trajectories both start and end there. The place of known concentration behaves both as a source and a sink. These behaviors are quite different and cannot be described by a single boundary condition. We needed an equation that allowed two boundary conditions. The corresponding Fokker–Planck equation (really, the Smoluchowski equation) showed that the distribution of velocity was in its steady state equilibrium configuration, namely Maxwellian. Then, we could solve or simulate the problem for a wide range of behaviors at the channel mouth, indeed, we could simulate it for any behavior that could be defined in the formalism of the Langevin equations.

That equation was easy to provide, once we realized it was needed. The full Langevin equation itself provides the extra boundary condition. It involves both the velocity and the location of a particle and describes trajectories of particles with a distribution of both. We had to use the full Langevin equation and deal with it, even when friction dominates movement of ions, as in the classical diffusion theory. Evidently, the distribution of velocities is **not** Maxwellian, even in the high friction limit, even to first approximation, in the non-equilibrium case of interest to us, when flux flows through channels with low barriers.[t]

At first, the prospect of dealing with the full Langevin equation seemed daunting. Neither the full Langevin equation nor the full Fokker–Planck equation is noted for its mathematical simplicity. Analytical solutions are rarely available, and even the general properties of the solutions are not known.[39,92,93] Imagine my pleasure then when we found identities, valid for potential barriers of any shape or size, describing the flux from one concentration to another, e.g., through a channel.

$$J_{\text{net}} = J_{\text{in}} - J_{\text{out}} = \frac{1}{\sqrt{2\pi\varepsilon}}\left(C_L \cdot \text{Prob}\{R|L\} - C_R \cdot \text{Prob}\{L|R\}\right). \quad (4)$$

ε is dimensionless temperature, named to be consistent with the literature of the Langevin equation (see precise definitions in Ref. 44). It is not a small number despite its name! C_R and C_L are the concentrations on the right and left, respectively. The probabilities $\text{Prob}\{R|L\}$ and $\text{Prob}\{L|R\}$ are the solutions of precisely defined stochastic problems (described in Ref. 44) that can solved analytically in some simple cases and simulated in all others.[35]

This formula is closely related to the traditional flux formula of the Nernst–Planck equation.

[t]The analytical solution to the problem demonstrates this point explicitly. The distribution of velocities deviates from the Maxwellian whenever flux flows.

$$J_{\text{net}} = J_{\text{in}} - J_{\text{out}} = \frac{C_L e^{\phi(0)/\varepsilon}}{\int_0^1 \frac{e^{\phi(\zeta)/\varepsilon}}{D(\zeta)} d\zeta} - \frac{C_R e^{\phi(1)/\varepsilon}}{\int_0^1 \frac{e^{\phi(\zeta)/\varepsilon}}{D(\zeta)} d\zeta} . \tag{5}$$

But now the integrals could be derived in terms of conditional probabilities of specific well-defined stochastic problems. The probabilities Prob $\{R|L\}$ and Prob $\{L|R\}$ of Eq. (4) are just the integrals of Eq. (5), it the dynamics are dominated by friction. If the dynamics are more complex, the conditional probabilities can be evaluated by simulations of Langevin equations or molecular dynamics.[u]

The analysis of the stochastic problem does more than provide a stochastic interpretation of the well-known Nernst–Planck flux formula. It also reveals the role of *cis* trajectories, namely *LL* and *RR*, previously unknown (in the physiological literature, at least to me) perhaps because they do not contribute to the mean steady-state fluxes of isotope customarily measured in experiments. *LL* trajectories do not contribute current (if the patch electrode is on just the outside, which we call the right), but they have a large effect on channel properties because they occupy the channel a substantial fraction of the time and so block other trajectories (in channels with interactions). *RR* trajectories block the channel in a similar way. Also they contribute directly and substantially to the variance (i.e., open channel noise) even though they contribute zero mean current. When an ion flows from the electrode into the channel it produces a short blip of current in one direction; when it flows back into the electrode, completing its *RR* trajectory, it contributes a blip in the other direction. These (the *RR* trajectories) significantly complicate analysis because they (unlike the others) do not form a stochastic process that can directly be analyzed by renewal theory.[116] If one of the ions in the electrode or in the bathing solution has a small diffusion constant (i.e., a"slow" ion, or a blocker), the *RR* and *LL* trajectories may determine most of the behavior observed experimentally. In any case, they are important determinants of experimental behavior, however, and so must be analyzed, not ignored. The *cis* trajectories *RR* and *LL* are likely to be important under normal conditions; indeed,

[u]Dynamics can be complex for many reasons, e.g., (1) if the ions and channel are described by molecular dynamics; (2) if ions interact with each other; (3) if they significantly modify the shape of the electric field; or (4) if friction or potential is "frequency dependent" so a generalized Langevin equation is needed to describe ionic motion.

they may dominate the normal open channel noise in at least some channels.[115]

The stochastic analysis also is closely related to traditional results on average fluxes:

- The flux law (5) is identical (in that special case of mean fluxes in the steady-state) to the integrated Nernst–Planck equation, the flux law assumed for radioactive isotopes by physiologists ever since the time of Ussing and Hodgkin.[116]
- The probabilistic analysis yields the familiar formula[108]

$$\text{Flux} = \frac{\text{Content}}{\text{Mean First Passage Time}} \tag{6}$$

  provided the flux, mean first-passage time MFPT, and contents are the appropriate conditional quantities (see Ref. 44 for precise definitions).
- The flux law is identical (for the mean fluxes in the steady-state) to the flux law derived previously with the adjoint method[35] and their simple statistical formulae turn out just to be the "best" estimators of the conditional probabilities Prob $\{R|L\}$ and Prob $\{L|R\}$ of the flux law (4)!

9. Chemical Reactions

Our stochastic model of the open channel has a general significance that caught me by surprise: it permits permeation (through the open channel) to be described **exactly** as a chemical reaction, moving ions by rate constants between well-defined states, namely definite concentrations at specific locations.

$$\boldsymbol{L} \underset{k_b}{\overset{k_f}{\rightleftarrows}} \boldsymbol{R}\,. \tag{7}$$

Such reactions have been written qualitatively, or in terms of vaguely defined states L and R for many years. But now we can define the rate constants accurately, for barriers of any shape and height, and the meaning of L and R is clear and unambiguous.

$$k_f \equiv \text{Prob}\{R|L\}\,; \qquad k_h \equiv \text{Prob}\{L|R\} \tag{8}$$

with the probabilities defined by the integrals in Eq. (5) above if dynamics are simple, or, if they are complex, by simulations of the underlying stochastic problems specified as Ref. 44.

This probabilistic interpretation of a chemical reaction with prescribed concentrations at its boundaries is not confined to channels and channel permeation. Rather, it applies to any chemical reaction in which the concentrations L and R are maintained fixed, for example, it applies to any chemical reaction in a continuous flow system[118] even if the reaction occurs over low potential barriers.

We were surprised to find a probabilistic model of a chemical reaction exactly valid for any shape potential barrier. Most theories of a chemical reaction assume a particular shape barrier, in particular a high barrier and I thought that was a necessary assumption.[39,59,92,93,98,99] It is certainly a reasonable assumption for the typical chemical situation involving distinct chemical species; after all, in a certain sense it is the high barrier that defines the species and makes them distinct. Assuming high barriers simplifies the mathematics dramatically, and I had thought that assumption necessary if simple results (like Eqs. (7) and (8)) were to be derived. In our stochastic model, however, we have a simple result true for any shape barrier. Assuming a high barrier would not significantly simplify the expressions for the rate constants and might obscure their physical meaning. The high barrier also has a profound effect on the role of boundary conditions in the problem. The high barrier ensures that flux is determined (asymptotically) by diffusion just over the top of the barrier. Boundary conditions do not occur at that location, however, and so (in a certain too vivid sense) the usual analysis hides boundary conditions behind its high barriers.

In the analysis of channels, barriers cannot be assumed large, as we have seen and so boundary conditions emerge into the prominence they usually have in physical problems. Inclusion of boundary conditions often complicates a problem, and adds terms to its solution, and so we were surprised to find such simple expressions (8) & (5) for the rate constants. These arise because of the boundary conditions we use. If the same chemical reaction[v]

[v]i.e., the chemical reation (7), with the same reactants, with the same dynamics over the same low potential barrier, with the same diffusion constant and other parameters.

took place in a different experimental set-up (with concentrations not held constant and so with different boundary conditions), the reaction would **not** follow exponential kinetics (in time) and so could not be described by simple forms of the law of mass action.

10. What Was Wrong?

It is natural to wonder what is different about the present analysis from other previous analyses (including my own) of the high friction case. The difficulty in earlier analyses was in the way friction increased in the limiting process. The limit of high friction was taken while keeping the overall driving force fixed (i.e., keeping fixed the difference in electrochemical potential, or electrical potential, or concentration, depending on the precise problem). Thus, the flux necessarily went to zero as the friction increased, implying that the system approached equilibrium in that high friction limit. The limit desired and used in the present analysis (i.e., in Eisenberg, Klosek, and Schuss[44]) is quite different. We study high friction, but with a fixed amount of flux (independent of the friction), a flux corresponding to that observed in experiments.

Subtleties often occur in limiting processes when multiple parameters are present: indeed, singular perturbation theory[174] was designed to deal with such situations. In general, a limiting process ought to correspond to the precise experimental situation of interest. Otherwise, the result of the limiting process is unlikely to describe the experimental situation, even qualitatively. In the present case, the limit of high friction and fixed driving force is inappropriate because it implies an equilibrium (or tiny perturbation from equilibrium) quite different from the situation in real open channels. In real channels, the limit of high friction occurs in a system far from equilibrium in which flux (of at least one ion) is nearly always substantial. Limiting a diffusion system too close to equilibrium is particularly dangerous according to (Hoover,[163] p. 266): "The change from equilibrium to non-equilibrium is *not a simple one that can be described by (regular, as opposed to singular) perturbation theory. The phase-space geometry is fundamentally changed in a way which frustrates traditional analyses based on Taylor series expansions.* Away from equilibrium only rare (fractal) states can correspond to nonequilibrium steady-states" (italics are in original).

11. Back to the Hierarchy

In this way, we have learned how to include diffusion in a stochastic model, how to predict the effects of concentration gradients. This step on the staircase of the our hierarchy is secure. But it is an isolated landing, standing by itself because the stochastic analysis assumed a potential function $\Phi(x)$; it did not compute it. This model is not self-consistent as we have defined it. The analysis is only valid under a range of conditions in which the potential function does not change, i.e., under a range of conditions that do not change the various types of charge described previously. Such conditions exist and can be studied experimentally, but they are quite restrictive, and certainly do not span the domain of biological interest.

Clearly we need to compute the potential in a self-consistent stochastic model from the physical description of the charges and the Langevin description of dynamics. That is an essential step in the construction of the hierarchy needed to analyze the atomic physics of channels, linking atomic structure to biological (macroscopic) function. But we do not know how to do that calculation (although, of course we have ideas, or should I say hopes?). For now, we can compute the electric field and the concentration coupled together only when we give up the discrete description of ions. To determine coupling, we have, at least up to now, described the electric field as a fluid, specified by the Poisson equation, and the concentration of each type of ion also as fluids, specified by the Nernst–Planck equations, all with appropriate boundary conditions. With that departure from reality, the entire problem can be analyzed, with charges producing potentials, potentials producing charges, both producing concentrations, and all producing flux. That approach has proven most fruitful in the analysis of semiconductor devices[6] and we will apply it now to channels. It is the highest, but most remote level of the hierarchy dealt with here.

To summarize, in our over-worked metaphor: the continuum model is the attic of our house of hierarchy, giving a good view of an attractive horizon, but too musty and remote from our foundations (of molecular dynamics) to inhabit any longer than we have to!

12. Macroscopic Models of the Continuum

Macroscopic models are rather unfashionable in molecular biology, particularly in the study of single ionic channels, where they seem faintly archaic and out of place. This unpopularity is more than understandable, given that measurements are made every day of current flow through single channel molecules. Nonetheless, the resolution of single channel measurements may make life seem more complicated than it is. General principles may be obscured by atomic detail, just as too much magnification makes a photograph too large and objects hard to recognize. A photograph showing each molecule of pavement will not show the road. The Navier–Stokes equations can be hard to see even in the Boltzmann transport equation let alone in the trajectories of molecular dynamics. A theory of an ensemble of channels might be simpler and more closely related to the underlying physics of channels than a stochastic theory of one channel!

Too little resolution is, of course, even worse than too much, because it can blur two different mechanisms into a hybrid very different from either. Consider a mechanism we all know first hand, reproduction. If human anatomy were studied with too little resolution, say by observers from another planet, males and females might be lumped together into one composite structure, a homogeneous blur of a heterogeneous population. In that blur the mechanism of reproduction would be lost. Until the *functionally* relevant populations were distinguished, and separately studied, the mechanism of reproduction would be entirely misunderstood. Indeed, until this error is corrected, more experimental work will lead to more confusion, not knowledge! Until we know where in the hierarchy of models evolution has made its adaptations, we must study all levels. We will not know which models are over-magnified and which are too blurred until we are finished.

13. Levels of the Hierarchy: Occam's Razor Can Slit Your Throat

If a complex system is studied or analyzed on only one level of the hierarchy, the (most understandable, necessary, but illogical) human need to solve something (to feel we have at least some control over our uncontrollable world) can lead us to accept the "simplest" explanation, using Occam's razor to justify our analysis, even if it is irretrievably blurred by low resolution of essential structures or events. If a theory ignores or does not well resolve the length scale relevant for a particular function of the system, it will not describe it properly, even if the theory is simple and fits some set of experimental data. No theory of reproduction that ignores the distinct morphology of sexes can be correct, no matter how well it fits a set of experimental data. Or (to choose an example closer to hand), no theory of the action potential in nerve fibers that lumps together Na^+ and K^+ currents will resolve the properties of the individual channel molecules. Occam's razor slits your throat if theories and experiments do not have enough resolution, if they blur together essentially separate components of the underlying mechanism, if they describe the wrong level of the hierarchy of models.

Macroscopic models seemed blurred and inappropriate descriptions of channels because ions, water molecules, and the channel's pore are about the same size. Ions and water moving through a channel clearly form a single file system (for example) and it seems ridiculous to represent or analyze such systems as a continuum. Nonetheless, macroscopic models describe some properties of channels quite well, particularly those that depend on long range phenomena like the electric field.

Continuum models have an important role because *they are consistent analyzed* and can be completely analyzed. Indeed, at the present time, they are the only models that can treat the electric field self-consistently. Only in continuum models have all the sources of the electric and concentration fields — that is to say, all the charges present and all the sources and sinks of matter — and all boundary conditions been included and their effects analyzed by mathematics, with minimal additional assumptions. Furthermore, interactions appear automatically from the mathematics (without any *ad hoc* assumption or analysis) if a complete set of field equations and boundary conditions are specified. The different types of interactions do not have to be specified, or at least enumerated so explicitly, as in discrete models, and subtle and high-order interactions which are consequences of fundamental laws arise automatically from the mathematics and will be found in the output of the calculations, even if the interactions are not manifested in the differential equations and boundary conditions that define the problem, even if the existence of interactions was not anticipated by the authors of the model, and even if the physical basis of the interactions is not understood *a priori*.

The importance of this mathematical power is great. Nonlinear coupled partial differential equations behave in complex ways often unanticipated even by researchers who have studied them for decades. Beforehand (before the theory was formulated, solved and computed) the behavior is often an unanticipated mystery. Later, after the complex behavior of a system is understood physically, it may seem obvious, natural, often inevitable. In nonlinear coupled systems, rarely is the "obvious" or "inevitable" predicted until after it is computed. The behavior may be simple and reproducible, and most useful technologically (e.g., a mechanical switch) but if it arises from essentially nonlinear coupling between underlying equations, it is nearly impossible to predict *a priori* by analysis, verbal or mathematical.

Thus, the reasons for making macroscopic self-consistent models of the open channel are:

- The sources of the electric field are understood.
- The sources and sinks of concentration are completely understood.
- The boundary conditions are (fairly) evident and quite well understood.
- A wide range of interactions are predicted automatically by the mathematics, even if they are not explicit in the original formulation of the problem.
- A complete analysis of a wide range of experimental conditions can be made using only mathematics, without making additional physical or biological assumptions. A complete consistent model can be made (solved and computed) of a hypothetical macroscopic system, an open channel filled with a continuous electric fluid, a fluid with charge density and concentration, not discrete ions.

We suspect that all effects found in such a continuum model are present in a real channel filled with discrete charges. Certainly the physics that is important in a continuum model should be present in a more realistic discrete reworking of the theory. If we find important effects of one type of charge or one type of boundary condition in a continuum model, both should be included in a discrete model. If we find important interactions in a continuum model, the physics producing them must be included in a discrete model, and must be presented in a way that can be analyzed.

In fact, we have found, much to our surprise, that the interactions in a continuum model of an open channel can account for a wide range of complex behavior, behavior that previously had been thought impossible unless the system were discrete. The richness in behavior occurs because the electric field changes the concentrations and also vice versa. Thus the electric field and concentration are coupled and neither can be analyzed without the other.

Most experimental interventions change the shape of the electric field and thereby have a large effect on permeation. Most previous theories of the open channel assume that the electric field is independent of experimental conditions, or varies in a simple pre-ordained way, and so miss these first-order effects, which can dominate even the qualitative behavior of the channel.

Changes in the shape of the electric field are used in semiconductor devices. They perform most of their functions that way. Semiconductor devices perform most of their functionsby changes in the shape of the electric field. They amplify, switch ON and OFF (i.e., gate), and do logic as the potential (or current flow) at one boundary changes the electric field everywhere. If the electric field were to assume a constant shape, transistors and integrated circuits could not be designed or understood. Unable to formulate or solve a discrete model, physicists have for nearly fifty years designed semiconductor devices using a consistent macroscopic model[6–8,97,119–122] similar to a continuum model of open channels. They too would dearly love to formulate and solve a consistent discrete model; but the continuum model has been good enough to support the exponential growth in capability, speed, and complexity that makes much of our modern technology possible.

14. Poisson–Nernst–Planck (PNP) Model

In a continuum model of an open channel, each ionic species is represented as a fluid, a concentration that is a function of time and location. The sources of the fluid are the electrodes far away at infinity. The fluids — the concentrations of ions — can flow under a variety of forces. In channels, the main fluxes are thought to be diffusion and drift in gradients of concentration and electrical potential. Convection is conventionally ignored and the system is supposed to be uniformly isothermal on all length scales. A thorough analysis demonstrating the universal insignificance of convection and temperature would be comforting!

The equation governing flow is the Nernst–Planck equation which relates the flux J_j ($\mathrm{mol} \cdot \mathrm{cm}^{-2} \cdot \mathrm{sec}^{-1}$), independent of location in this steady-state situation of species j with charge z_j to the diffusion constant $D_j(x)$ ($\mathrm{cm}^2\ \mathrm{sec}^{-1}$) at location x, with constant RT, the thermal energy per mole at room temperature, where R is the gas constant and T is the absolute temperature, F the Faraday constant.

$$J_j = -D_j(x)\left[\frac{dC_j}{dx} + \frac{z_j F}{RT} C_j \frac{d\Phi}{dx}\right]. \tag{9}$$

The driving forces of gradients of concentration C_j and electric potential Φ are often written as shown without their list of arguments, but *this notation is seriously misleading.* Because the concentration(s) of ions help determine the potential (through Poisson's equations) and the potential helps determine the concentration (through the Nernst–Planck equations), both variables depend on the entire properties of the system everywhere, not just at one location. Thus, $\Phi_j\left(x; C_j\left[\tilde{x}\right]\right)$ might be an appropriate notation, showing that the potential depends not only on location x, but also on the electric field *at every other location* $\tilde{x}$, as well. Similarly, the concentration of a given species $C_j(x)$ may also depend on the concentration $C_k\left(\hat{x}\right)$ of each other species $k \neq j$ at every location $\hat{x}$, and also on the potential $\Phi_j\left(\breve{x}\right)$ at every location $\breve{x}$, as well. The appropriate notation is then something like $C_j\left(x;\ \Phi\left[\breve{x}\right];\ C_{k\neq j}\left[\hat{x}\right]\right)$. Of course, both the concentration and the electric field depend on the diffusion constants, dimensions of the channel, and all other parameters in the system. While it is understandable that a simplified nomenclature is often used, and I use it myself as much as possible, it is regrettable that the underlying complexity of the physics is sometimes forgotten.

The Nernst–Planck equation can be written in terms of the electrochemical potential $\tilde{\mu}_j\left(x\right)$ and either the electrical potential or the concentration

$$\tilde{\mu}_j(x) \equiv RT\ \ln\tilde{C}_j(x) + z_j F\Phi(x)\,, \tag{10}$$

by neglecting the effect of one ion on the free energy of another (i.e., the mixed ion effects mentioned previously). Note that the potential, set to zero, is measured with respect to a reference point to be chosen, the solution outside the channel on the right-hand side, in the usual convention. Similarly, the (dimensionless) concentration $\tilde{C}_j\left(x\right)$ is measured with respect to a reference concentration $\bar{C}$, a standard state, so

$$\tilde{C}_j(x) \equiv \frac{C_j(x)}{\bar{C}}\,. \tag{11}$$

Later we will measure the flux with respect to the same standard concentration, introducing the flux $\tilde{J}_j$, measured in units of $\bar{C}$, defined by

$$\tilde{J}_j \equiv \frac{J_j}{\bar{C}}\,. \tag{12}$$

The Nernst–Planck equation can be written in terms of the electrochemical potential $\tilde{\mu}_j\,(x)$, the electrical potential $\Phi\,(x)$, and the dimensionless flux $\tilde{J}_j$.

$$\tilde{J}_j = -D_j(x)\left[\exp\left\{-z_j F\Phi(x)/RT\right\}\right]\frac{d}{dx}\left[\exp\left\{\tilde{\mu}_j(x)/RT\right\}\right]\,, x_1 \ge x \ge x_2\,. \tag{13}$$

Remember that flux is independent of location and time in this steady-state model of an open channel and so this expression (13) can be explicitly integrated between any two locations x_2 and x_1 of its domain of validity.

$$\tilde{J}_j \int_{x_1}^{x_2} \frac{\exp\{z_j F\Phi(x)/RT\}}{D_j(x)}dx = \exp\left\{\tilde{\mu}_j(x_1)/RT\right\} - \exp\left\{\tilde{\mu}_j(x_2)/RT\right\} \tag{14}$$

giving what we call the *Integrated Nernst Planck* equation:

$$\overbrace{\tilde{J}_j}^{\text{Net Flux}} = \overbrace{\frac{\exp\left\{\tilde{\mu}_j(x_1)/RT\right\}}{\int_{x_1}^{x_2}\frac{\exp\{z_j F\Phi_j(x;C_j[x])/RT\}}{D_j(x)}dx}}^{\text{Unidirectional Efflux}} - \overbrace{\frac{\exp\left\{\tilde{\mu}_j(x_2)/RT\right\}}{\int_{x_1}^{x_2}\frac{\exp\{z_j F\Phi_j(x;C_j[x])/RT\}}{D_j(x)}dx}}^{\text{Unidirectional Influx}}\,. \tag{15}$$

The numerators of the integrated Nernst–Planck Eq. (15) depend only on the electrochemical potential in the baths, where the concentrations and electrical potentials are known and under experimental or biological control. The numerator is entirely independent of the electric field and is uncoupled, despite the nonlinear coupled nature of the overall system.[w]

[w]It would be interesting to try to recover the intergrated flux equation from a discrete model, particularly to see how the various interaction and couplings disappear as the macroscopic model arises from its more realistic discrete cousin.

The denominator of the integrated Nernst–Planck Eq. (15) depends only on the electric field and is independent of the concentration of a tracer ion in the bath (present in negligible concentrations — when measured in moles/liter, not disintegrations per second). For this reason, the ratio of the terms in Eq. (15), namely the flux ratio, is particularly useful, as we discuss later in this paper. The flux ratio (originally introduced by Ussing,[117] see Refs. 123–128) should be independent of the shape of the electric field within the channel. Indeed, the ratio of fluxes should be given by (the exponential of) the difference in electrochemical potential on the two sides of the channel, quite independent of what goes on in the channel.

$$\frac{\textbf{Unidirectional Efflux}}{\textbf{Unidirectional Influx}} = \exp\left\{\left[\tilde{\mu}_j(0) - \tilde{\mu}_j(1)\right]/RT\right\} \equiv \textbf{Ideal Flux Ratio}. \qquad (16)$$

Here, $\tilde{\mu}_j(0) - \tilde{\mu}_j(1)$ is the difference in electrochemical potential between the inside and the outside bath, the overall driving force for electrodiffusion determined by the difference in electrical potential and the difference in (the logarithm of) the concentrations (really free energies per mole, namely activities).

The unidirectional fluxes of the main species can easily be estimated from measurements of the total fluxes of radioactive tracers (indeed, that is why Ussing defined them!). Each term can be experimentally identified by measurements of the total flux of radioactive isotope, when isotope is present only on one side of the channel (in trace amounts, nonetheless easily detected by its radioactivity), and is absent from the other (see Refs. 5 and 123 for details). The flux ratio determined from Eq. (15) gives such powerful results — a ratio quite independent of what goes on inside the channel or of the shape of the electric field in the channel — because the trace amount of radioactive isotope used to estimate macroscopic unidirectional fluxes does not change the denominator of Eq. (15).

The meaning of unidirectional fluxes and their ratio became clearer to me when I tried (unsuccessfully) to define them in another way, when I realized that the Nernst–Planck equation can be written in terms in terms of electrochemical potential and concentration (as well as in the traditional form shown in Eq. (13)). Thus,

$$J_j = \frac{1}{RT} D_j(x) C_j(x) \frac{d\tilde{\mu}_j}{dx}; \qquad x_1 \leq x \leq x_2 . \qquad (17)$$

This expression can be integrated between any two locations x_1 and x_2 in its domain of validity.

$$J_j = \frac{\tilde{\mu}_j(x_1)}{RT\int_{x_1}^{x_2}\frac{1}{D_j(x)C_j(x)}dx} - \frac{\tilde{\mu}_j(x_2)}{RT\int_{x_1}^{x_2}\frac{1}{D_j(x)C_j(x)}dx}\,. \tag{18}$$

The numerators of this expression (18) depend only on the electrochemical potential in the baths, and it was instructive to try to define each term as a unidirectional flux, and the ratio of terms as a unidirectional flux ratio, as we have done previously with Eq. (15). Such a definition is quite different from the traditional one and gives quite different results.

The apparent paradox is easily resolved by applying the classical (e.g., Refs. 5 and 123) analysis of unidirectional and tracer flux to Eq. (18). Then, we see that holding the concentration of tracer (nearly) zero on the *trans* side of the channel (which is the key step in relating the net flux of tracer to the unidirectional flux of the main species) does not set either of the terms in Eq. (18) to zero. Furthermore, the denominator of Eq. (18) is different for each species of ion[x] and so the denominator does not cancel if the flux of one species (e.g. tracer isotope) is divided by the flux of another (e.g., the main species). For these reasons the flux of isotope does **not** estimate one of the terms in Eq. (18), nor is the ratio of terms in Eq. (18) robust and revealing.

Returning now to the classical analysis (the flux ratio (16)) we notice that it, the ratio, is uncoupled from the electric field, that is to say, independent of the potential profile $\Phi(x)$ within the channel, although the fluxes themselves are not. The terms in the traditional integrated Nernst–Planck Eq. (15) are coupled because the denominators of that equation depend on the potential $\Phi(x)$ everywhere within the channel, and thus on all the properties of the channel, but the ratio of terms does not.

The potential within the channel (the function $\Phi(x)$ in the denominator of Eq. (15)) is generally unknown, except as the solution of Poisson's equation, which in turn depends on the Nernst–Planck equation. The denominator may be the same for every ion, but it is the same unknown value that depends on the potential everywhere $\Phi(x)$, which also depends on the concentration $C_j(x)$ of every ion everywhere: $\Phi(x)$ depends on almost

[x]Strikingly different, if one of the species is held at zero concentration on one side, as the tracer species customarily is.

every parameter and variable in the system. For that reason, the integrated Nernst–Planck Eq. (15) cannot be used until the electric field is specified. The electric field is specified by Poisson's equation

$$\frac{d^2\Phi(z)}{dz^2} = \overbrace{-2\tilde{\omega}_0(z)}^{\text{Permanent Charge}} - \overbrace{\tilde{q}(z)}^{\text{Channel Contents}} - \overbrace{\underbrace{2\tilde{\varepsilon}\Delta(1-z)}_{\text{Constant Field}} - 2\tilde{\varepsilon}\Phi(z)}^{\text{Induced Charge: Dielectric Effect}}, \tag{19}$$

where Δ is the trans-membrane potential and $\tilde{q}(z)$ is the sum of the concentration of all ions at location z. The channel extends from $z = 0$ to $z = 1$ (see Refs. 4 and 5 for precise definitions).

This one-dimensional version of the Poisson equation was derived from the full three-dimensional system and boundary conditions[20,129,130] where the system of dimensionless units is described. Briefly, the permanent charge is the charge density created by the chemical bonds of the protein, the channel contents are the ions within the channel, and the induced charge is the rearrangement in charge (in channel and pore) created by the normalized dielectric constant $\tilde{\varepsilon}$ and the deviation $[\Delta(1-z) - \Phi(z)]$ in potential from the (spatially) constant field, namely, $\Delta(1-z)$, which has been assumed so widely in traditional physiology and channology.

The coupling between concentration and potential is most clearly seen if a single integro-differential equation is written combining the Nernst–Planck and Poisson equations. Just as the Nernst–Planck equations can be integrated once to give an expression for the flux, so they can be integrated to give an expression for the concentration of ions in the channel (Eq. (5.7) of Ref. 131) which can then be substituted in the Poisson equation to give a single integro-differential equation for the potential, an equation uncoupled to any other. I present this expression to emphasize the complex nonlinear dependence of the potential profile on the contents of the channel, hoping thereby to discourage unanalyzed simplifications and approximations.

$$\frac{d^2\Phi(z)}{dz^2} + \overbrace{2\tilde{\varepsilon}\,[\Delta(1-z) - \Phi(z)]}^{\text{Induced Charge: Dielectric Effect}}$$

$$= \overbrace{-\,2\tilde{\omega}_0(z)}^{\text{Permanent Charge}} - \overbrace{\lambda^2 \sum_j z_j \frac{C_j(L)e^{z_j\Delta}\int_z^1 e^{z_j\Phi(\zeta)}d\zeta + C_j(R)\int_0^z e^{z\Phi(\zeta)}d\zeta}{e^{2_t\Phi(z)}\int_0^1 e^{z_j\Phi(\zeta)}d\zeta}}^{\text{Channel Contents}} . \tag{20}$$

It would be a mathematical miracle if the potential profile $\Phi(z)$, namely the solution of this equation, was a linear function of z (for example) or was independent of the trans-membrane potential Δ or bath concentrations $C_j(L)$ and $C_j(R)$. The flux and all other properties of the channel are a strong function of the potential profile, so every property of the channel is expected to be a complex function of experimental conditions.

Approximations to the electric field, or other properties of the channel, may be possible, under certain restricted conditions, but they are inherently unlikely, and therefore need to be justified analytically and numerically under the range of experimental conditions in which they are used. For example, Chen, Barcilon, and Eisenberg,[20] analyzed five special cases (in which the permanent charge density was zero!) and showed that in four of them the traditional constant field expression[1] can be derived. Whether the constant field expression or any other simple expression can arise in the presence of permanent charge is not known.

The fact that the shape of the potential profile changes with experimental interventions of many types restricts the validity of many of the traditional canons of membrane physiology. Most obviously, a single profile of potential vs distance cannot be used to describe a channel under a range of experimental conditions, whatever theory (Eyring rate theory, electrodiffusion, etc.) is used at whatever level of the hierarchy to convert that profile into a prediction of current flow. Furthermore, it is not possible to guess how the potential profile changes shape as the trans-membrane potential changes. Perhaps we can (under some conditions) assume that the trans-membrane potential only changes the slope of a (spatially) linear component of the total potential, but one does not know beforehand when that assumption is correct. Indeed, the *assumption is likely to be incorrect just when the phenomena become most interesting!* Many theories of the open channel are subject to this criticism. Theories of gating are also subject to this criticism if they are sensitive to the shape of the potential profile as well as the difference in potential from one side of the channel to the other.

The sensitivity of the potential profile to experimental conditions has profound effects for the analysis of all membrane transporters, not only channels. For example, it has long been believed (Refs. 1, 124 and 125; reviewed in Refs. 126 and 132) that the unidirectional efflux of an ion (measured with radioactive tracer placed on only one side of the membrane, as described above) is independent of the *trans* concentration of the main (i.e., non-radioactive) species in continuum electrodiffusion. Yet, if the *trans* concentration modifies the shape of the electric field, the unidirectional efflux will vary with *trans* concentration. The *PNP* model[4] shows this effect clearly and in a tantalizing way: if a channel has a uniform charge density of one sign, the unidirectional efflux decreases as *trans* concentration is increased, showing a characteristic sign of "single filing", even though the model is of a continuum fluid. If the sign of the permanent charge is changed, the unidirectional efflux increases as *trans* concentration is increased, showing a characteristic sign of mediated transport, even though the structure involved is a channel of unchanging conformation.[y]

I think any consistent model of ion transport (continuum, discrete, stochastic, even a simulation of molecular dynamics) is likely to produce a unidirectional efflux that depends on *trans* concentration. The contents of the channel will vary with experimental conditions, e.g. *trans* concentration. Thus, the net charge will vary with experimental conditions (and vary significantly because the number of ions in the pore is more or less equal to the charge on the channel's wall). Because flux depends steeply (often exponentially) on the potential, the current through a channel *in any consistent model* will depend strongly on experimental conditions; in particular, the unidirectional efflux will depend on *trans* concentration.

14.1. *Flux Ratios*

Membrane transporters have been classified by the properties of unidirectional fluxes, as just described, but even more by their ratio. The *PNP* model just presented allows only one ratio of unidirectional fluxes, an ideal ratio quite independent of channel properties, as we have seen. However,

---

[y]It is interesting that the dependence of flux on *trans* concentration was observed in the output of a numerical integration of the *PNP* model long before its meaning, origin, or significance were understood. The numerical analysis of the complete model was needed to uncover this property of the system, although it could have been anticipated, had we been wise enough.

a wide range of flux ratios is found experimentally, and this discrepancy between experiment and theory has been viewed as fundamental, as a main justification of the usual state models of ion transport in membranes and channels.

Membrane transport has been divided into two classes, channel and mediated, ever since papers of Ussing[117] and, Hodgkin and Keynes (Refs. 123 and 124)[z] depending on the value of the flux ratio, for nearly fifty years. Channel transport is single-file, so the flux ratio is "less" than ideal. Mediated transport involves flux coupling and so is "more" than ideal. The cited references and Ref. 4 give precise definitions of experimental conditions and the words "less" and "more".

Classical active transport (e.g., the coupled movement of Ca^{++} and Na^+ ions,[133] , or the coupled movement of Na^+ and K^+ ions produced by the hydrolysis of ATP in the Na/K pump,[126,127,134,135] are examples of mediated transport. The sodium conductance (i.e., voltage activated sodium channel of nerve membrane,[1]) is an example of channel transport.

Any successful theory of transport needs to predict the flux ratios seen in experiment. And continuum theories have been (understandably) rejected in the literature (e.g., Ref. 1) because they predict only ideal flux ratios (see the extensive investigations of Bass' group[136–138]).

Chen and Eisenberg[5] show how a simple modification of the *PNP* model produces non-ideal flux ratios. If a phase boundary potential δ_j (a jump in electrochemical potential) is placed in series with the ends of a channel, the flux ratio is not ideal. In that case the integrated flux expression becomes

$$\overbrace{\tilde{J}_j}^{\text{Net Flux}} = \overbrace{\frac{\exp\left\{\left[-\delta_j(0) + \tilde{\mu}_j(0)\right]/RT\right\}}{\int_{x_1}^{x_2} \frac{\exp\{z_j F\Phi_j(x;C_j[x])/RT\}}{D_j(x)}dx}}^{\text{Unidirectional Efflux}} - \overbrace{\frac{\exp\left\{\left[-\delta_j(1) + \tilde{\mu}_j(1)\right]/RT\right\}}{\int_{x_1}^{x_2} \frac{\exp\{z_j F\Phi_j(x;C_j[x])/RT\}}{D_j(x)}dx}}^{\text{Unidirectional Influx}} \tag{21}$$

where $\delta_j(0)$ and $\delta_j(1)$ are phase boundary potentials thought to arise during the dehydration and resolvation process at the outside and inside of the

[z]More recent literature is reviewed in Refs. 1, 126 and 127.

channel, although the physical basis of the phase boundary potentials is not known and was not specified in Chen and Eisenberg's theory.
The flux ratio of the system with phase boundary potentials is then

$$\frac{\text{Unidirectional Efflux}}{\text{Unidirectional Influx}} = \exp\left\{\left[-\delta_j(0) - \delta_j(1) + \tilde{\mu}_j(0) - \tilde{\mu}_j(1)\right]/RT\right\}$$
$$= \textbf{Ideal Flux Ratio} \times \left[\exp\left\{\left[-\delta_j(0) - \delta_j(1)\right]/RT\right\}\right] . \qquad (22)$$

The flux ratio here can have any (positive real) value, depending on the difference in the values of $\delta_j(0)$ and $\delta_j(1)$. That is the main result of Chen and Eisenberg,[5] quite independent of the specific description of δ_j implemented there. Indeed, the underlying mechanism of the phase boundary potentials is not known or postulated in the theory, although some properties of the mechanism and potential are clear.

For example, phase boundary potentials of this sort disappear when flux disappears. Otherwise, current could flow in the absence of an overall driving force and thus violate thermostatics. Therefore, phase boundary potentials cannot be equilibrium phenomena independent of flux through the channel. However, nonequilibrium potential drops are well-known to most scientists, e.g., the ohmic potential drop across electrical resistors, and are found at many interfaces and phase boundaries where ions undergo large changes in state (e.g., hydration). For example, nonequilibrium potential drops are found at the interface of gases and solids,[23,139] and are described by an expression closely related to the description of $\delta_j(0)$, as predicted by Chen and Eisenberg. Nonequilibrium potential drops called over-potentials are found at electrodes when applied current drives them away from equilibrium. Over-potentials are described by the Butler–Volmer equation[60] which is nearly the same as the analogous expression of Chen and Eisenberg or of statistical rate theory.

Chen and Eisenberg elaborated their theory to allow the flux of one species of ion to be driven by a gradient of another. If the phase boundary potentials of the two species fall in a certain broad range, this cross-coupling is compatible with irreversible thermodynamics (because in that broad range the local total entropy production, i.e., local change in temperature, is positive). The flux ratio in that system is determined by asymmetry in the phase boundary potentials. The interaction of the different ions in the dehydration/resolvation step at the ends of the channel provides the driving force for the "active" transport.

In bulk solutions, it is well-known that one ion can drag another by electrophoretic and relaxation effects, allowing the electrochemical potential gradient of one ion can drive the movement of another ion. Flux coupling is a commonplace phenomena in ordinary ionic solutions. It is "... always apparent when pronounced differences exist between the mobility of similarly charged ions", (Ref. 55, p. 211). In fact, as previously mentioned, ions in bulk solution can in this way be induced to move against their own concentration gradient (Ref. 45, p. 377). Thus, one can easily imagine how a similar mechanism might apply at the selectivity filters at the ends of channels. There, one can imagine that the mobility of different ions is quite different (which is how selectivity might work); the electrophoretic and relaxation effects on different ions are different (with the channel protein taking the place of water; Nitzan and Ratner,[156]); and energy is transferred from the gradient of one ion to the movement of another. Despite these current speculations, I must clearly say that the physical basis of the coupling postulated by Chen and Eisenberg[5] is not known; an analysis with atomic resolution is dearly needed to see if the phenomena of mediated (non-ATP driven) transport might occur in a channel protein of one conformation.

14.2. *Pumping by Field Coupling*

The *PNP* model described up to now is what engineers call a two-terminal device, a structure in which the path for current flow has no branches. Branches are needed, however, to give many semiconductor devices (that use the same physical laws as the *PNP* theory) their useful functions, for example, field-effect transistors (FETs), bipolar transistors, indeed integrated circuits in general.[8,9,78,140] It is not surprising then to find that the *PNP* model is capable of rich behavior when it is extended to branched geometries.

In particular, the *PNP* model (of a channel of just one conformation) can also produce ATP driven active transport, say of Na^+ and K^+, if it is extended to the branched structures found in some channels.[141–145]

Imagine a **Y** shaped channel, with the top (i.e., the two-sided end) of the **Y** inside the cell and the bottom (i.e., one-sided end) outside. Imagine that an ATPase is present at the end of the channel in the upper left branch[aa] of the $^*\mathbf{Y}$ on the inside of the cell membrane. If that branch $^*\mathbf{Y}$ of the channel were anion selective, the phosphate ions produced by the ATPase would flow down into the channel, driven both by the concentration of phosphate (created by the ATPase by the hydrolysis of ATP) and by the local electric field. Imagine that the other upper branch of the channel (the right branch of the $\mathbf{Y}^*$) were nonselective. Then, the phosphate would come back out of the channel on the same (in)side it was created, moving through the nonselective upper right hand branch of the channel. If the bottom part of the channel (the vertical stroke of the $\underset{*}{\mathbf{Y}}$) were cation selective, the movement of the phosphate (in the upper right branch $\mathbf{Y}^*$, the nonselective branch of the channel) would move Na^+ and K^+ because phosphate would be the major determinant of the electric field within that part of the channel. In this way, the energy of ATP hydrolysis would drive the movement of cations without changes in the conformation of the protein. The electric field would couple the energy produced by the hydrolysis of ATP to the movement of cations. In particular, the electric field driving the movement of cations would be (mostly) determined by the phosphate concentration, in turn determined by the ATPase.

The Na/K pump, of course, moves Na^+ and K^+ in a definite ratio (Refs. 126, 127, 134, 135) and the model suggested so far would not produce a definite ratio of Na^+ efflux to K^+ influx. That would arise if phase boundary potentials δ_j were present as described previously in this paper (see Ref. 5 for details). The ratio of the δ_j's at the ends of the channel (here at the bottom of the vertical stroke of the $\underset{*}{\mathbf{Y}}$ and at the top of right hand stroke of the $\mathbf{Y}^*$) would determine the flux ratio. The flux ratio could have any value (in a wide range) including the values usually found in experiments.

In this way, a structure of one conformation could pump ions, with energy derived from ATP without changing conformation. The requisite change in shape would be in the electric field. Many of the conformations postulated in traditional analysis would correspond to different shapes in the electric field, as ATP is hydrolyzed, phosphate enters the channel, etc. The occlusion phenomena so extensively studied (e.g., Ref. 146) would represent changes in the shape of the electric field within the channel produced by experimental interventions. Nonelectrolytes could be moved by

[aa]Marked with an asterisk *

the mechanism of dielectrophoresis in which uncharged objects move in a nonuniform electric field.[30]

One interesting feature of this model is the location and properties of the ATPase itself. In this model there are no special demands on the ATPase. Any one that converts ATP to a high concentration of phosphate will do, and changes the local electric field in the channel. In particular, the ATPase need not have a complex structure capable of "swallowing" one ion, disgorging it in response to another ion, then turning to swallow the other ion, etc.

Models of the Na/K pump involving a complex cycle of conformation changes have always posed a certain evolutionary (and perhaps conceptual) difficulty. Because membranes are imperfect and leak Na^+, all cells have to pump Na^+ out of their cytoplasm. Otherwise, the cells will soon accumulate Na^+, and, even worse, water, thus increasing in volume until they burst. The incompressibility of water and the fragility of a membrane (only two lipid molecules thick) means that bursting occurs quickly if the Na/K pump is blocked, in seconds in the typical sized cell (say 10 μm in diameter). For a cell to exist, it needs an active transport system, a pump, usually a Na^+ pump.

It is not clear how the earliest cells in evolutionary history could have evolved an enzyme as complex as that postulated in traditional models of the Na/K ATPase. Yet the earliest cells must have had ionic pumps so they could exist long enough to reproduce. The *PNP* model of the Na/K pump avoids this issue. In it almost any (nonspecific) ATPase will do, as long as it creates a high local concentration of phosphate. The special requirements are in the channel and its boundary conditions, not the ATPase itself. Of course, this kind of evolutionary argument is more interesting than decisive. The enzymes involved in DNA replication perform complex essential tasks and must have existed early in the history of life. They may also undergo complex conformation changes as they function but it is easier to imagine[ab] how this might have happened than it is to imagine how an ATPase might have evolved the complex set of conformation changes (of the traditional state model of the Na^+/K^+ pump) before a membrane had evolved to define a cell. Clearly, the three dimensional structure of the Na^+/K^+

[ab]Of course, this lack of imagination may reflect my limitations more than those of the creator's.

ATPase must be known before speculations of the sort I have indulged in can become real science. Nonetheless, the properties of branched channels are worth pursuing, because they exist, because they are capable of rich behavior, and because they are used for that very purpose in analogous systems in semiconductors. Indeed, branched geometries form the solid state technology of contemporary electronics.

15. Gating in Channels of One Conformation

So far, I have only discussed properties of the open channel. Little has been said of how the channel opens and closes. And little can be said in general, if the channel opens and closes by a conformation change, that is to say, by a substantial movement of many atoms of the channel protein. Conformation changes cannot be understood without structural data (including measurements of structure as a function of time) and so they cannot be analyzed by *a priori* physical theory. They cannot be reduced to atomic biology, unless the laws of motion of proteins are known. However, not only are such laws not known, it is not clear that they exist in any form simpler than the molecular dynamics of all the atoms of the protein. Different conformation changes may require different descriptions. Macroscopic rules may not be able to describe even one conformation change, under a range of conditions let alone all the conformation changes of all proteins.

Open channels of one conformation have certain properties that hint of gating, however, and it is important to mention and discuss these, *albeit* in a speculative manner.[ac] Current flow through an open channel can be controlled in a way that does not involve a conformation change at all (a change in shape of the channel protein) but involves a change in the shape of the electric field within the protein's pore.[131]

If the shape of the electric field in a channel protein changes, so will the current flow through the open channel. If the shape of the electric field changes a little, the current flow can change a lot. For example, if a barrier height increases by a few kT, the current flow would decrease substantially and the channel would appear to close, at least no detectable current flows through it. As we have already discussed, the shape of the electric field

[ac]The speculation can be turned to facts only through the path of theory, confirmed by experiments. That theory must eventually be specific, starting with structure and ending with predictions of experimental data. We cannot do that yet.

depends on many variables, for example, the concentration of ions in the bath, trans-membrane potential, and the potential across the membrane far from the channels pore, or even away from the channel protein itself. Thus, changes in any of those variables could close or open the pore of a channel, just by changing the shape of the electric field, without changing the conformation of the channel protein, just as the potential on the gate of a field-effect transistor changes the shape of the electric field in its conducting channel and thus changes its conduction. In a similar way, gating of a channel could be produced by a change in membrane potential, or by the binding of an agonist to a nearby receptor (which changes the far-field), all without a change in the conformation of the protein (as that word is ordinarily understood). Of course, something is changing shape when the electric field changes shape, namely the distribution of charge and, if one wished, one could call this a conformation change. But that is not at all what enzyme kineticists have in mind when they speak of conformation change: they mean a change in shape of the protein, a geometrical or steric effect, not an electrical one. And changes in geometrical shape and changes in electric field are not equivalent, particularly when constructing theories. A conformation change of an electric field is much more tightly constrained by known physical laws than a conformation change in the steric sense. Changes in the electric field are described by specific known physical laws (the laws that relate potential to charge: in continuum models this is Poisson's equation; in discrete models, this is Coulomb's law summed over all charges) whereas a shape change in a protein cannot be described in anything like such a specific way. Calling a change in shape of an electric field a "conformation change" might make the idea more palatable by connecting it with an already accepted notion, but it would obscure the essential difference between two quite different mechanisms. I reserve "conformation change" for its original use.

The kind of gating just described, resulting from a change in shape of the electric field, but not a change in shape of the distribution of permanent charge, is well-known in semiconductor physics. In semiconductor devices, like transistors or thyristors, changes in current are produced by a change in the electric field.[6,7,97,120–122] They are never produced by a change in structure, that is to say, by a change in what we call the distribution of permanent charge or what semiconductor physicists call the doping function.

Changes in the electric field produce the enormous richness in behavior seen in semiconductor circuits. Indeed, a computer can be built entirely from (three terminal) devices made of semiconductors obeying (two dimensional) drift-diffusion equations, and so changes in the electric field can produce all the behavior a computer can produce, *including all the behavior that can be coded in a semiconductor memory.* In this way, a two dimensional *PNP* equation with complex distribution of permanent charge can reproduce all the information and behavior human beings can code into words or numbers! No wonder a general theory of the properties of the *PNP* equation (with arbitrary charge distribution) looks hard to derive.[6]

Channels are thought to be essentially one-dimensional devices, with the exception of the branched channels as we have discussed. Putting those aside for the moment, although they might have interesting gating behavior as we discuss later, the question arises whether simple unbranched one-dimensional channels of one conformation, described by the *PNP* model, are capable of the richness of behavior seen in semiconductor devices. In particular, can such simple structures gate the way thyristors do[8,9]? Can channels open and close (conduct or block current) as semiconductor devices do, simply by changing their electric field without changing their arrangement of permanent charge?

I would dearly love to answer these questions, but I cannot. We know that the *PNP* equations have this property[131] as will be discussed later in this paper, and we know that channel gating (or some types of channel gating) *might* work this way, but we do not know if it actually does. It seems likely that some of the phenomena of gating arise this way, either subconductance states, flickers, or the spontaneous or controlled opening of channels. But it is not evidence that force this conclusion, only intuition (e.g., prejudice) motivates it. Until we can construct a discrete *PNP* theory, in which charges are discrete but the potential profiles are computed (not assumed), we cannot tell whether charge rearrangement might produce some or all of the phenomena of gating seen in single channels.

Some general conclusions can be reached without a discrete theory. The sensitivity of the electric field in the channel to the contents of the channel's pore has some interesting consequences that will survive the details and analysis of any consistent theory. For example, it is clear that the contents of an open channel have something like the same total charge as the wall of the channel, otherwise the unneutralized charge would create large fields. These would either be large enough to destroy the three-dimensional structure of the channel protein (which is after all held together by much the same electrical forces that act between permeating ion and channel wall), or, in a less extreme case, they would create insurmountable barriers to current flow, potential barriers that would produce the closed channel seen experimentally, even though the pore remained physically open.

Indeed, the sensitivity of the electric field to channel contents is so great that one can reach a rather important general conclusion about single channel currents just from their existence, as rectangular events with one height (but variable duration). *Single channel currents of (much) the same amplitude probably arise from channels with (nearly) the same potential profile and thus with (much) the same contents.*

If the contents of the channel changed in any material way, the current flow would change much more. The change in contents would produce a change in the net charge of the channel protein (i.e., a change in the sum of the channel contents plus the charge on the channel wall) which would produce a change in the electric field. That change in field would produce a larger change in measured current because flux is a sensitive, usually exponential function of barrier height. The fact that single channel currents do not change from opening to opening (more than the noise in their measurement), nor do they drift during the duration of their opening, implies that the (time averaged) shape of the electric field and the (time averaged) contents of the channels do not drift during an opening, or from opening to opening. If the channel contents did change significantly, the electric field would change shape, most likely producing a large barrier somewhere. A barrier anywhere is enough to prevent conduction (since the channel is a series device). Thus, *any significant change in contents of a channel is likely to close it!*[ad]

15.1. *Gating by Field Switching*

An intriguing clue to a possible mechanism of gating is given by the mechanism of switching in a thyristor.[8,9] The thyristor is a semiconductor device with (typically) two positive and two negative regions of doping (what we

---

[ad] Some changes in contents would not change the shape of the electric field. Examples would include (1) the exchange of one Na^+ for another; (2) the exchange of one Na^+ for another isotope of Na^+; or (3) the exchange of one Na^+ for another ion with indistinguishable interactions with the channel protein, if such exists.

call permanent charge), more or less two transistors back to back, not quite a two terminal device. The thyristor functions by switching between conducting and non-conducting states. The thyristor has two different current-voltage relations, one conducting and the other non-conducting, namely isolating or insulating. Both current-voltage relations are stable, that is to say, once the device has switched from one state to the other, it can stay put (in conducting or non-conducting state) over a wide range of conditions.

It is intriguing to examine unbranched channels with charge distributions similar to those of a thyristor (Chen and Eisenberg, personal communication). It is a particularly seductive line of thought because Numa has suggested[147] that the classical voltage activated sodium channel also has a shockingly similar distribution with four maxima in its distribution of permanent charge, although evidently he was unaware of the analogy with thyristors. If the *PNP* model is used to compute the current-voltage relation of Numa's distribution of charge, the current-voltage relation is not unique.[131] There are (at least) two current-voltage relations, one with a tiny average slope (i.e., low conductance), and the other with a substantial slope, of something like 10 pS. These different states of the channel look quite like closed and open sodium channels.

The "closed" channel has a distribution of potential (computed from the Numa arrangement of permanent charge) that has much the same shape as the distribution of permanent charge, with four maxima. In the closed state, almost no mobile (ionic) charge is present within the channel; thus, the permanent charge is not neutralized and the potential barriers are large. Current cannot flow because the channel contains little mobile charge and that charge cannot cross the barriers. As one might expect, the current-voltage relation of the closed state has very small average slope (some femtosiemens or less), but the relation is quite nonlinear. If such a state exists, it might contribute to the so-called background (or seal or leakage) current found in patch clamp experiments. One might expect part of that *background* current to be removed by pharmacological agents (i.e., by channel blockers). Effects of this sort have in fact been observed in the closed porin channel and in background currents in preparations of potassium channels (Fred Quandt, personal communication).

The "open" channel computed from the Numa charge distribution has very different properties. Its distribution of potential is monotonic, showing no signs of the maxima of the permanent charge distribution, or of the closed channel, because the open channel contains substantial amounts of neutralizing charge, namely something like one or two ions (smeared into a continuum in the PNP model, regrettably but inevitably). This neutralizing charge screens the permanent charge of the channel, producing a potential profile without barriers. The resulting current-voltage curve is quite linear (because the potential profile is monotonic) and with relatively large slope, (i.e., conductance). The channel is filled with charges and those charges have no barriers to cross.

In this way, a single unbranched structure of permanent charge could account for both the open and closed channel properties of a single channel, without any change in conformation in the normal sense of the word. Of course, the electric field in such a model changes shape dramatically as the channel opens and it is that change which allows current through the channel. The charge that changes the shape of the electric field includes the charge in the channel, the charge induced in the channel protein, and the (nonlinear) induced charge at the boundaries of the system. The very first charge that enters the channel after a step change in voltage "fights its way" against the potential barriers of the closed channel. This takes a long time,[ae] as one might expect, but eventually enough charge enters the channel to screen and shield the permanent charge, lowering the potential barriers, allowing more current to flow, creating the open channel. The ON response (from an empty, unshielded, low conductance state, to a filled, shielded, high conductance state) is thus S-shaped, starting slowly; interestingly the OFF response (from a filled, shielded, high conductance open state to an empty, unshielded, low conductance closed state) is monotonic, starting very quickly. This time course is quite like current through an ensemble K channels of squid nerve, although the time scale we compute is too fast.

What is not clear is whether this kind of field switching actually occurs in biological channels, as well as in the *PNP* theory. If field-switching does occur, however, it could produce many of the phenomena of gating (e.g., the voltage or agonist driven time dependent gating analyzed by Hodgkin and Huxley,[83] or Katz,[91]) the opening and closing of single channels, or their flickering and switching into subconductance states, as Duan Pin Chen and I realized as soon as we constructed the *PNP* model.[131] But, to be fair, I

[ae]Microseconds in the simulations of Chen and Eisenberg (personal communication) of the time dependent *PNP* model containing univalent ions only.

must say that we do not know if such field-switching occurs at all in real biological channels, as appealing as the idea is, at least to me.

15.2. *Gating Current*

The changes in contents of channels just described are not electrically silent. If the charges inside a channel move, an exactly equal movement of charge occurs in the surrounding solution (according to Maxwell's equations) and the external charge movement can be collected in electrodes and recorded by electronics, if they are faithful enough. If a rectangular pulse of voltage is applied to such a channel, the charge flowing on the ON of the pulse will equal the charge flowing on the OFF, because the channel contents switch from one state to another and then back to the original state. A plot of this charge vs voltage will be nonlinear, because the dependence of channel contents on potential is nonlinear.

Nonlinear displacement current with these characteristics was postulated by Hodgkin and Huxley[83] and discovered by Schneider and Chandler,[147] Armstrong and Bezanilla,[148] and also Keynes and Rojas.[149] They attributed the current to the movement of a voltage sensor, presumably a part of the channel protein (more recent references in Hille,[1]). This kind of gating current might be called allosteric current or conformation current, because it results from the change in shape of the protein.

Changes in the contents of a channel might mimic some or all of the properties of an allosteric current, of nonlinear gating charge. The voltage sensor might be just the contents of the channel; in any case the contents of the channel are likely to be part of the voltage sensor. Clearly only a few ions can fit into a channel (at one time), and so one would expect a gating charge of only a few elementary charges in this model of gating current, as is found experimentally. This finding of small nonlinear gating charge in a channel protein is hard to understand if the current arises from a large scale conformation of the protein, as is commonly believed. The channel protein contains thousands of polar chemical bonds with partial charges of 0.1–0.6 elementary charges,[12] and so it is hard to see how a conformation change involving so many (electrically charged polar) chemical bonds can produce as **little** (nonlinear) gating charge as is seen experimentally.

On the other hand, a field-switching theory of gating current has serious difficulties dealing with some of the other properties observed experimentally. The slow time course of gating current might not be expected from a change in the contents of a channel. The time dependent Nernst–Planck equation changes contents in nsec, if it is analyzed assuming a constant electric field,[150–152] independent of trans-membrane potential and time. If the channel supplies the charge to modify the field, as in the time dependent *PNP* model, the channel contents change much more slowly, but still in micro-seconds at the longest (Chen and Eisenberg, personal communication[af]) compared to the milli-seconds seen experimentally.

The field-switching theory has at least one more difficulty. It implies that the shape of gating current should depend on the species of ion which enters and then exits the channel, thereby creating the (equal) ON and OFF charge movement. Different ions should penetrate the channel differently even if they do not permeate. Experiments, however, show much the same gating current no matter what impermeant ions are present.

In view of these significant problems, I believe that change in the charge within an unbranched channel cannot account for the gating current seen experimentally, probably because that kind of charge moves too quickly to be resolved experimentally.

15.3. *Gating in Branched Channels*

The discussion of gating in this paper, and all the literature I know, has assumed that the classical voltage sensitive channels have an unbranched structure. It is worthwhile speculating briefly how a branched, **Y** shaped channel might behave.[141–145] If one branch of the channel were unselective, say the left branch[ag] of the $^*\mathbf{Y}$, but separated by a high potential barrier from the rest of the channel, it would help control the potential in the main branch of the channel. The contents of the left branch in that case would be the gating charge that moves in response to changes in the trans-membrane potential, thereby controlling the field and permeation in the main branch of the channel. If the **Y** were sufficiently broad, the voltage sensor branch $^*\mathbf{Y}$ would sense the membrane potential, quite independently of the potential within the main branch $\underset{*}{\mathbf{Y}^*}$ of permeation. Thus, permeation would be

---

[af] Only univalent ions were present in the theory we calculated. It is possible that Ca^{++} ions would have as large an effect on the theory as they do on real channels and so the theory must be reformulated to include them and then recalculated.

[ag] Marked with an asterisk .

controlled by the membrane potential (chiefly) and not be sensitive to the current flow and potential within the permeating path itself. The fact that permeation through voltage sensitive channels is a function of the trans-membrane potential (mainly) and is quite insensitive to current flow through the channel has been known for some time[83,157] and was one of the motivations for the original design of the voltage clamp. Any satisfactory model of gating, with branched or unbranched channel, will predict gating phenomena quite insensitive to the amount of current flowing through the channel. But anything that changes the current flow through the channel is likely to change the shape of the electric field within the channel's pore. And anything that changes the shape of the electric field in the pore is likely to change gating that occurs there (e.g. gating by field switching in the pore). Thus, a successful theory of an unbranched channel will somehow isolate the properties of gating from the current flowing in the channel's pore, probably by making gating quite insensitive to the shape of the electric field within the pore.

A branched channel, as just described, would have properties quite similar to a field-effect transistor, in which (gate) potential (mainly) controls current flow. The branched channel described previously as a Na/K pump would have properties quite similar to a bipolar transistor, in which both (base) current and potential control current flow. It is appealing, but wildly speculative at this stage of knowledge, to imagine that evolution might have constructed two devices much like the two main types of transistors, using them to perform its functions as we use them to make our machines.

15.4. *Blocking*

The effects of ions within a channel are hardly a surprise to workers who have studied the blocking effect of slow ions (i.e., weakly permeable ions) on open channel currents (Hille,[1] summarizes this large area). They know that a channel can be closed (i.e., blocked) by what appears to be the entry of a single ion into the channel's pore. Perhaps the entering ion simply changes the electric field within the channel, without necessarily changing the shape of the protein (although of course that might happen as well). The ion need make a rather small change in the shape of the electric field to block current flow because the flow tends to be an exponential function of peak barrier height. A change of barrier height of 4.6 kT would change flux by a factor of 100 and produce a "blocked" channel.

In a way, what I suggest is simply a renaming of a "blocked channel" to a "channel with an altered electric field". The unspecified blocked state is now identified with a state of a different distribution of charge and thus potential. But more is involved than renaming. The distribution of charge and potential are governed by specific physical laws[ah] and so the new description can be specific, predictive, and falsifiable. It is also consistent. None of those features characterize state models of the usual type.

The entry of an additional ion, adding charge to the usual contents of the channel, clearly should change the potential profile substantially in any consistent model, likely producing a blocking event. But what if one type of ion exchanges with an ion of another type, say a monovalent cesium replaces a monovalent potassium? The charge does not change, so why should the field change?

Clearly if the ions are indistinguishable, the field will not change. If the channel is nonselective between cesium and potassium, it cannot tell them apart and the field will not change, nor will permeation, nor will the single channel current. The open channel current will be maintained and there will be no obvious experimental signature of the new contents of the channel.

But cesium and potassium are not the same. They have quite different atomic radii and so quite different electric field strengths at their edge where they interact with water and (say) carbonyls of the channel protein. Their interactions with other atoms are described by quite different "potentials" in the simulations of molecular dynamics and overall energies of interaction will differ by several kT.

Thus, the various properties of the permion would be expected to change. The diffusion constant, the potential barrier, the effective mass would all change. A *PNP* theory of this kind of blocking would require additional information describing the modified conformation of the channel, that might come from calculations of molecular dynamics. Analysis of this kind of blocking would require an explicit link between the hierarchies of models because the conformation of the channel protein is expected to be different

[ah]Poisson's equation and the Nernst–Planck equations in the *PNP* model; Langevin equation in the stochastic models; Coulomb's law plus the molecular dynamics of protein, water, and permeating and blocking ions in the highest resolution models.

when it is occupied by a cesium rather than a potassium. The movement of the atoms of the channel protein will change the distribution of permanent charge, even though no covalent bonds have been changed. The existing permanent charge will be in different places, in different densities (if one chooses the continuum point-of-view). Induced charge will be different as well. In this case, one would expect the channel protein to be in a different state, as it is usually said to be. That state can be specified in the physical language of atomic biology by theoretical or (better) experimental analysis. Structural determination by X-ray crystallography or NMR can (in principle) give all the atomic coordinates, but those techniques cannot yet be routinely applied to channels. Numerical simulations of molecular dynamics can give a theoretical prediction of the new structure (with cesium in the channel's pore) and that is what we will work with for now until structures can be experimentally known.

16. Back to the Hierarchy: Linking Levels

The consequences of contents/field coupling in the macroscopic *PNP* model are so profound that we have nearly lost sight of where we started earlier in this paper, in a hierarchy of models of which the *PNP* is the most lofty, but with lowest resolution and therefore most in the air, most remote from the solid foundation of molecular dynamics! The properties of that model are important and exciting, particularly as they reveal the physics needed in any consistent model at any level of resolution, but the *PNP* model needs to be related to models with higher resolution, those resting more solidly on the foundation of a hierarchy, models with atomic resolution, like molecular dynamics.

The proper way to proceed when linking levels of the hierarchy is to actually derive mathematically the dynamics, conservation laws, and boundary conditions, of the low resolution model from the high(er) resolution model. That derivation will show explicitly where the laws come from, what the functions and parameters mean, and how the boundary conditions relate to the experimental situation.

Such derivations are rarely possible. In the problems I have worked on, we could only do this in one case[44]: there, it took nearly eight years to derive the flux laws of the Nernst–Planck equation from the stochastic properties of the underlying diffusion trajectories. And there it was necessary to make assumptions about the electric field and consider only conditions in which the electric field does not change. So, it is not clear how long it will take to derive a coupled macroscopic model — a *PNP* equation — from a description of stochastic motion of individual ions, let alone from molecular dynamics in atomic detail.

Until then, it is difficult to discuss even the meaning (in terms of parameters of discrete high resolution models) of the effective parameters of the smoothed low resolution models. But at least one important thing can be said about one of the macroscopic variables of the *PNP* model. The effective concentration of an ionic species in a channel depends as much on the duration of time a channel is occupied by the ion as it does on the number of ions that can fit in (a given volume of) the channel.

16.1. *Occupancy*

A channel is not always occupied by the same ions. It may even spend a significant fraction of time occupied by no ions. The time that the channel is occupied by each ion (or a particular combination of ions) is variable, and dependent on all the parameters of the problem and so the effective concentration averaged over all times depends strongly on the duration of each occupancy state of the channel. Thus, a theory that links stochastic theory with the *PNP* model must determine the duration of each occupancy state, and how that duration varies with parameters and experimental conditions. Indeed, in a time dependent problem, the theory must predict the time dependence of the occupancy states as well, e.g., how they vary after a step in potential.

Some of this theory can be avoided if the single channel current represents the only conducting occupancy state, because, as argued previously, any other contents of the channel are likely to produce a large internal potential barrier that blocks current flow. If we accept that argument or better yet, if molecular dynamics or experimentation shows it to be correct the various statistics of the open state (e.g., the open probability, histograms of open times, and so on) directly estimate the properties of a single occupancy state, making the theory quite specific by dramatically reducing the number of unknown parameters. Furthermore, in this situation occupancy and gating become intimately linked, even identical (in some reduced cases). Changes in occupancy produce individual channel opening and closing; changes in occupancy are produced by random motions of ions over the pre-existing potential barrier (that are present just before the ion enters the channel and changes the occupancy state). The pre-existing potential barrier would be a function of trans-membrane potential, ionic conditions and so on, and in that way the qualitative properties of gating might arise as a result of changes in occupancy. Unfortunately, however, the measured statistics of the open channel are not in themselves enough to imply a theory of occupancy let alone gating. They do not say what is in the channel during its open time or how much of it is there, so a full analysis is not possible, without further experimentation to determine the number and type of ions n the channel, and, of course, a consistent theory of occupancy (i.e., a theory of ion entry, ionic movement within the channel, and ion exit, on one side or the other). The theory must predict (not assume) the dependence of occupancy on the various parameters of the system if it is to be useful.

Changes in occupancy may produce many of the more complex phenomena of channels: flickering and subconductance states, both are easy to explain as the consequence of a change in occupancy. Flickering would result from a channel occupied by an ion that raises potential barriers enough to cut off current flow. Subconductance states would come from channels occupied by an ion that hardly changes the potential profile, that is to say, that changes it just enough (say a few kT) to make a small (say 10%) change in conductance.

Occupancy affects each of the parameters of macroscopic smoothed theories like the *PNP* model. Each needs microscopic interpretation. Even parameters like the density of permanent charge are likely to depend on channel occupancy (Ref. 4: Discussion section) and general analysis is difficult because the various effective parameters (potential, friction, permanent charge, mobile charge, etc.) are not averaged the same way. They do not depend on occupancy in the same way. It seems unlikely that verbal analysis will suffice. A theory is needed.

The problems just described occur when we try to link the macroscopic with the microscopic, smoothed with stochastic. But similar problems arise whenever one level of the hierarchy is connected to another.

Every theory has its effective parameters. The two body "potential functions" of molecular dynamics are the output of quantum chemical considerations and calculations. The "potential functions" of the Langevin theory are the output of molecular dynamics and perhaps macroscopic theories as well (because of the long range nature of the electric field discussed previously). The input of one theory is the output of the other. Linking input and output is an important part of building a hierarchy extending from atomic structure to biological function. I only wish we knew how to do it.

So we proceed as best we can, inventing methods as we proceed, stumbling on generalizations by working on the particular. Peeking through the keyhole of an open channel, we gaze at the horizon of atomic biology in general.

16.2. *Is There a Theory?*

With all the complexity of the hierarchy and the difficulty of linking its levels, it is easy to become discouraged. After all, how can I be so sure there is a theory simpler than the molecular dynamics of all the atoms of the protein that describes the function of a channel?

The reason I am so sure is evolution. It is hard to believe that life could exist if its main molecular functions did not follow reproducible, reasonably general laws. An engineer requires his amplifier or logic element to follow simple deterministic rules under a range of conditions. Evolution requires its molecules to perform the same way. They function reproducibly under varying conditions to perform life's functions. The engineers' amplifiers are **internally** governed by a hierarchy of models at least as daunting as the hierarchy of channel models, but **externally** they follow the simple phenomenological rules of circuit elements. Otherwise, the engineer would not know what they would do next and he could not design a functioning system.

What are the biological equivalents of these linear amplifiers and nonlinear logic elements? What are the building blocks of channel function? What are the phenomenological rules that determine the biological function of a channel, or protein, for that matter? We do not know. Indeed, finding these rules is one of the main goals of modern biology. We can be confident a simple rule exists (for the natural function of a biological molecule), but we do not know at what level of the hierarchy it will emerge. The biological function might be the result of atomic interactions (e.g., selectivity) and so need to be described at the subatomic level of molecular orbitals. The function might be the result of twists in the permeation path, and so need to be described at the level of molecular dynamics. The function might be the result of interactions of individual ions and so need to be described at the level of stochastic dynamics. Or it might be the result of the macroscopic electric field and be at the smoothed level of continuum theories. We do not know at what level of the hierarchy to work until we find the simple phenomenological law that describes a natural function. But then it will involve a great deal of (seemingly unnecessary) work[ai] on "wrong" levels of the hierarchy. At what level will the adaptation be found? It is unfortunately necessary to examine every level of the hierarchy of models linking structure and function until we find where and how evolution created an adaptation to solve its problem, to enhance reproduction, to ensure fitness, or survival. If we examine only one level of the hierarchy, we will not be able to observe or resolve some adaptations and phenomena at all, i.e., adaptations that are best understood on other levels of the hierarchy, not clearly seen from the one level we are examining. For example, at the level of continuum models we cannot describe adaptations and phenomena that depend on discrete atoms; on the level of molecular dynamics we cannot compute, and can hardly describe or observe the macroscopic electric field. Only when we look at the appropriate level of the hierarchy (the one at which the adaptation occurs) can we expect simplicity. If we observe on only one level, and that happens to be a wrong one (a level that does not reveal and resolve the adaptation) then we will be confused. We will most likely find too many phenomena that we cannot quite explain or we can explain, but not uniquely. It will be easy to become confused, discouraged, or even worse to be misled that knowledge is greater than it really is.

For these reasons, work is needed on all levels of the hierarchy until we learn where each biological phenomena fits. On that level, each biological function will be best (and most simply) described. Different adaptations will be best described on different levels, but each will follow a simple law when described on its own level, I believe.

16.3. *Simplicity, Evolution, and Natural Function*

Simple laws can be found on some level of the hierarchy, of course, only if they exist. The argument that simple rules can be found is essentially evolutionary and depends on life's need for simple, reproducible reasonably general function in the face of a varying environment. That is a homeostatic argument on a molecular scale.

[ai]Perhaps practical discoveries will arise serendipitously from the "unnecessary" work, as they usually have in the past.

The argument does not apply, however, to all properties of channels or proteins, only to those that are directly involved in the natural function of the molecule, the function that evolution adapted and selected. Properties that are not natural functions may be complex special cases, quite changeable as the environment changes. An engineer can construct a limiter (i.e., threshold detector) from a linear amplifier by putting it in an "unnatural" state, by changing its connections and supply voltages. The jury-rigged limiter will work, but it is likely to have hysteresis and sensitivity to noise that make it hard to use and study. Devices designed as limiters in the first place will be easier to use and study.

For the same reason, the natural functions of a protein will probably be (much) easier to study than unnatural ones. Studying unnatural function is likely to be an unrewarding task. Studying voltage dependent gating in a channel that never sees a changing voltage in life (if such exists) is likely to reveal an experimental mess, a mass of hard-to-reproduce phenomena sensitive to the details of the experimental conditions and source of the preparation. Studying that channel is unlikely to reveal the general laws of voltage gated channels.[aj] I believe phenomenological laws (general rules of natural function) exist describing natural functions under the range of natural conditions, but they will not apply to every experimental condition we can create.

Seeking and understanding the general rules of natural function is the task of atomic and molecular biology. Hopefully, the reader will not be too discouraged to continue what we have barely begun. Our successors will be able, I trust, to compute the natural function of channel proteins from their structure, thus creating an atomic theory of ionic channels, a small, perhaps initial piece of the eventual atomic biology of proteins.

17. Afterword

Biology is a descriptive science because it studies the particular outcome of a chaotic process, and is essentially a historical science studying an evolving process at just one time. Molecular biology should be seen in that light, as (for the most part) the identification, description, manipulation and control of the proteins life uses. Most biologists are needed for those tasks. Their work will change our lives more than we can easily imagine, by improving

[aj] Unless the channel had evolved from (or into) a voltage dependent channel and has a vestigial (or nascent) voltage sensor and/or voltage controlled gate!

our technology in general and our health in particular. Their work will soon give us staggering control over the human life cycle.

This paper describes another kind of biology, the analytical kind, and arises from the reductionist tradition in which I was trained, by John Edsall and Andrew Huxley, Alan Hodgkin (mostly by example) and Julian Cole, among many others. I hope I have not misunderstood their lessons.

Acknowledgment

A hierarchy of models is much more than anyone can build or master. The builder responsible for each level of the hierarchy requires an expertise in the techniques, traditions, successes, and failures of the corresponding physical science. I have had the enjoyment and honor of working with many such builders, sharing the excitement of the hierarchy with them: Victor Barcilon, Ron Elber, Joe Jerome, Mark Ratner, Zeev Schuss, with the continuing help of Duan Pin Chen, who has worked with me on so many levels.

Victor Barcilon, Joe Blum, Tom DeCoursey, Truman Esmond, Lane Niles, and Mark Ratner were kind enough to make vigorous criticisms of the manuscript. I thank them for their interest, time, and help!

The steadfast encouragement and support of Andrew Thomson and the National Science Foundation made this work possible. To them I am most grateful.

This paper was written in 1994, mostly in January. It has not been significantly revised since then.

References

1. B. Hille, *Ionic Channels of Excitable Membranes*, 2nd edition, (Sinauer Associates Inc., MA, 1992) pp. 1–607.
2. R. S. Eisenberg, *J. Memb. Biol.* **115**, 1 (1990).
3. A. H. Soukhanov, *The American Heritage Dictionary of the English Language*, 3rd edition (Houghton Mifflin Co., New York, 1992).
4. D. P. Chen and R. S. Eisenberg, *Biophys. J.* **64**, 1405 (1993).
5. D. P. Chen and R. S. Eisenberg, *Biophys. J.* **65**, 727 (1993).
6. J. W. Jerome, *Mathematical Theory and Approximation of Semiconductor Models* (Springer-Verlag, New York, 1994).
7. M. Shur, *Physics of Semiconductor Devices* (Prentice Hall, New Jersey, 1990).
8. S. M. Sze, *Physics of Semiconductor Devices* (John Wiley & Sons, New York, 1981).

9. P. D. Taylor, *Thyristor Design and Realization* (John Wiley & Sons, New York, 1987).
10. W. Panofsky and M. Phillips, *Classical Electricity and Magnetism*, 2nd edition (Addison Wesley, 1962) pp. 1–494.
11. E. W. Hobson, *The Theory of Spherical and Ellipsoidal Harmonics* (Chelsea Publ. Co., New York, 1965).
12. G. E. Schultz and R. H. Schirmer, *Principles of Protein Structure* (Springer-Verlag, New York, 1979).
13. M. K. Gilson and B. Honig, in *Conformations and Forces in Protein Folding*, eds B. T. Nall and K. A. Dill (Am. Assoc. for the Advancement of Science, Washington, 1988) pp. 1–17.
14. T. E. Creighton, *Proteins: Structures and Molecular Properties* (W. H. Freeman & Co., New York, 1983)
15. A. Fersht, *Enzyme Structure and Mechanism* (W. H. Freeman & Co., New York, 1985).
16. C. Walsh, *Enzymatic Reaction Mechanisms* (W. H. Freeman Co., San Francisco, 1979).
17. T. L. Hill, *Cooperativity Theory in Biochemistry* (Springer-Verlag, New York, 1985).
18. T. L. Hill, *Free Energy Transduction in Biology* (Academic Press, New York, 1977).
19. T. L. Hill, *Statistical Mechanics* (Dover Publ., New York, 1977).
20. D. P. Chen, V. Barcilon and R. S. Eisenberg, *Biophys. J.* **61**, 1372 (1992).
21. A. J. Baid and L. R. Faulkner, *Electrochemical Methods* (John Wiley & Sons, New York, 1980).
22. C. A. Ward, *J. Chem. Phys.* **79**, 5605 (1983).
23. C. A. Ward and M. Elmoselhi, *Surface Science* **176**, 457 (1986).
24. J. D. Jackson, *Classical Electrodynamics*, 2nd edition (Wiley, New York, 1975).
25. J. A. Stratton, *Electromagnetic Theory* (McGraw-Hill, New York, 1941) p. 1–615.
26. H. Fröhlich, *Theory of Dielectrics*, 2nd edition (Oxford Univ. Press, London, 1958).
27. V. V. Daniel, *Dielectric Relaxation* (Academic Press, New York, 1967).
28. C. J. F. Böttcher, *Theory of Electric Polarization*, 2nd edition (Elsevier, New York, 1973).
29. P. Hedvig, *Dielectric Spectroscopy of Polymers* (John Wiley & Sons, New York, 1977).
30. H. A. Pohl, *Dielectrophoresis* (Cambridge Univ. Press, New York, 1978).
31. E. H. Grant, R. J. Sheppard and G. P. South, *Dielectric Behavior of Biological Molecules in Solution* (Oxford Univ. Press, London, 1978).
32. J. R. Macdonald, *Impedance Spectroscopy* (John Wiley & Sons, New York, 1987).

33. A. C. Eringen and G. A. Maugin, *Electrodynamics of Continua I. Foundations and Solid Media* (Springer-Verlag, New York, 1990).
34. G. D. Mahan and K. R. Subbaswamy, *Local Density Theory of Polarizability* (Plenum Press, New York, 1990).
35. V. Barcilon, D. Chen, R. Eisenberg and M. Ratner, *J. Chem. Phys.* **98** 1193 (1993).
36. E. M. Purcell, *Am. J. Phys.* **45**, 3 (1977).
37. H. C. Berg, *Random Walks in Biology* (Princeton Univ. Press, New Jersey, 1983).
38. D. R. Cox and H. D. Miller, *The Theory of Stochastic Processes* (Chapman and Hall, 1965).
39. C. W. Gardiner, *Handbook of Stochastic Methods: For Physics, Chemistry and the Natural Sciences* (Springer-Verlag, New York, 1985).
40. J. C. Crank, *The Mathematics of Diffusion* (Clarendon Press, New York, 1975).
41. E. Zauderer, *Partial Differential Equations of Applied Mathematics* (John Wiley & Sons, 1983).
42. R. Ghze, *A Primer of Diffusion Problems* (John Wiley & Sons, New York, 1988).
43. J. Kevorkian, *Partial Differential Equations* (Wadsworth & Brooks/Cole Advanced Books, California, 1990).
44. R. S. Eisenberg, M. M. Klosek and Z. Schuss, Stochastic trajectories of diffusion between two concentrations: A model of an open ionic channel (1994), submitted.
45. H. J. V. Tyrrell and K. R. Harris, in *Butterworths, Monographs in Chemistry* (Butterworths, Boston, 1984).
46. J. E. Mayer and M. G. Mayer, *Statistical Mechanics* (John Wiley & Sons, New York, 1940).
47. S. Chapman and T. G. Cowling, *The Mathematical Theory of Non-Uniform Gases*, 3rd edition (Cambridge Univ. Press, Cambridge, 1970).
48. H. L. Anderson and R. H. Wood, in *Water*, ed. F. Franks, (Plenum Press, New York, 1973).
49. F. Franks, *Water* (Plenum Press, New York, 1973).
50. H. S. Harned and B. B. Owen, *The Physical Chemistry of Electrolytic Solutions*, 3rd edition (Reinhold Publishing Corporation, New York, 1958).
51. R. A. Robinson and R. H. Stokes, *Electrolyte Solutions*, 2nd edition (Butterworths Scientific Publ., London, 1959).
52. H. L. Friedman, *Ionic Solution Theory* (Interscience Publ., New York, 1962).
53. B. E. Conway, J. O. Bockris and E. Yeager, *Comprehensive Treatise of Electrochemistry* (Plenum Press, New York, 1983).
54. W. C. Bray and F. L. Hunt, The conductance of aqueous solutions of sodium chloride, hydrochloric acid, and their mixtures, *J. Am. Chem. Soc.* **33**, 781 (1911).

55. G. Kortüm, *Treatise on Electrochemistry*, 2nd edition (Elsevier, New York, 1965).
56. J. O. Hirschfelder, and C. F. Curtis, *Molecular Theory of Gases and Liquids* (John Wiley & Sons, New York, 1954).
57. J.P. Hansen and I. R. McDonald, *Theory of Simple Liquids*, 2nd edition (Academic Press, New York, 1986).
58. C. G. Gray and K. E. Gubbins, *Theory of Molecular Fluids*, Vol. 1 (Clarendon Press, Oxford, 1984).
59. A. Nitzan, in *Evolution of Size Effects in Chemical Dynamics*, Part 2, eds. I. Prigogine and S. A. Rice (John Wiley & Sons, New York, 1988) pp. 489–555.
60. J. Bockris and A. Reddy, *Modern Electrochemistry* (Plenum Press, New York, 1970) pp. 1–1432.
61. J. N. Israelachvili, *Intermolecular and Surface Forces* (Academic Press, New York, 1985).
62. C. L. Brooks, M. Karplus and B. M. Pettitt, *Proteins: A Theoretical Perspective of Dynamics, Structure and Thermodynamics* (John Wiley & Sons, New York, 1988).
63. J. A. McCammon and S. C. Harvey, *Dynamics of Proteins and Nucleic Acids* (Cambridge, New York, 1987).
64. T. S. Kuhn, *The Structure of Scientific Revolutions*, 2nd edition (Univ. of Chicago Press, Chicago, 1970).
65. S. G. Brush, *The Kind of Motion We Call Heat* (North Holland, New York, 1986).
66. F. R. W. McCourt, J. J. M. Beenakker, W. E. Kohler and I. Kuscer, *Nonequilibrium Phenomena in Polyatomic Gases. Dilute Gases.*, Vol. 1 (Oxford Science Publi., 1990).
67. G. A. Martynov, *Fundamental Theory of Liquids* (Adam Hilger, New York, 1992).
68. E. R. Davidson, *Chem. Rev.* **93**, 2337 (1993).
69. C. E. Dykstra, *Chem. Rev.* **93**, 2339 (1993).
70. G. C. Maitland, M. Rigby, E. B. Smith and W. A. Wakeham, *Intermolecular Forces* (Oxford Science Publi., Oxford, 1987).
71. M. Rigby, E. B. Smith, W. A. Wakeham and G. C. Maitland, *The Forces Between Molecules* (Oxford Science Publ., London, 1986).
72. J. M. Haile, *Molecular Dynamics Simulation* (John Wiley & Sons, New York, 1992).
73. M. P. Allen and D. J. Tildesley, *Computer Simulation of Liquids* (Clarendon Press, Oxford, 1989).
74. H. Bai-Lin, *Chaos* (World Scientific, Singapore, 1984).
75. F. C. Moon, *Chaotic Vibrations* (John Wiley & Sons, New York, 1987).
76. S. K. Scott, *Chemical Chaos* (Clarendon Press, Oxford, 1991).
77. P. Brumer, in *Evolution of Size Effects in Chemical Dynamics. Part 1*, eds. I. Prigogine and S. A. Rice (John Wiley & Sons, New York, 1988) pp. 365-439.

78. P. Horowitz and W. Hill, *The Art of Electronics*, 2nd edition (Cambridge Univ. Press, Cambridge, 1989).
79. B. Sakmann and E. Neher, *Single Channel Recording* (Plenum Press, New York, 1983).
80. T. G. Smith, H. Lecar, S. J. Redman and P. W. Gage, *Voltage and Patch Clamping With Microelectrodes* (Am. Physiological Society, Maryland, 1985).
81. L. Kelvin, *Phil. Mag.* **11**, 146 (1856).
82. L. Kelvin, *Proc. R. Soc.* **7**, 382 (1855).
83. A. L. Hodgkin and A. F. Huxley, *J. Physiol.* (London) **117**, 500 (1952).
84. M. S. Ghausi and J. J. Kelly, *Introduction to Distributed-Parameter Networks* (Holt, Reinhart and Winston, New York, 1968).
85. R. W. P. King, *Transmission-Line Theory* (Dover Publications, New York, 1965).
86. J. J. B. Jack, D. Noble and R. W. Tsien, *Electric Current Flow in Excitable Cells* (Oxford, Clarendon Press, 1975).
87. L. Arnold, *Stochastic Differential Equations: Theory and Applications.* (John Wiley & Sons, New York, 1974).
88. Z. Schuss, *Theory and Applications of Stochastic Differential Equations* (John Wiley & Sons, New York, 1980).
89. T. C. Gard, *Introduction to Stochastic Differential Equations* (Marcel Dekker, Inc., New York, 1988).
90. J. L. Doob, *Stochastic Processes* (John Wiley & Sons, New York, 1953).
91. B. Katz, *The Release of Neural Transmitter Substances* (C. Thomas Publ. Co., Illinois, 1969).
92. P. Hänggi, P. Talkner and M. Borokovec, *Rev. of Modern Phys.* **62**, 251 (1990).
93. H. Risken, *The Fokker-Planck Equation: Methods of Solution and Applications* (Springer-Verlag, New York, 1984).
94. R. Elber, D. Chen, D. Rojewska and R. S. Eisenberg, Sodium in gramicidin: An example of a permion (1994), submitted.
95. C. Kittel, *Introduction to Solid State Physics*, 5th edition (John Wiley & Sons, New York, 1976).
96. K. Seeger, *Semiconductor Physics*, 5th. ed. (Springer-Verlag, New York, 1991).
97. N. K. Rogers, in *Prediction of Protein Structure and The Principles of Protein Conformation*, ed. G. D. Fasman, (Plenum Press, New York, 1990) pp. 359-389.
98. J. T. Hynes, in *Theory of Chemical Reactions*, ed. M. Baer (CRC Press, Florida, 1986).
99. J. I. Steinfeld, J. S. Francisco and W. L. Hase, *Chemical Kinetics and Dynamics* (Prentice Hall, New Jersey, 1989).

100. P. Billingsley, *Probability and Measure*, 2nd edition (John Wiley & Sons, New York, 1986).
101. K. Cooper, E. Jakobsson and P. Wolynes, *Prog. Biophys. Molec. Biol.* **46**, 51 (1985).
102. E. Jakobsson and S. W. Chiu, *Biophys. J.* **52**, 33 (1987).
103. E. Jakobsson and S. W. Chiu, *Biophys. J.* **54**, 751 (1988).
104. K. E. Cooper, P. Y. Gates and R. S. Eisenberg, *Quarterly Rev. of Biophys.* **21**, 331 (1988).
105. K. E. Cooper, P. Y. Gates and R. S. Eisenberg, *J. Membr. Biol.* **109**, 95 (1988).
106. S. W. Chiu and E. Jakobsson, *Biophys. J.* **55**, 147 (1989).
107. P. Y. Gates, K. E. Cooper and R. S. Eisenberg, *Ion Channels*, Vol. 2, ed. T. Narahashi (Plenum Press, New York, 1990) pp. 233–281.
108. T. Naeh, M. M. Klosek, B.J. Matkowsky and Z. Schuss, *Siam J. Appl. Math.* **50**, 595 (1990).
109. G. J. Szekely, *Paradoxes in Probability Theory and Mathematical Statistics* (D. Reidel Publ. Co., 1986)
110. J. P. Romano and A. F. Siegel, *Counterexamples in Probability and Statistics.*(Wadsworth & Brooks/Cole Advanced Books & Software, California, 1986).
111. M. H. Jacobs, *Diffusion Processes* (Springer-Verlag, New York, 1935).
112. S. Karlin and H. M. Taylor, *A First Course in Stochastic Processes* (Academic Press, New York, 1975).
113. H. R. Lerche, in *Lecture Notes in Statistics*, eds. D. Brillinger, S. Fienberg, J. Gani, J. Hartigan and K. Krickerberg (Springer-Verlag, New York, 1986).
114. A. Gut, *Stopped Random Walks* (Springer-Verlag, New York, 1988).
115. A. H. Hainsworth, R. A. Levis and R. S. Eisenberg, Origins of open-channel noise in the large potassium channel of sarcoplasmic reticulum (1994), submitted.
116. D. R. Cox, *Renewal Theory* (Chapman and Hall, 1962).
117. H. H. Ussing, *Acta Physiol. Scand.* **19**, 43 (1949).
118. E. B. Nauman and B. A. Buffman, *Mixing in Continuous Flow Systems* (John Wiley & Sons, New York, 1983).
119. W. Shockley, *Electrons and Holes in Semiconductors with Applications to Transistors Electronics* (D. Van Nostrand Co. Inc., New York, 1950).
120. M. S. Mock, *Analysis of Mathematical Models of Semiconductor Devices* (Boole Press, Dublin, 1983).
121. P. A. Markowich, *The Stationary Semiconductor Device Equations* (Springer-Verlag, New York, 1986).
122. K. Hess, *Advanced Theory of Semiconductor Devices* (Prentice Hall, New Jersey, 1988).
123. J. A. Jacquez, *Compartmental Analysis in Biology and Medicine* (Univ. of Michigan Press, Michigan, 1985).
124. A. L. Hodgkin and R. D. Keynes, *J. Physiol.* (London) 128, 28 (1955).

125. A. L. Hodgkin and R. D. Keynes, *J. Physiol.* (London) **128**, 61 (1955).
126. S. G. Schultz, *Basic Principles of Membrane Transport* (Cambridge Univ. Press, New York, 1980).
127. B. Hille, in *Textbook of Physiology,* Vol. 1, 21st edition, eds. H. D. Patton, A. F. Fuchs, B. Hille, A. M. Scher and R.D. Steiner, (WB Saunders Co., Philadelphia, 1989) pp. 24–47.
128. O. Sten-Knudsen, in *Membrane Transport in Biology: Concepts and Models,* Vol. 1, ed. D. C. Tosteson (Springer-Verlag, New York, 1978) pp. 5–113.
129. V. Barcilon, *SIAM J. Appl. Math.* **52** 1391 (1992).
130. V. Barcilon, D. P. Chen and R. S. Eisenberg, *SIAM J. Appl. Math.* **52** 1405 (1992).
131. D. P. Chen and R. S. Eisenberg, *J. Gen. Physiol.* **100**, 9a (1992).
132. A. L. Hodgkin, *Biol. Rev.* **26**, 339 (1951).
133. M. P. Blaustein, R. DiPolo and J. P. Reeves, *Sodium-Calcium Exchange,* Annals of the New York Academy of Sciences, Vol. 639, New York.
134. J. F. Hoffman and B. Forbush, in *Structure, Mechanism, and Function of the Na/K Pump,* Vol. 19 (Academic Press, New York, 1983).
135. P. Läuger, *Electrogenic Ion Pumps* (Sinauer Associates, Massachussetts, 1991).
136. L. Bass and A. J. Bracken, *Mathematical Biosci.* **66**, 87 (1983).
137. L. Bass, A. Brackens and J. Hilden, *J. Theor. Biol.* **118**, 327 (1986).
138. L. Bass and A. McNabb, *J. Theor. Biol.* **133**, 185 (1988).
139. C. A. Ward, *J. Chem. Phys.* **67**, 229 (1977).
140. D. J. Roulston, *Bipolar Semiconductor Devices* (McGraw-Hill Publ. Co., New York, 1990).
141. C. Miller, *J. Gen. Physiol.* **79**, 869 (1982).
142. H. Matsuda, *J. Physiol.* (London) **397**, 237 (1988).
143. E. A. Richard and C. Miller, *Science* **247**, 1208 (199).
144. M. S. Weiss, T. Wacker, J. Weckesser, W. Welte and G. E. Schulz, *FEBS Lett.* **267**, 268 (1990).
145. M. S. Weiss, U. Abele, J. Weckesser, W. Welte, E. Schiltz and G. E. Schulz, *Science* **254**, 1627 (1991).
146. B. Forbush, *J. Biol. Chem.* **262**, 11116 (1987).
147. M. Noda, S. Shimizu, T. Tanabe, T. Takai, T. Kayano, T. Ikeda, H. Takahashi, H. Nakayama, Y. Kanaoka, N. Minamino, K. Kangawa, H. Matsuo, M. A. Raftery, T. Hirose, S. Inayama, H. Hayashida, T. Miyata and S. Numa, *Nature* (London) **312**, 121 (1984).
148. C. M. Armstrong and F. Bezanilla, *J. Gen. Physiol.* **63** 533 (1974).
149. R. D. Keynes and E. Rojas, *J.Physiol.* **239**, 393 (1974).
150. H. Cohen and J. W. Cooley, *Biophys. J.* **5**, 145 (1965).
151. K. S. Cole, *Physiol. Rev.* **45**, 340 (1965).
152. K. S. Cole, *Membranes, Ions, and Impulses* (Univ. of California Press, Berkeley, 1968) p. 569.

153. K. S. Cole, *Four Lectures on Biophysics* (Univ. of Brazil, Rio de Janiero, 1947).
154. J. S. Rowlinson and F. L. Swinton, *Liquids and Liquid Mixtures* (Butterworth Scientific, New York, 1982).
155. L. Onsager and R. M. Fuoss, Irreversible processes in electrolytes, *J. Phys. Chem.* **36**, 2689 (1932).
156. A. Nitzan and M. A. Ratner, Conduction in polymers: dynamic disorder transport, *J. Phys. Chem.* **98**, 1765 (1994).
157. B. M. Deb, *The Force Concept in Chemistry* (Van Nostrand Reinhold, New York, 1981).
158. R. F. W. Bader, *Atoms in Molecules. A Quantum Theory* (Oxford, New York, 1990).
159. L. Szasz, *The Electronic Structure of Atoms* (John Wiley, New York, 1992).
160. R. G. Parr and W. Yang, *Density-Functional Theory of Atoms and Molecules* (Oxford, 1989).
161. P. Hohenberg and W. Kohn, Inhomogeneous electron gas, *Phys. Rev.* **136**, B864 (1964).
162. R. A. Kumpf and D. A. Dougherty, A mechanism for ion selectivity in potassium channels: computational studies of cation-p interactions, *Science* **261**, 1708 (1993).
163. W. G. Hoover, *Computational Statistical Mechanics* (Elsevier, New York, 1991).
164. A. Gray, *Lord Kelvin* (1908; reprinted 1973, Chelsea Publishing, New York).
165. C. Smith and M. N. Wise, *Energy and Empire. A Biographical Sketch of Lord Kelvin* (Cambridge Univ. Press, New York, 1989).
166. C. B. Boyer, *The History of Calculus and its Conceptual Development* (Dover Publications, New York, 1949).
167. F. Cajori, *A History of Mathematics*, 3rd edn. (Chelsea Publishing, New York, 1980).
168. J. Grabiner, *The Origins of Cauchy's Rigorous Calculus* (MIT Press, Cambridge, MA, 1981).
169. R. Czerminski and R. Elber, Reaction path study of conformational transitions in flexible systems: applications to peptides, *J. Chem. Phys.* **92**, 5580 (1990).
170. W. Nowak, R. Czerminski and R. Elber, Reaction path study of ligand diffusion in proteins: application of the self-avoiding walk (SPW) method to calculate reaction coordinates for the motion of CO through hemoglobin, *J. Amer. Chem. Soc.* **113**, 5627 (1991).
171. S. Karlin and H. M. Taylor, *A First Course in Stochastic Processes* (Academic Press, New York, 1975).
172. H. R. Lerche, *Boundary Crossing of Brownian Motion* (Springer-Verlag, New York, 1986).
173. A. Gut, *Stopped Random Walks* (Springer-Verlag, New York, 1988).
174. J. Kevorkian and J. D. Cole, *Perturbation Methods in Applied Mathematics* (Springer-Verlag, New York, 1981) pp. 1–558.

CHAPTER 6

THE STATISTICAL MECHANICAL BASIS OF SEQUENCE ALIGNMENT ALGORITHMS FOR PROTEIN STRUCTURE RECOGNITION

RICHARD A. GOLDSTEIN, ZAIDA A. LUTHEY-SCHULTEN
and PETER G. WOLYNES

School of Chemical Sciences, University of Illinois,
Urbana, IL 61801, USA

Contents

1. Introduction

Statistical mechanical considerations enter into protein structure prediction in two different ways. Protein folding can occur *in vitro*.[1] We can then imagine predicting protein structure through simulation of the folding event. This physical process resembles a phase transition in a finite system. Thus, the various *ab initio* methods of structure determination based on free energy minimization need to cope with the statistical-mechanical description of the underlying phase transition. Complementing these *ab initio* methods of structure prediction are a variety of approaches based on the empirical notion that similar sequences fold to similar structures. This maxim underlies methods based on sequence alignment and homology modeling[2,3] and is also ultimately at the heart of statistical analyses,[4–8] expert system methods[9–11] and neural-network based approaches.[12–17] These methods rely on the database of information adduced from X-ray diffraction and NMR experiments. These semi-empirical methods then present us with problems of statistical inference and information theory. The formal similarity of these two different approaches makes the use of many concepts from statistical mechanics fruitful. An example of this is how the theoretical methods originally developed to analyze the statistical mechanics of spin glasses has proven useful in analyzing both neural networks in general[18] and information-based approaches to protein structure prediction in particular.[19–22,53]

Mean-field theory has given important insight into the phase transition aspects of protein folding. These mean-field theories resemble those of first-order transitions like crystallization and exhibit all-or-none cooperativity. A significant conceptual development was made by Reva and Finkelstein, who showed how an appropriate mean-field theory was an intellectual basis of an iterative sequence-structure alignment scheme for optimally threading a sequence onto a template structure.[23] Their study used a single idealized lattice representation of a β-barrel as a template to illustrate the idea. Later, Bowie *et al.* introduced an energy-based approach for threading sequences onto templates that was usable for a variety of real protein structures.[24] Their scheme, based on a one-body potential that encoded amino acid preferences for each residue's environment, did not fully utilize the self-consistency central to the Reva-Finkelstein approach. Other approaches to energy functions have combined the simple one-body and multi-body potentials with these mean-field approaches in the attempt to recognize protein structures.[25–27,53]

The mean-field theories that lead to sequence-structure alignment algorithms to date have not focused on finding the basis for discriminating against alternative structures. A single template at a time is assumed, and the best alignment to this template is obtained. Presumably, the winner among several such templates is the best guess for a structure. This discrimination issue is the one which we have focused on previously in the context of free-energy minimization based approaches.[20–22] The crucial notion here is that if folding into the correct structure is prevented, a phase transition can still occur, but this transition resembles those seen in spin-glasses which have an extremely-rough many-dimensional free-energy surface. Finding the ground state for such a system is NP complete and requires a search that is exponentially difficult as in Levinthal's paradox. This spin-glass view of misfolding has been buttressed by numerous detailed calculations modelling proteins as completely random heteropolymers, which lack the features of cooperativity required for kinctically folding into organized structures.[28-31]

The spin-glass transition into a frozen misfolded state is characterized by a temperature T_g, the glass transition temperature. The [optimal energy function] for discriminating between folded and misfolded structures is one that maximizes the ratio of the folding to the glass transition temperature for most sequences. The success of this optimization scheme for associative memory Hamiltonians when predicted structures were found using free-energy minimization methods was illustrated in a previous paper.[20]

In this paper, we discuss the issue of the optimal energy function for sequence-structure alignment algorithms. Perhaps unsurprisingly, the optimal energy function for sequence alignment is closely related to that for free-energy minimization by Monte Carlo methods. The reason for this is that when a mean-field alignment is carried out with an incorrect template, the alignment algorithm is also approximating the solution to a free-energy minimization problem on a rough energy landscape. The probability distribution of deep incorrect minima can be calculated from spin-glass theory. The typical depth of a minimum is related to the energy at the glass-transition temperature. Maximizing the energy-gap between correct structures and incorrect structures therefore gives a criterion for optimal discrimination similar to the dynamic one. This result can be made more precise using a statistical analysis that takes into account the probability distribution of both correct minima and of incorrect template minima. The probability of a given alignment being correct varies as a Boltzmann distribution of the free-energies, at a temperature related to a spin-glass transition temperature for an underlying random energy landscape. This analysis also allows us to deal with one of the most problematic features of alignment algorithms, the assignment of gap penalties. Gap penalties arise from a chemical potential-like term determined by the logarithm of probability distribution of gaps in alignments against correct templates as compared to the probability distribution of gaps in alignments against random templates, as determined self-consistently. This contrasts with standard methods for determining gap penalties, which either involve adjusting these penalties in an ad-hoc manner,[32,33] or by looking at the statistical distribution of gaps in correct alignments only.[34,35]

We discuss the statistical-mechanical formulation in Sec. 2 of this paper. We first review how the Hamiltonian for molecular dynamics can be optimized based on concepts derived from spin-glass theory. We then develop a statistical approach to the optimization of the Hamiltonian for sequence-structure alignment, including the choice of gap penalties. Such an approach, while developed here for sequence-structure alignments, could be used for any alignment methodology, including more traditional sequence-sequence alignments. Following this, we develop the analysis for a particular form of Hamiltonian based on local interactions, and discuss the mean-field approach to alignment in the presence of multi-body terms. We finally present a variety of results from this method from the resulting algorithm, showing in a series of anecdotal and statistical studies how the predictions which are arrived at are extremely good for a set of known examples, with a fairly complete database of templates. Application of this method to proteins of unknown structure, such as sequences in the genetic databases that only have information about a local functional signature and the hormone-binding domain of a set of nuclear-receptor proteins, including the estrogen, progesterone, thyroid, and glucocorticoid receptors, is presented elsewhere.[36,53]

2. Theory: Optimal Hamiltonians for Molecular Dynamics

In contrast to other methods which attempt to find a "potential of mean force" to capture the salient aspects of the "true" Hamiltonian in terms of a reduced representation of the protein chain,[37–39] we have previously tried to engineer an energy function so as to stabilize the correct protein structure while avoiding the multiple-minima problem. The dynamics of the system involves a competition between the interactions favoring the correctly folded state, and other random, frustrated interactions. As discussed in earlier papers, the results of the dynamical simulation involves the competition between two possible transitions. When the native-like interactions dominate, there is a transition to the correctly-folded state at a temperature T_f. Conversely, when the random interactions dominate, the roughness of the resulting energy landscape causes a transition to a non-ergodic glassy state at a temperature T_g, from which the correctly folded state is kinetically inaccessible. The glass transition then represents the situation where the dynamics is dominated by the multiple-minima problem, suggesting that an appropriate measure of the ability of any particular form of the Hamiltonian to avoid the multiple-minima problem is the ratio T_f/T_g. By maximizing this ratio, it is possible to generate an optimal Hamiltonian of any given functional form. This analysis has been briefly described earlier[20–22]; we take the opportunity in this paper to present these ideas in more depth.

The estimates of T_f and T_g are based on a random energy approximation, where the thermodynamics are defined by the distribution of energy levels in that state, and correlations between different states are neglected.[40] In this model, there is a set of liquid-like states with a Gaussian energy distribution centered at $E = E_l$, with width Γ_l, so $P(E)\delta E$, the probability of a state having an energy between E and $E + \delta E$, is given by:

$$P(E)\delta E = \frac{1}{\sqrt{2\pi\Gamma_l{}^2}} \exp\left\{-\frac{(E-E_l)^2}{2\Gamma_l^2}\right\}\delta E\,. \tag{1}$$

If Ω_o is the number of conformational states of the polymer, then $\rho(E)$, the density of conformational states, is $\rho(E) = \Omega_o P(E)$, so the entropy S as a function of energy is:

$$S(E) = \log\Omega_o - \frac{(E-E_l)^2}{2\Gamma_l^2} - \log\left(\frac{\sqrt{2\pi}\Gamma_l}{\delta E}\right), \tag{2}$$

where δE is large relative to the spacings between energy levels but small relative to Γ_l. Using

$$\frac{1}{T} = \frac{\partial S}{\partial E} \tag{3}$$

yields

$$E - E_l = -\frac{\Gamma_l^2}{T}\,. \tag{4}$$

We can provide an appropriate value of δE in the thermodynamic limit by assuming $\delta E \sim N^{1/2}$, where N is the total number of residues. As Γ_l should also scale by the same factor, $\frac{\Gamma_l}{\delta E}$ becomes a constant. Since we are interested in the tails of the distribution, we can neglect $\log\left(\frac{\sqrt{2\pi}\Gamma_l}{\delta E}\right)$:

$$S = S_o - \frac{\Gamma_l^2}{2T^2} \tag{5}$$

where S_o, the configurational entropy of the chain, is given by $S_o = \log \Omega_o$. The free-energy, as a function of temperature, is then:

$$F(T) = E_l - \frac{\Gamma_l^2}{2T} - TS_o\,. \tag{6}$$

To a first approximation, we can neglect the entropy of the folded state so that its free energy is equal to its internal energy, E_f. Equating the free-energy of the folded and liquid-like states at the folding temperature, yields an expression for the folding temperature T_f:

$$T_f = \frac{\Delta E_{lf} + \sqrt{\Delta E_{lf}^2 - 2S_o\Gamma_l^2}}{2S_o} \tag{7}$$

where $\Delta E_{lf} = E_l - E_f$, the energy difference between the average of the liquid-like states and the folded state.

The glass transition occurs at the temperature where there are too few states available for interconversions between different minima to take place, and the system remains locked in one of the local minima. This is characterized by an entropy crisis, where $S(T_g) = 0$:

$$T_g = \frac{\Gamma_l}{\sqrt{2S_o}}\,. \tag{8}$$

Solving for T_f/T_g:

$$\frac{T_f}{T_g} = \sqrt{\Lambda} + \sqrt{\Lambda - 1} \tag{9}$$

where

$$\Lambda = \frac{1}{2S_o}\left(\frac{\Delta E_{lf}}{\Gamma_l}\right)^2 \tag{10}$$

which predicts, as mentioned in previous papers, that T_f/T_g is maximized when $\Delta E_{lf}/\Gamma_l$ is maximized.

In general, this presents a nonlinear optimization problem with the possibility of multiple maxima in parameter space, that can be solved using standard techniques. It is, however, possible to find a closed-form solution for any Hamiltonian that it can be expressed as a linear function of a set of parameters, $\{\gamma_i\}$. In this case, ΔE_{lf} is linear in γ_i while Γ_l^2 is quadratic in γ_i. For instance, if we can express the energy of protein μ in its native state as $E_T^\mu = \Sigma_i \xi_{Ti}^\mu \gamma_i$, and in misfolded state ζ as $E_\zeta^\mu = \Sigma_i \xi_{\zeta i}^\mu \gamma_i$ for some coefficients ξ_{Ti}^μ and $\xi_{\zeta i}^\mu$, ΔE_{lf} and Γ_l can be written in the form:

$$\Delta E_{lf} = \mathbf{A}\gamma \tag{11}$$

$$\Gamma_l^2 = \gamma\mathbf{B}\gamma \tag{12}$$

for vector **A** and matrix **B**, where

$$A_i = \xi_{Ti}^\mu - \left\langle \xi_{\zeta i}^\mu \right\rangle_\zeta \tag{13}$$

$$B_{ij} = \left\langle \xi_{\zeta i}^\mu \xi_{\zeta j}^\mu \right\rangle_\zeta - \left\langle \xi_{\zeta i}^\mu \right\rangle_\zeta \left\langle \xi_{\zeta j}^\mu \right\rangle_\zeta \tag{14}$$

and $\langle\rangle_\zeta$ indicates an average over misfolded states. Taking advantage of the fact that multiplying γ by any arbitrary constant does not change T_f/T_g, the value of γ that maximizes T_f/T_g is given by:

$$\gamma = \mathbf{B}^{-1}\mathbf{A}\,. \tag{15}$$

In practice, **B** and **A** are averaged over a set of target proteins of known structure, spanning a range of structural motifs. We have previously shown the ability of Hamiltonians optimized in this way to generalize, rather than simply memorize the properties of the target proteins, so that the Hamiltonian can be used to correctly select the appropriate configuration of a protein dissimilar in sequence and structure from any in the training set.[21]

In the most general form, the energy contributions of a particular type of interactions, such as a given pair of residues in contact with each other, are the adjustable parameters. The coefficient $\xi_{\zeta i}^\mu$, for instance, then represents the number of interactions of type i present when sequence μ is in configuration ζ. It is interesting to contrast the optimization formula of Eq. 15 with the more common potential of mean force approach of, for instance, Sippl.[38,39] With his approach, the energy contributions of particular interactions, the adjustable parameters γ_i^{mf}, are given by:

$$\gamma_i^{mf} = -T \log \left(\frac{\xi_{Ti}^\mu}{\left\langle \xi_{\zeta i}^\mu \right\rangle_\zeta} \right). \tag{16}$$

This assumes that there are no correlations between different interactions. Making the same assumption in the optimization approach, and also assuming no correlation exists between the interactions in the liquid-like states, allows us to write **B** as a diagonal matrix, simplifying Eq. (15) to:

$$\gamma_i = \frac{\xi^{\mu}_{Ti} - \left\langle \xi^{\mu}_{\zeta i} \right\rangle_{\zeta}}{\left\langle (\xi^{\mu}_{\zeta i})^2 \right\rangle_{\zeta} - \left\langle \xi^{\mu}_{\zeta i} \right\rangle^2_{\zeta}} . \tag{17}$$

Assuming a Poisson distribution of interactions of each type, $\left(\left\langle (\xi^{\mu}_{\zeta i})^2 \right\rangle_{\zeta} - \left\langle \xi^{\mu}_{\zeta i} \right\rangle^2_{\zeta} \right)$ is proportional to $\left\langle \xi^{\mu}_{\zeta i} \right\rangle_{\zeta}$, leading to

$$\gamma_i \propto \frac{\xi^{\mu}_{Ti}}{\left\langle \xi^{\mu}_{\zeta i} \right\rangle_{\zeta}} - 1 \tag{18}$$

which is proportional to γ_i^{mf} in Eq. (16) in the limit of $\xi^{\mu}_{Ti} / \left\langle \xi^{\mu}_{\zeta i} \right\rangle_{\zeta}$ close to 1.

3. Probabilistic Analysis of Alignment Search Algorithms

In general, energy-based protein prediction methodologies must address two separate problems. The first is the construction of an appropriate energy function, the second the search for either the global minimum or the lowest accessible minima of either the free energy or enthalpy. In previous papers, we have developed ideas of energy landscaping, of engineering an energy function to try to reduce the multiple minima problem, and then used either molecular dynamics with simulated annealing, or a screening method to evaluate the energy of particular structures. While molecular dynamics provides the most rigorous test of an energy-based scheme, and can be performed in reasonable amounts of computer time using optimized Hamiltonians, such approaches are not amenable to surveys of more than a small fraction of the known protein sequences.

One generalization of the screening method is to take advantage of the fact that there are both experimental and theoretical arguments indicating that the universe of possible protein structural motifs is somewhat limited.[24, 41–45] As graphically illustrated in Fig. 1, the space of folded protein configurations is considerably smaller than the space of all possible protein configurations. The structures of particular proteins whose structure has been solved by X-ray crystallography or multi-dimensional NMR

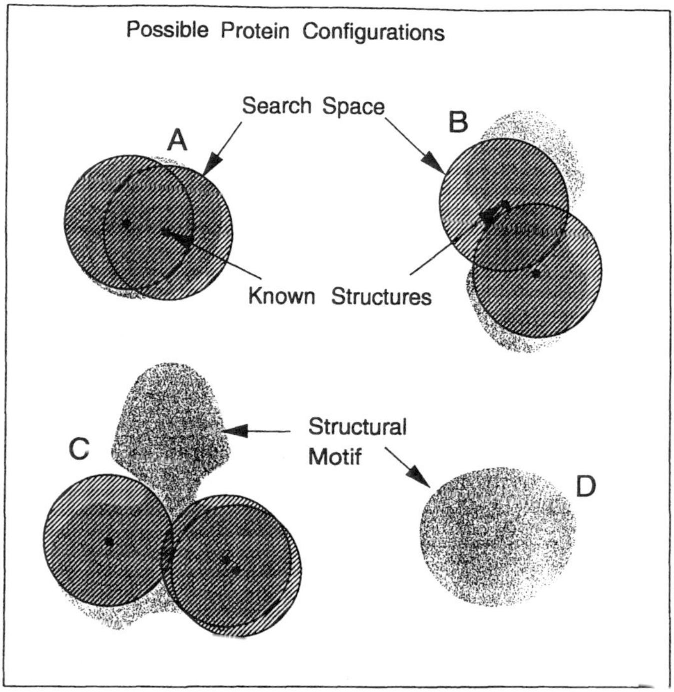

Fig. 1. Graphical representation of the search strategy described in the text. Instead of trying to find the minimum energy configuration in the space of all possible protein configurations, the search is instead focused on limiting the search to the possible structural motifs, represented in the figure as irregular shaded regions. These motifs are defind based on the set of protein structures determined experimentally using X-ray crystallography or multi-dimensional NMR. These individual protein structures are identified as points in the figure. Various alignments of the target protein to these known sample structures are generated with a gap penalty that penalizes the alignments by an amount representing how much the resulting structures differ from the known sample. This restricts the search of possible configurations to areas surrounding the sample, represented in the figure as the circular hatched regions. The choice of the optimal gap penalty can be calculated from statistical considerations using Eq. (33). Some structurel motifs, such as A in the figure, are well represented by the samples. Some structural motifs are probably not as well represented, such as B and C in the figure, while some existent structural motifs have to await observation, such as D.

exists as points in this configuration space. These points can also be considered as defining more generalized "structural motifs", subspaces of configurations that have certain similarities with each other and the known structure, shown as the shaded regions in Fig. 1. While some structural motifs such as that labeled **D** in the figure may be currently unobserved, at some future time it is likely that most of the points in the space of folded protein configurations will correspond to structural motifs associated with known structures. Even if this is not strictly true for all sequences, it might be true for a significant fraction of the biologically relevant proteins. In this case, the structure prediction problem can be posed as determining into what previously observed structural motif a particular protein sequence folds.

The perspective that comes from the finite nature of the set of possible structural motifs allows us to rephrase the structure prediction problem as the problem of choosing the predicted structure out of a set of possible structural motifs. We are then interested in making a decision about what the predicted structure is and in having some indication about the probable correctness of the prediction. The optimal Hamiltonian in this sense refers to the Hamiltonian that maximizes our chances of deciding correctly. As mentioned above, it is not *a priori* obvious that the optimal Hamiltonian for molecular-dynamics minimization is also optimal for this decision-based sequence-structure alignment algorithms. Here we use a probabilistic analysis to optimize the cost function for choosing between a set of possible structures. Following this, we extend this analysis to include alignment of the sequence to the structural motifs, including generating the optimal choice of gap penalties.

We start with a target protein whose structure is to be predicted, a particular Hamiltonian, and an ensemble of configurations $\{k\}$, some of which are correct, with correctness defined in some way such as some proportion of the pairwise distances between residues in the predicted structure and the actual structure matching within some tolerance. The sequence information of the target protein is combined with the Hamiltonian and the coordinates of the configurations to compute the energy of the target protein in each configuration. Because there is an ensemble of structures considered adequately accurate, there will be a corresponding distribution of energies of correct structures. We are interested in the the ratio of the conditional probability that a configuration with energy E_k is correct compared with the conditional probability that it is incorrect. We write these probabilities as $P(\mathcal{C}|E_k)$, and $P(\mathcal{R}|E_k)$, where $\mathcal{C}$ represents the correctness of the configuration, and $\mathcal{R}$ represents that the configuration is random and incorrect. As long as our ensemble is sufficiently large, we can express this ratio as the ratio of $\Omega(E_k, \mathcal{C})$, the number of configurations that are correct and have energy E_k, divided by $\Omega(E_k, \mathcal{R})$, the total number of incorrect configurations with energy E_k. Since the number of configurations should increase rapidly with the energy, it is easier to consider the log probabilities, and write:

$$\log \frac{P(\mathcal{C}|E)}{P(\mathcal{R}|E)} = \log \frac{\Omega(E,\mathcal{C})}{\Omega(E,\mathcal{R})} . \tag{19}$$

We are primarily interested in evaluating these probabilities to determine the correct structure, that is, when $E_k \sim E_c$, where E_c is the most probable energy for the correct structures.

Expanding around this value leads to:

$$\begin{aligned} \log \frac{P(\mathcal{C}|E_k)}{P(\mathcal{R}|E_k)} = \log \frac{\Omega(E_c,\mathcal{C})}{\Omega(E_c,\mathcal{R})} \\ +(E_k - E_c) \left.\frac{\partial \log \Omega(E,\mathcal{C})}{\partial E}\right|_{E=E_c} \\ -(E_k - E_c) \left.\frac{\partial \log \Omega(E,\mathcal{R})}{\partial E}\right|_{E=E_c} . \end{aligned} \tag{20}$$

Because E_c is a maximum of $\Omega(E, \mathcal{C})$, the first of these partial derivatives is zero. The second of these partial derivatives is just the reciprocal of the temperature for this probabilistic analysis, T_p. If we approximate the distribution of energies of the random configurations as a Gaussian centered at $E = E_r$ with standard deviation Γ_r, we can write:

$$\frac{P(\mathcal{C}|E_k)}{P(\mathcal{R}|E_k)} = \frac{\Omega(E_c,\mathcal{C})}{\Omega(E_c,\mathcal{R})} e^{-(E_k - E_c)/T_p} \tag{21}$$

where

$$T_p = \frac{\Gamma_r^2}{\Delta E_{rc}} \tag{22}$$

$$\Delta E_{rc} = E_r - E_c . \tag{23}$$

In the case of one true exactly-correct structure, we can solve for T_p in terms of the various quantities in the random-energy model. Assuming that the random states are equated with the liquid-like states, so that $\Delta E_{rc} = \Delta E_{lf}$ and $\Gamma_r = \Gamma_l$, substitution and simplification yields:

$$T_p = 2 \left(\frac{R}{1 + R^2} \right) T_g \tag{24}$$

where $R = T_f/T_g \gtrsim 1$, indicating that the temperature of the alignment is somewhat lower but approximately equal to the glass transition temperature.

The value of $\frac{\Omega(E_c,\mathcal{C})}{\Omega(E_c,\mathcal{R})}$ can be estimated by assuming that the energy of the correct configurations can be represented by a Gaussian centered at $E = E_c$ with width Γ_c:

$$\frac{\Omega(E_c,\mathcal{C})}{\Omega(E_c,\mathcal{R})} = \frac{\Gamma_r}{\Gamma_c}\frac{\Omega(\mathcal{C})}{\Omega(\mathcal{R})}e^{(E_r^2-E_c^2)/2\Gamma_r^2}, \tag{25}$$

where $\Omega(\mathcal{C})$ and $\Omega(\mathcal{R})$ represent the total number of correct and incorrect configurations respectively. It is interesting to note that, as long as the ratio of Γ_r to Γ_c is approximately constant, the probability that the correct structure has energy E less than this lowest energy random state is a monotonically increasing function of $\Delta E_{rc}/\Gamma_r$. The optimal Hamiltonian for use with the probabilistic analysis corresponds to the Hamiltonian optimized using spin glass theory presented in the previous section.

The above analysis is for the case where we have a set of possible structures, including one that is correct. In general, we do not start with such a set of possible structures, but rather start with a set of known protein structures, each a member of a known structural motif. We want to search through all of the possible configurations consistent with these structural motifs, while avoiding the space of configurations outside that defined either by the motif or by the space of possible folded configurations. Generally, this involves making a mapping between the residues in the target protein whose structure is to be determined, and in the known structure that defines that motif, allowing insertions and deletions in the sequence. The set of possible structures are then represented by the set of possible protein scaffolds combined with these alignments. Obviously some flexibility is required, yet allowing all possible insertions and deletions would explore a wider range of conformation space than is necessary to have a reasonable chance of including the correct conformation. Generally, some type of penalty is included in the cost function for the addition of gaps, restricting the configuration space explored in the search, as shown in Fig. 1, so that less consideration is given to configurations with unlikely combinations of insertions and deletions. This corresponds to the notion of importance sampling in Monte-Carlo simulations. Here we extend the above analysis to account for insertions and deletions.

A configuration is represented by a choice of scaffold protein, $\mathcal{S}_k$, and by the corresponding alignment of the target protein to this scaffold. We characterize the alignment by specifying the number and placement of gaps, where a gap is defined as any place in the alignment where, if target residue i is aligned to scaffold residue i', target residue $i+1$ is not aligned to scaffold residue $i'+1$. We allow for an increased vocabulary of gaps by not only allowing insertions or bulges, where target residue j is aligned to scaffold residue $i'+1$, with $j > i+1$, and deletions or skips, where target residue $i+1$ is aligned with scaffold residue $j', j' > i'+1$, but considering every possible alignment with target residue j aligned with scaffold residue j', with $j - i \geq 1$ and $j' - i' \geq 1$, with no intervening paired residues between (i, i') and (j, j'). Insertions and deletions are then the two limiting cases of these more-general gaps.

We categorize a gap by the number of residues in the target residue spanning the gap $(j - i)$ and the distance in the scaffold protein spanned by the gap $(r_{i'j'})$, combined with whatever other parameters we consider relevant. Relevant characterizations might include where in the scaffold protein the gaps occurred. One must bear in mind, however, that the finer the characteristics, the more difficult it is to ensure an adequate statistical sampling from a database of known structures. A gap of type 0 corresponds to no gap with $j = i + 1$ and $j' = i' + 1$. We can then make a list $\{G\}$ of all the gaps in the target sequence in order (e.g. $\{0, 0, 0, 3, 0, \ldots\}$ specifies that residue 4 starts a gap of category 3). Because a gap cannot start in a region that is already an inserted region, this list will contain $N - \sum_m (\lambda_m) n_m$ elements, where N is the total number of residues in the target protein, n_m is the number of gaps of category m, and λ_m is the number of residues inserted in a gap of category m.

We consider the conditional probability that this given configuration k, specified by $\mathcal{S}_k$ and $\{n_l^k\}$, with calculated energy E_k, is correct. The Hamiltonian has to include as part of its definition how to treat the energies of residues that are in insertions. In this work, we consider the energy of any residue in an insertion to have the average energy of that residue in the liquid-like states. We write this probability as $P(\mathcal{C} \mid E_k, \{n_l^k\}, \mathcal{S}_k)$. As in the previous section, this probability relative to the probability that the configuration is incorrect can be written in terms of $\Omega(E_k, \{n_l^k\}, \mathcal{S}_k, \mathcal{C})$, the number of configurations in the ensemble that are correct and have energy E_k and number of gaps $\{n_l^k\}$ using scaffold $\mathcal{S}_k$, divided by $\Omega(E_k, \{n_l^k\},$

$\mathcal{S}_k, \mathcal{R})$, the number of incorrect configurations based on scaffold $\mathcal{S}_k$ with energy E_k and number of gaps $\{n_l^k\}$. Again, we can expand $\log \Omega(E_k, \{n_l^k\}, \mathcal{S}_k, \mathcal{C})$ and $\log \Omega(E_k, \{n_l^k\}, \mathcal{S}_k, \mathcal{R})$ around the most probable energy for the correct configuration, E_c, and the most probable numbers of each type of gap, $\{n_l^c\}$: Expanding around these values leads to:

$$\begin{aligned}\log \frac{P(\mathcal{C}|E_k, \{n_l^k\}, \mathcal{S}_k)}{P(\mathcal{R}|E_k, \{n_l^k\}, \mathcal{S}_k)} &= \log \frac{\Omega(E_c, \{n_l^c\}, \mathcal{S}_k, \mathcal{C})}{\Omega(E_c, \{n_l^c\}, \mathcal{S}_k, \mathcal{R})} \\ &+ (E_k - E_c) \frac{\partial \log \Omega(E, \{n_l\}, \mathcal{S}, \mathcal{C})}{\partial E}\bigg|_{\substack{E=E_c \\ \{n_l\}=\{n_l^c\}}} \\ &- (E_k - E_c) \frac{\partial \log \Omega(E, \{n_l\}, \mathcal{S}, \mathcal{R})}{\partial E}\bigg|_{\substack{E=E_c \\ \{n_l\}=\{n_l^c\}}} \\ &+ \sum_l (n_l^k - n_l^c) \frac{\partial \log \Omega(E, \{n_l\}, \mathcal{S}, \mathcal{C})}{\partial n_l}\bigg|_{\substack{E=E_c \\ \{n_l\}=\{n_l^c\}}} \\ &- \sum_l (n_l^k - n_l^c) \frac{\partial \log \Omega(E, \{n_l\}, \mathcal{S}, \mathcal{R})}{\partial n_l}\bigg|_{\substack{E=E_c \\ \{n_l\}=\{n_l^c\}}} .\end{aligned} \tag{26}$$

Again, because $\Omega(E, \{n_l\}, \mathcal{S}, \mathcal{C})$ is a maximum at $E = E_c$ and $\{n_l^k\} = \{n_l^c\}$, its partial derivatives equal zero. $\frac{\partial \log \Omega(E, \{n_l\}, \mathcal{S})}{\partial E}$ is still $1/T_p$ as in the previous section. All that remains to be evaluated is $\frac{\partial \log \Omega(E, \{n_l\}, \mathcal{S}, \mathcal{R})}{\partial n_l}$, equal to μ_l/T_p, where μ_l is the chemical potential for a gap of type l.

In order to calculate the dependence of $\Omega(E, \{n_l\}, \mathcal{S}, \mathcal{R})$ on $\{n_l\}$, we need to calculate $P(\{n_l\}, \mathcal{R})$, the probability of a given set of $\{n_l\}$ in random alignments. Assuming that the gaps are independent of each other, this is equal to the product of the probabilities that that sequence of gaps and non-gaps exists in the random configurations, times all of the arrangements n_l gaps of type l and $N - \sum_l (\lambda_l + 1) n_l$ nongaps:

$$P(\{n_l\}, \mathcal{R}) = \frac{(N - \sum_l \lambda_l n_l)!}{(N - \sum_l (\lambda_l + 1) n_l)! \prod_l n_l!} p_{r_0}^{(N - \sum_l (\lambda_l + 1) n_l)} \prod_l p_{r_l}^{n_l} \tag{27}$$

where p_{r_l} is the probability of a gap of type l beginning at any location in the sequence, and $p_{r_0} = 1 - \sum_l p_{r_l}$. Multiplying by the total number of configurations, taking the logarithm of both sides, using Stirling's approximation for the factorials, and differentiating with respect to n_l yields:

$$\frac{\partial \log \Omega(E, \{n_l\}, \mathcal{S}, \mathcal{R})}{\partial n_l}$$
$$= \log \left\{ \frac{(N - \sum_m \lambda_m n_m)^{\lambda_l}}{n_l (N - \sum_m (\lambda_m + 1) n_m)^{\lambda_l + 1}} \right\} + \log \frac{p_{r_l}}{p_{r_0}^{\lambda_l + 1}} . \tag{28}$$

In order to express the first term in terms of the probabilities of gaps occurring in correct alignments, we can make a similar expansion for $\frac{\partial \log \Omega(E, \{n_l\}, \mathcal{S}, \mathcal{C})}{\partial n_l}$:

$$\frac{\partial \log \Omega(E, \{n_l\}, \mathcal{S}, \mathcal{C})}{\partial n_l}$$
$$= \log \left\{ \frac{(N - \sum_m \lambda_m n_m)^{\lambda_l}}{n_l (N - \sum_m (\lambda_m + 1) n_m)^{\lambda_l + 1}} \right\} + \log \frac{p_{c_l}}{p_{c_0}^{\lambda_l + 1}} \tag{29}$$

where p_{c_l} is the probability of a gap of type l beginning at any location in correct alignments, and $p_{c_0} = 1 - \sum_l p_{c_l}$. As this partial derivative is equal to zero at $\{n_l\} = \{n_l^c\}$,

$$\log \left\{ \frac{(N - \sum_m \lambda_m n_m^c)^{\lambda_l}}{n_l^c (N - \sum_m (\lambda_m + 1) n_m^c)^{\lambda_l + 1}} \right\} = \log \frac{p_{c_0}^{\lambda_l + 1}}{p_{c_l}} \tag{30}$$

which defines the normalization of the probabilities. Since Eq. (26) is evaluated at $\{n_l\} = \{n_l^c\}$, we can substitute the result from Eq. (30) into Eq. (29) to yield:

$$\frac{\partial \log \Omega(E, \{n_l\}, \mathcal{S}, \mathcal{R})}{\partial n_l} = \log \left\{ \frac{p_{r_l} p_{c_0}^{\lambda_l + 1}}{p_{r_0}^{\lambda_l + 1} p_{c_l}} \right\} \tag{31}$$

suggesting that when making and comparing alignments with the scaffold proteins, the quantity that should be minimized is the free energy F:

$$F = E + \sum_l n_l \mu_l \tag{32}$$

where, using the expression in (22) for T_p,

$$\mu_l = \frac{\Gamma_r^2}{\Delta E_{rc}} \log \left\{ \frac{p_{r_l} p_{c_0}^{\lambda_l + 1}}{p_{r_0}^{\lambda_l + 1} p_{c_l}} \right\} . \tag{33}$$

Equation (33) represents one of the principal results of this paper. This formula, given the assumptions made in its derivations, explicitly determines a unique choice of optimal gap penalties for any energy or cost-based alignment scheme. It is important to note that these gap penalties are a function of the distribution of gaps in random as well as correct alignments, and will be therefore Hamiltonian, or cost-function dependent.

The values of E_c and p_{c_l} can be evaluated by considering an ensemble of structurally-similar, correctly-aligned proteins. E_r, Γ_r, and p_{r_l} are evaluated in the region around E_c and p_{c_l}. We make the assumption that, as the region of correct alignments correspond to low values of F, the most accurate estimates of E_r, Γ_r, and p_{r_l} in the region of $E = E_c$ and $\{n_l\} = \{n_l^c\}$ correspond to values of these parameters obtained by looking at alignments corresponding to free-energy minima. Pairs of structurally dissimilar proteins are aligned in such a way as to minimize F. These alignments are used to calculate E_r, Γ_r, and p_{r_l}. As these values are themselves used in the calculation of the form of F, a self-consistent solution is searched for, where the free energy function used to find the minima is the same as that calculated using those minima.

In principle, the Hamiltonian energy parameters should also be determined in a self-consistent manner, by using minima in the alignments between random pairs of proteins to create the set of random structures, allowing for the possibilities of gaps. In practice, the values of these parameters optimized as described in the previous section perform well for the various test cases. Further refinement of these parameters are, however, currently in progress.[53]

4. Methodology

A Hamiltonian based on local interactions, similar in form to those developed earlier,[21] was used for this work. It can be expressed as:

$$\mathcal{H} = \sigma\Bigg[\sum_i \gamma_p(A_i, C_i) - \langle\gamma_p(A_i)\rangle_\zeta + \sum_\rho \sum_{i<j} \left\{\gamma_c^\rho(A_i, A_j)\, u(r_c^\rho - r_{ij}) - \langle\gamma_c^\rho(A_i, A_j)\rangle_\zeta\right\}\Bigg]. \quad (34)$$

$\gamma_p(A_i, C_i)$ is the energy contribution for amino acid A_i to be in context C_i, where context is described as the combination of secondary structure (α-helix, β-sheet, turn, or coil, as defined using the algorithm of Kabsch and Sander[46] and surface accessibility (less than or greater than 15% of the sidechain surface area exposed to solvent, normalized by the sidechain surface area of Gly-X-Gly tri-peptides as calculated by Shrake & Rupley,[47] calculated using the rolling-sphere method of Richards[48] (as implemented in *Midas*). $\gamma_c^\rho(A_i, A_j)$ is the energy term for amino acids A_i and A_j to be in contact, where contact is defined by the C^β distance between the residues being closer than distance r_c^ρ. These contact potentials could be, in

Table 1. Optimized values for the contact ($\gamma_c^\rho(A_i, A_j)$) and profile ($\gamma_p(A_i)$) parameters for the Hamiltonian described in Eq. (34). The profile term for any residue in a buried coil is ser equal to zero. The Hamiltonian was optimized for a set of 42 sample proteins of a variety of structural motifs, using Eq. (15) as described in the text.

						$\gamma_c^\rho(A_i, A_j)$					
						$r_c = 5$ Å					
	ILE	PHE	VAL	LEU	TRP	MET	ALA	GLY	CYS	TYR	
	− 1.7	− 1.3	− 1.1	− 1.3	− 1.1	− 1.0	0.4	− 0.7	0.1	0.2	ILE
		2.7	− 0.5	− 0.6	1.8	− 0.7	0.7	− 0.7	− 0.9	0.7	PHE
ILE	1.0		− 0.9	0.1	0.6	− 0.2	1.5	− 0.1	− 1.6	− 0.8	VAL
PHE	1.0	1.2		0.2	− 0.3	0.1	− 0.1	0.0	− 0.8	− 0.8	LEU
VAL	0.7	0.7	0.4		− 0.3	1.1	− 0.4	− 0.7	2.8	− 1.1	TRP
LEU	0.9	0.9	0.5	0.8		3.4	− 0.6	− 0.4	2.7	− 0.8	MET
TRP	0.8	0.3	0.2	0.4	1.2		0.7	0.5	− 0.8	0.8	ALA
MET	0.6	0.6	0.3	0.3	0.7	0.1		0.2	0.1	0.2	GLY
ALA	0.2	0.1	0.1	0.2	0.0	0.2	0.2		0.0	1.0	CYS
GLY	− 0.2	− 0.2	− 0.1	− 0.1	0.1	− 0.1	− 0.1	0.2		1.0	TYR
CYS	0.9	1.0	0.8	0.4	0.4	0.4	0.1	0.1	1.1		PRO
TYR	0.2	0.4	0.1	0.3	0.9	0.2	− 0.1	0.0	0.4	0.2	THR
PRO	− 0.5	− 0.3	− 0.2	0.0	0.1	0.0	− 0.1	0.1	− 0.2	− 0.2	SER
THR	0.0	0.0	0.0	− 0.1	− 0.2	− 0.1	− 0.1	0.0	0.2	− 0.2	HIS
SER	− 0.3	− 0.2	− 0.1	− 0.2	− 0.1	0.2	0.0	− 0.1	− 0.1	0.1	GLU
HIS	0.2	0.4	0.1	0.1	− 0.2	0.8	− 0.3	0.2	0.2	0.4	ASN
GLU	− 0.2	− 0.2	− 0.1	− 0.4	− 0.1	− 0.4	− 0.1	0.0	0.1	0.2	GLN
ASN	− 0.2	− 0.3	− 0.1	− 0.2	0.1	0.0	− 0.2	− 0.1	0.1	− 0.2	ASP
GLN	− 0.3	− 0.3	− 0.2	− 0.2	− 0.5	0.0	0.1	− 0.1	− 0.2	0.1	LYS
ASP	− 0.2	− 0.5	− 0.3	− 0.3	− 0.2	− 0.4	− 0.1	0.1	− 0.3	− 0.1	ARG
LYS	− 0.2	− 0.3	− 0.1	− 0.1	− 0.1	− 0.3	− 0.1	0.2	− 0.2	0.1	
ARG	− 0.2	− 0.1	− 0.1	− 0.1	− 0.2	0.2	− 0.1	0.1	− 0.4	− 0.3	
	ILE	PHE	VAL	LEU	TRP	MET	ALA	GLY	CYS	TYR	

Table 1. (Continued)

	PRO	THR	SER	HIS	GLU	ASN	GLN	ASP	LYS	ARG	
	– 1.1	– 1.0	– 0.3	0.4	– 0.4	– 1.0	– 1.3	– 0.6	– 1.0	– 0.8	ILE
	– 0.2	– 0.4	– 0.3	1.3	– 0.4	– 0.5	– 0.4	– 0.8	– 1.0	– 0.3	PHE
ILE	– 1.0	– 0.8	– 0.1	– 0.6	– 0.5	– 0.3	– 0.1	– 0.2	– 0.5	– 0.5	VAL
PHE	– 0.6	– 0.9	– 0.6	0.3	– 0.6	– 0.3	– 0.6	– 1.0	– 0.6	– 1.0	LEU
VAL	0.4	– 1.0	0.6	– 2.8	– 0.5	– 1.5	– 1.0	– 2.0	0.7	– 0.3	TRP
LEU	– 0.1	– 0.3	– 0.7	3.4	0.2	– 0.5	– 1.2	– 0.1	– 0.8	– 1.4	MET
TRP	0.5	0.7	0.7	0.2	0.1	0.6	0.4	0.3	0.2	0.3	ALA
MET	0.3	0.2	0.6	0.6	0.0	0.2	– 0.8	0.7	– 0.5	0.1	GLY
ALA	– 1.0	0.1	– 0.1	3.2	– 0.1	0.1	– 0.4	– 0.6	– 1.3	– 1.4	CYS
GLY	0.0	– 0.5	– 0.2	– 1.0	– 0.4	0.4	0.1	– 0.8	– 0.5	– 0.6	TYR
CYS	0.3	0.1	0.1	– 1.6	– 0.6	0.2	0.2	– 0.4	– 0.4	0.2	PRO
TYR		– 0.3	0.6	0.0	0.2	0.4	0.1	0.3	– 0.3	– 0.8	THR
PRO	0.4		1.3	0.7	0.4	1.4	– 0.1	1.9	0.5	– 0.2	SER
THR	– 0.1	0.1		0.4	0.3	– 0.4	– 0.3	– 1.1	– 0.3	– 0.1	HIS
SER	0.0	0.0	– 0.1		– 0.6	0.1	– 0.1	0.1	0.5	– 0.1	GLU
HIS	0.1	0.1	– 0.3	1.1		0.2	0.5	0.3	– 0.1	– 0.2	ASN
GLU	0.2	0.0	– 0.2	0.1	0.1		– 0.4	– 0.8	0.0	0.4	GLN
ASN	0.1	0.1	0.0	0.2	0.0	0.0		– 0.2	0.4	0.6	ASp
GLN	– 0.1	0.1	0.2	– 0.2	0.2	– 0.1	0.3		– 0.3	– 0.3	LYS
ASP	0.0	0.1	– 0.1	0.3	0.1	0.1	0.3	0.3		– 0.5	ARC
LYS	0.2	0.0	0.1	– 0.3	0.1	0.0	0.0	0.2	– 0.1		
ARG	0.0	0.2	0.1	– 0.1	0.2	0.2	0.0	0.1	– 0.1	0.1	
	PRO	THR	SER	HIS	GLU	ASN	GLN	ASP	LYS	ARG	

$r_c = 12$ Å

principle, dependent upon the context of the residues, C_i and C_j. In these calculations a summation was made over two values of r_c^ρ, with $r_c^1 = 5$ Å and $r_c^2 = 12$ Å. All energetic contributions are relative to the average contribution for the corresponding residues in the liquid-like states, so that $E_l = 0$. The alignments resulting from use of this methodology are independent of any scaling of the Hamiltonian. In order to ensure self-averaging of the various distributions and averages, the scale factor σ was chosen so that Γ_l, a characteristic energy parameter, is equal to one.

The local interaction Hamiltonian of Eq. (34) was optimized for the same set of 42 example proteins used by Goldstein *et al.*[21] selected from the PDB data base,[49,50] representing a range of tertiary folds, including

Table 1. (Continued)

	$\gamma_P(A_i, C_i)$						
	Buried			Surface			
Residue	**Helix**	**Sheet**	**Turn**	**Helix**	**Sheet**	**Turn**	**Coil**
ILE	0.7	0.6	− 0.1	− 0.3	− 0.4	− 0.2	0.1
PHE	0.2	0.1	0.1	− 1.1	0.1	− 0.3	0.2
VAL	0.2	0.9	− 0.5	0.0	− 0.2	− 0.4	− 0.2
LEU	0.4	0.2	0.1	− 0.3	− 0.4	− 0.2	− 0.2
TRP	− 1.2	− 0.6	− 0.2	− 1.5	− 1.4	− 0.9	− 1.5
MET	1.4	− 0.4	− 0.4	− 0.7	− 0.2	− 0.6	0.2
ALA	0.5	0.0	− 0.6	0.3	− 0.2	− 0.3	− 0.2
GLY	− 0.5	− 0.2	0.8	− 0.5	− 0.2	0.8	0.0
CYS	− 1.6	− 1.3	− 1.1	− 0.3	− 1.1	− 0.44	0.0
TYR	− 0.1	0.8	− 0.2	− 0.2	0.2	− 0.3	− 0.2
PRO	− 1.8	− 1.7	− 1.4	− 1.9	− 1.9	− 0.7	− 0.5
THR	0.2	0.3	− 0.2	0.8	2.0	0.3	1.0
SER	0.4	− 0.2	− 0.4	0.5	− 0.4	0.4	0.5
HIS	− 0.2	− 0.4	− 0.5	− 0.7	0.2	− 0.3	0.0
GLU	0.2	0.3	− 0.6	1.3	0.9	0.7	0.5
ASN	− 0.4	− 0.3	− 0.8	− 0.3	0.0	0.6	0.4
GLN	0.1	0.0	− 0.3	0.9	0.9	0.6	0.7
ASP	− 0.1	0.3	− 0.1	1.0	0.4	1.0	0.9
LYS	− 0.3	0.0	0.4	1.4	1.4	1.1	1.2
ARG	0.1	0.5	0.1	1.9	1.1	0.7	1.7

α-helical proteins (e.g. sperm whale myoglobin, 5MBN), β-sheet proteins (e.g. mouse Fab fragment heavy chain, 3HFM(H)), and mixed proteins (e.g. chicken dehydrofolate reductase, 8DFR). We again used the X-ray coordinates of a set of proteins between 10 and 50 percent larger than the training protein, including all possible translations along the sequence, to model the liquid-like states.[20,21,51] **A** and **B** were averaged over the set of training proteins, and used to calculate optimal γ's, presented in Table 1.

The covalent nature of the disulfide bonds was incompatible with the mean-field alignment algorithm using the local interaction Hamiltonian, in that the alignment procedure could cluster more than two cysteines in close proximity, resulting in excessive stability of those structures. For this reason, the interaction between cysteines closer than 5 Å was set to zero. Information about cysteine bonds could be added to the Hamiltonian as independent distance constraints, or used *a posteriori* as a check of the final predicted structure.

Gaps were divided into three different categories, examples of which are shown in Fig. 2: insertions ($j-i > 1$ with $r_{i'j'} = 3.8$ Å, the distance between successive C_α atoms, illustrated by example **A** in Fig. 2), deletions ($j-i = 1$ with $r_{i'j'} > 3.8$ Å, illustrated by example **B**), and more generalized gaps (($j - i > 1$ and $r_{i'j'} > 3.8$ Å, illustrated by example **C**). These three

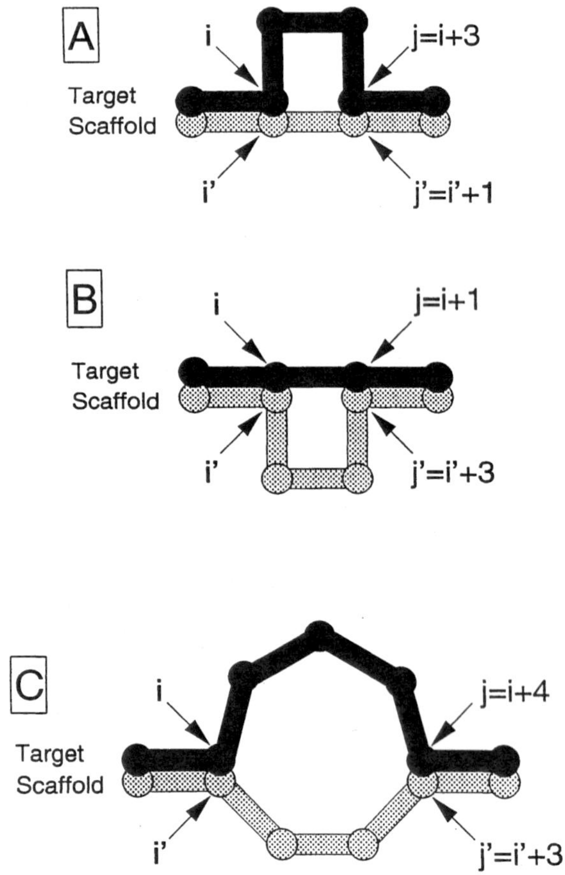

Fig. 2. Examples of the three categories of gaps, including: (A) Insertions of target residues where, if residue i on the target protein is pinned to residue i' on the scaffold protein, target residue j, the next pinned target residue, is pinned to scaffold residue j', with $j > i + 1$ but $j' = i' + 1$, (B) deletions of target residues, where $j' > i' + 1$ but $j = i + 1$, (C) generalized gaps, where both $j > i + 1$ and $j' > i' + 1$.

categories were treated independently, with simple functional forms used to describe their respective statistics:

$$p_l = \begin{cases} A_1 \exp\left(-k_1(j-i)\right) & j-i>1, r_{i'j'}=3.8 \\ A_2 \exp\left(\frac{(r-r_0)^2}{2D_2^2}\right) & j-i=1, r_{i'j'}>3.8 \\ A_3\{D_3(j-i)\}^{-3/2} r^2 & j-i>1, r_{i'j'}>3.8 \\ \times \exp\left(-k_3(j-i)+\frac{r^2}{D_3(j-i)}\right) & \end{cases} . \tag{35}$$

The set $\{A_{1r}, A_{2r}, A_{3r}, k_{1r}, k_{3r}, r_{0r}, D_{2r}, D_{3r}\}$ describes the distribution of gaps in alignments of random pairs of proteins, while $\{A_{1c}, A_{2c}, A_{3c}, k_{1c}, k_{3c}, r_{0c}, D_{2c}, D_{3c}\}$ describes the corresponding distribution in homologous pairs. These distributions resulted in gap penalties of the form:

$$\mu_l = \frac{\Gamma_r^2}{\Delta E_{rc}} \times \begin{cases} C_{11} + C_{21}(j-i) & j-i>1, r_{i'j'}=3.8 \\ C_{12} + C_{22} r_{i'j'} + C_{32} r_{i'j'}^2 & j-i=1, r_{i'j'}>3.8 \\ C_{13} + C_{23}(j-i) + C_{33} r_{i'j'}^2/(j-i) & j-i>1, r_{i'j'}>3.8 \end{cases} \tag{36}$$

where:

$$C_{11} = \log\left(\frac{A_{1r}}{A_{1c}}\right) - \log\left(\frac{p_{r0}}{p_{c0}}\right) \tag{37}$$

$$C_{21} = -(k_{1r} - k_{1c}) \tag{38}$$

$$C_{12} = \log\left(\frac{A_{2r}}{A_{2c}}\right) - \left(\frac{{r_{0r}}^2}{D_{2r}} - \frac{{r_{0c}}^2}{2D_{2c}}\right) - \log\left(\frac{p_{r0}}{p_{c0}}\right) \tag{39}$$

$$C_{22} = \frac{r_{0r}}{{D_{2r}}^2} - \frac{r_{0c}}{{D_{2c}}^2} \tag{40}$$

$$C_{32} = -\left(\frac{1}{{D_{2r}}^2} - \frac{1}{{D_{2c}}^2}\right) \tag{41}$$

$$C_{13} = \log\left(\frac{A_{3r}}{A_{3c}}\right) - \frac{3}{2}\log\left(\frac{D_{3r}}{D_{3c}}\right) - \log\left(\frac{p_{r0}}{p_{c0}}\right) \tag{42}$$

$$C_{23} = -(k_{3r} - k_{3c}) \tag{43}$$

$$C_{33} = -\left(\frac{1}{D_{3r}} - \frac{1}{D_{3c}}\right) . \tag{44}$$

As $\frac{\Gamma_r^2}{\Delta E_{rc}}$ is proportional to the scaling parameter σ in Eq. (34), $\frac{\Gamma_r^2}{\sigma \Delta E_{rc}}$ would be expected to be more self-averaging than $\frac{\Gamma_r^2}{\Delta E_{rc}}$ itself. In order to account both for this dependence and for possible correlations between Γ_r and ΔE_{rc}, $\frac{\Gamma_r^2}{\sigma \Delta E_{rc}}$ was averaged over a set of target proteins as aligned to homologous and random proteins, and $\sigma \langle \frac{\Gamma_r^2}{\sigma \Delta E_{rc}} \rangle$, with the value of σ appropriate for that target protein, was used in Eq. (33) as the best approximator to $\frac{\Gamma_r^2}{\Delta E_{rc}}$.

In order to generate statistics on gaps between homologous proteins, 182 pairs of structurally-similar homologous proteins were aligned so as to maximize the pair's q score, defined as[20]:

$$q_{ab} = [N(N-1)]^{-1} \sum_{i \neq j} \theta(r_{ij}^a - r_{i'j'}^b) \,, \tag{45}$$

where θ is a Gaussian function of width 1–2 Å depending upon $j - i$, and r_{ij}^a and $r_{i'j'}^b$ are distances between residues i and j of protein a and corresponding residues i' and j' of protein b, respectively. These structures were then annealed by allowing the alignment to shift, until the lowest energy structure with $q > 0.35$, indicating adequate structural similarity, was found. Gaps were not allowed within structural elements such as an α-helix or β-sheet strands. Unphysical gaps, where $r_{i'j'}$ exceeded $3.8(j-i)$ Å by more than 5 Å when $j - i > 1$, or exceeded 7.5 Å in the case of $j - i = 1$, were also prohibited. The total number of gaps of each category, the average value of $j - i$ and the average and second moment of $r_{i'j'}$ were used to calculate the best estimators for $\{A_{1c}, A_{2c}, A_{3c}, k_{1c}, k_{3c}, r_{0c}, D_{2c}, D_{3c}\}$.

An iterative method was used to obtain a self-consistent set of gap penalties. 382 pairs of unrelated proteins were aligned, using the optimized local-interaction Hamiltonian and an estimated set of gap penalties, $\{\mu^n\}$. As with the correct alignments, unphysical gaps and gaps within structural elements were prohibited. Again, the total number of gaps of each category, the average value of $j - i$ and the average and second moment of $r_{i'j'}$ were used to calculate the best estimators for $\{A_{1r}, A_{2r}, A_{3r}, k_{1r}, k_{3r}, r_{0r}, D_{2r}, D_{3r}\}$, while the resulting distribution of energies for the random pairs of proteins compared with the distributions for the same targets in alignments between homologs furnished $\langle \frac{\Gamma_r^2}{\sigma \Delta E_{rc}} \rangle$. These values were used to calculate $\{\mu'^n\}$ using Eqs. (36) to (44). The next estimated set of gap penalties, $\{\mu^{n+1}\}$, was calculated using:

$$\{\mu^{n+1}\} = \{\mu^n\} - \epsilon\{f\} \tag{46}$$

where $f_l(\{\mu^n\}) \equiv \mu_l'^n - \mu_l{}^n$ and ϵ is a small constant. In practice, various values of $\epsilon \sim 0.05$ were used to generate a set of values of $\{\mu^{n+1}\}$, with the values chosen that minimized the largest values of $\left|\frac{f_l}{\mu_l}\right|$.

Exact self-consistency would be achieved when $\{\mu_l{}^n\} = \{\mu_l'^n\}$, or

$$f_l(\{\mu_l{}^n\}) \equiv \mu_l'^n - \mu_l{}^n = 0 \quad \forall\, l\,. \tag{47}$$

There are subtleties, however, involving the use of a somewhat limited set of example proteins. The statistical theory involves expanding $\log \Omega(E_k, \{n_l^k\}, \mathcal{S}_k, \mathcal{R})$ and $\log \Omega(E_k, \{n_l^k\}, \mathcal{S}_k, \mathcal{C})$ as a function of the number of gaps around $\{n_l\} = \{n_l^c\}$. As small changes in the gap parameters can cause large changes in the alignments, especially of non-homologous pairs of proteins, these functions have some chaotic characteristics for individual proteins, and only become smooth and analytical in the thermodynamic limit. For a finite set of proteins, it will be possible to coarse-grain $\{\mu_l\}$ so that $\log \Omega(E_k, \{n_l^k\}, \mathcal{S}_k, \mathcal{R})$ and $\log \Omega(E_k, \{n_l^k\}, \mathcal{S}_k, \mathcal{C})$ are analytical, and the iterative method will converge at the resolution of the course-graining. Any attempt to locate the optimal values of $\{\mu_l\}$ on a finer scale will lead to a numerically ill-conditioned problem. The scale of the coarse-graining required will depend upon the number of pairs of proteins being used to determine the statistics. For the number of proteins defining our set of random pairs, it was possible to satisfy Eq. (47) within a final value of $\left|\frac{f_l}{\mu_l}\right|$ of approximately 10% for insertions and deletions, with a larger error range of about 20% for more generalized gaps, which were more poorly represented among the alignments.

Final values of the gap parameters were:

$$\begin{aligned}
C_{11} &= 0.1240\\
C_{21} &= 0.0685\\
C_{12} &= 0.9927\\
C_{22} &= -0.2929\\
C_{32} &= 0.0283\\
C_{13} &= 0.1680\\
C_{23} &= 0.0652\\
C_{33} &= 0.0063\,.
\end{aligned}$$

The final value of $\langle\frac{\Gamma_r{}^2}{\sigma\Delta E_{rc}}\rangle$ was 22.19. The value of $T_p = \sigma\langle\frac{\Gamma_r^2}{\sigma\Delta E_{rc}}\rangle$ will depend upon the target protein and the corresponding value of σ, which can be computed by calculating the width of the energy distribution of the target protein in a set of random configurations. For the proteins in the test set, the value of σ varied between approximately 9.0 to 16.0, resulting in values of T_p of between 1.4 to 2.4.

There are complications adapting dynamic programming techniques to the sequence-structure alignment problem in the presence of multi-body interactions. For sequence-sequence alignments, and energy functions that only involve one-body terms (such as the profile methods of Eisenberg and co-workers[24]) the dynamic programming algorithm of Smith and Waterman can be shown to generate the optimal alignment.[52] In the presence of two-body terms, such as in our Hamiltonian, the alignment of one residue depends upon the position of all of the other residues in the protein. We used the approach developed by Finkelstein and Reva,[23] later used by Thornton and Skolnick and their respective co-workers[25-27] where an initial alignment was constructed and further refinements made in an iterative fashion. This can be considered as an iterative form of mean-field approach which converges when a self-consistent result is obtained, and the alignment is optimal given the "field" at each residue produced by all of the other residues in their final alignments.

In the implementation of the dynamic programming algorithm two metrics are required, a scoring matrix **S**, whose elements $S_{ii'}$ represent the energetic contribution from residue i being pinned in location i' on the template protein, and the gap penalty functions μ_l calculated above. In our previous work the alignments were made using a scoring matrix based on the Dayhoff similarity matrix with the default penalty functions. In the work presented here, initial alignments were generated using a scoring matrix **S** based on our energy function in which the identity $A_{i'}^S$ and the context $C_{i'}^S$ of the amino acids in the template structure were used in evaluating the energy contributions.

$$\begin{aligned} S_{ii'}^0 &= \gamma_p(A_i, C_{i'}^S) - \langle \gamma_p(A_i) \rangle_\zeta \\ &\quad + \sum_\rho \sum_{j'} \gamma_c^\rho(A_i, A_{j'}^S) u(r_c^\rho - r_{i'j'}) \,. \end{aligned} \tag{48}$$

Refinements of the initial alignment were found in a meanfield iterative procedure in which the scoring matrix was updated by replacing the scaffold residue in location j' by the target residue j aligned to that location.

$$S_{ii'} = \gamma_p(A_i, C^S_{i'}) - \langle\gamma_p(A_i)\rangle_\zeta + \sum_\rho \sum_{\substack{j \\ \text{pinned}}} \left\{\gamma^\rho_c(A_i, A_j)u(r^\rho_c - r_{i'j'}) - \langle\gamma^\rho_c(A_i, A_j)\rangle_\zeta\right\} . \quad (49)$$

Interactions involving residues in the target protein not aligned to any corresponding residue in the scaffold protein were given the average interaction of the liquid-like states, equal to zero with the current Hamiltonian.

In order to provide smoother convergence, a relaxation method was employed, where a linear combination of the previous and the new values of $S_{ii'}$ were used for the subsequent value of $S^n_{ii'}$:

$$S^n_{ii'} = (1-\epsilon)S^{n-1}_{ii'} + \epsilon S_{ii'} . \quad (50)$$

The maximum similarity measure **M** is calculated according to a generalization of the standard dynamic programming algorithm:

$$M_{jj'} = S_{jj'} + \max\left\{M_{j-1,j'-1}, \max_{\substack{1<i<j \\ 1<i'<j'}} \{M_{i,i'} - \mu(j-i, r_{i'j'})\}, 0\right\} . \quad (51)$$

The segments of maximum similarity are determined by the usual traceback procedure beginning at the maximum element of **M**.

5. Results

The set of 42 sample proteins used to optimize the Hamiltonian was used to form the set of possible structural motifs. These represented well-defined structures of between 50 and 250 residues in length, with overly similar structures eliminated. A set of 17 proteins of known structure was used to create a test set. Each of the test proteins was structurally similar to one of the motif proteins, but with limited sequence similarity (17 to 40% sequence identity). In addition, the test protein itself was included in the set of structural motifs, in order to test self-recognition. The mean-field threading algorithm was used to align the test proteins to each of the structural motif proteins in turn, using the statistically-derived gap penalties in Eq. (33).

Table 2. Results of aligning 17 test proteins with a set containing test and the 42 sample proteins, at least one of which had a similar structure but dissimilar sequence. Listed for each of the test proteins are the free-energy of the protein aligned to itself combined with the respective q-scores, with $q = 1.0$ indicating an exactly correct alignment. Also shown is the alignment of lowest free-energy to the sample proteins excluding the test protein, with the free-energy and resulting q-score, and the sequence similarity of the target protein with the structural homolog in the sample set, as measured by percentage identity (% I). In all cases, the lowest free-energy structure was the test protein aligned to itself. With the exception of tuna cytochrome C, excluding the exactly correct structure from consideration, the alignment of lowest free energy was to a structural homolog, with generally accurate alignments.

Self-recognition				Prediction				
	Target	**F**	**q**		**Scaffold**	**F**	**q**	**% I**
3DFR	*L. casei* DHFR	− 10.70	0.95	8DFR	Chicken DHFR	− 9.49	0.45	31
1ALC	Baboon Alpha-lactalbumin	− 10.34	1.00	2LYZ	Hen Egg White Lysozyme	− 9.83	0.68	37
1FX1	*D. vulgaris* Flavodoxin	− 13.53	1.00	3FXN	*Clostridium MP* Flavodoxin	− 9.61	0.51	31
1MBA	Sea Hare Myoglobin	− 11.07	1.00	5MBN	Sperm Whale Myoglobin	− 9.04	0.62	28
2LHB	Sea Lamprey Hemoglobin	− 12.18	1.00	5MBN	Sperm Whale Myoglobin	− 8.84	0.52	26
2LH4	Lupin Leghemoglobin	− 9.84	1.00	5BMN	Sperm Whale Myoglobin	− 8.61	0.41	19
1FDH(G)	Human Fetal Hemoglobin (γ)	− 9.71	1.00	5MBN	Sperm Whale Myoglobin	− 8.12	0.65	23
3HFM(L)	Mouse Fab(L)	− 9.78	0.85	3HFM(H)	Mouse Fab (H)	− 8.09	0.27	26
4FAB(L)	Mouse FAB(L)	− 11.01	0.83	3HFM(H)	Mouse Fab (H)	− 7.93	0.30	21
4DFR(B)	*E. coli* DHFR	− 11.48	0.95	8DFR	Chicken DHFR	− 7.36	0.58	34
1hds(b)	Deer SC Hemoglobin (β)	− 8.70	1.00	5MBN	Sperm Whale Myoglobin	− 7.20	0.72	25
1F19(L)	Mouse Fab (L)	− 10.46	0.75	3HFM(H)	Mouse Fab (L)	− 7.12	0.32	24
2RHE	Human Bence-Jones (Variable)	− 7.50	1.00	3HFM(H)	Mouse Fab (H)	− 6.86	0.57	29
2HHB(A)	Human Homoglobin (α)	− 7.72	1.00	5MBN	Sperm Whale Myoglobin	− 6.48	0.64	26
5CYT	Tuna Cytochrome C	− 7.52	1.00	1RBB(A)	Bovine Ribonucleases (A)	− 5.94	0.17	40
1REI(A)	Human Bence-Jones (Variable)	− 8.63	1.00	1CD4	Human CD4	− 5.63	0.30	30
256B(A)	*E. coli* Cytochrome 562	− 8.45	1.00	2CCY(A)	*R. molischianum* Cytochtome C'	− 4.18	0.36	17

The structure of lowest F-score, excluding self-recognition, was taken as the predicted structure.

The sequence alignment method achieved correct self-recognition for all of the test cases, generally with exactly correct alignments. sometimes with minor shifts resulting in q-scores slightly smaller than 1.0. As shown in Table 2, when one leaves out the exactly correct template in the scaffold set but retains a structural homolog, a correct assignment was made in 16 out of 17 of the test cases. As shown in Fig. 3, a comparison of the distance plots for the correct (X-ray crystallography) structure and the structure based on the alignment to the structural motif shows that the alignments were generally quite accurate, even in cases that are generally difficult for standard sequence alignment methods (leghemoglobin with sperm whale

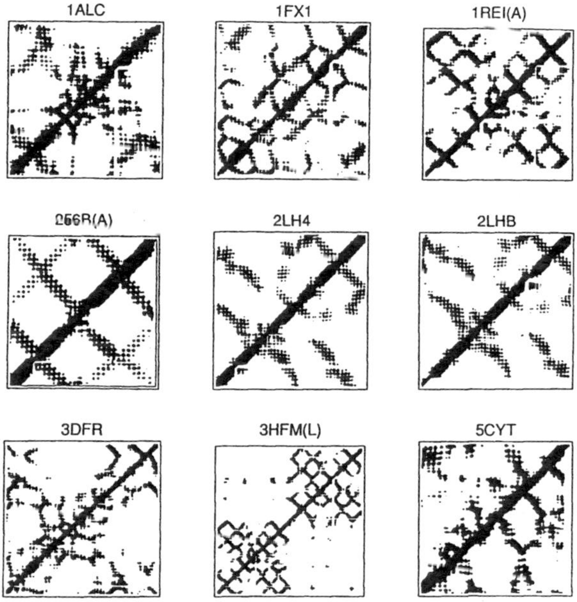

Fig. 3. Distance plots for some of the test proteins listed in Table. 2. A point on the distance map is shaded if the residues corresponding to that point are within 12 Å. For each protein, the distance plot above the diagonal refers to the known structure determined using X-ray crystallography, while the distance plot below the diagonal refers to the prediction based on the lowest free-energy alignment with the set of sample proteins excluding itself. A symmetric pattern indicates a successful prediction, as all of these are, with the exception of tuna cytochrome C, 5CYT.

myoglobin, Bence-Jones protein with a CD-4 fragment, and cytochrome 562 with cytochrome C').

The one true failure involved the prediction of the structure of tuna cytochrome C (5CYT). The scaffold protein with the lowest free-energy bovine ribonuclease B (1RBB(A)), although the correct homolog (3C2C), was only slightly higher in free-energy ($F = -5.31$ for 3C2C, vs $F = -5.94$ for 1RBB(A)). The alignment of 5CYT to 3C2C was accurate, with a q-score of 0.76. Although this is one of the easiest alignments in the test set for more standard sequence alignment methods, it is interesting to note that this protein was also a problem for the earlier studies using molecular dynamics with the Associative Memory Hamiltonian.[20] As with this other Hamiltonian, no account is taken of the interactions with the heme prosthetic group. Apo-cytochrome C is known to be much less stable than holo-cytochrome C; this suggests that including prosthetic groups explicitly in this model might be necessary for small proteins where interactions with these groups furnish much of the stabilization energy. The absence of such interactions in the model did not affect as much the predictions for larger prosthetic-group containing proteins, such as the myoglobins.

6. Conclusion

Major advances have often been made in the protein structure prediction problem when connections have been made with the techniques and approaches of statistical mechanics. In this paper we have developed a statistical approach to the problem of aligning a protein sequence with a set of known protein structures. The results indicate that the optimal Hamiltonian or cost function for the alignment is related to the optimal Hamiltonians that we have developed earlier based on ideas from spin-glass theory. As with this optimal Hamiltonian, the optimal gap penalties involve the statistics of distributions of not only correctly aligned proteins, but also the distributions of random, misaligned proteins. We show that by using this Hamiltonian and gap penalties, we are able to correctly predict the structure of a vast majority of sets of test proteins, including proteins that generally fail to be aligned with more standard sequence alignment methods. Simultaneous self-optimization of the energy and gap parameters has been recently carried out and shown to improve the discrimination between folded and misfolded states. These parameters appear to be successful in predicting structures with known local prosite signatures.[53]

This result is of general applicability, and should provide a statistical basis for alignment methodologies in general, including the more commonly used sequence–sequence alignment techniques.

Acknowledgments

This work was supported by NIH grant PHSIRO/GM44557-01. Computational resources were provided by the National Center for Supercomputing Applications in Urbana and by the FMC Corporation. Helpful discussions with D. Evensky are gratefully acknowledged.

References

1. C. Anfinsen, E. Haber, M. Sela, and F. White, *Proc. Nat. Acad. Sci. USA* **47**, 1309 (1961).
2. T. L. Blundell, B. L. Sibanda, M. J. E. Sternberg, and J. M. Thornton, *Nature* (London) **326**, 347 (1987).
3. A. Sali, J. P. Overington, M. S. Johnson and T. L. Blundell, *TIBS* **15**, 235 (1990).
4. P. Chou and G. Fasman, *Biochemistry* **13**, 222 (1974).
5. A. V. Finkelstein and O. Ptitsyn, *J. Mol. Biol.* **62**, 613 (1971).
6. J. Garnier, D. Osguthorpe and B. Robson, *J. Mol. Biol.* **120**, 97 (1978).
7. O. Gascuel and J. Golmard, *CABIOS* **4**, 357 (1988).
8. J. F. Gibrat, J. Garnier and B. Robson, *J. Mol. Biol.* **198**, 425 (1987).
9. F. Cohen, R. Abarbanel, I. Kuntz and R. Fletterick. *Biochemistry* **25**, 266 (1986).
10. V. Lim, *J. Mol. Biol.* **88**, 873 (1974).
11. B. Sibanda and J. Thornton, *Nature* (London) **316**, 170 (1985).
12. H. Bohr, J. Bohr, S. Brunak, R. M. J. Cotterill, B. Lautrup, L. Norskov, O. Olsen and S. Petersen, *FEBS Lett.* **241**, 223 (1988).
13. H. Bohr, J. Bohr, S. Brunak, R. M. J. Cotterill, B. Lautrup, L. Norskov, O. Olsen and S. Petersen, *FEBS Lett.*, **261**, 43 (1990).
14. L. H. Holley and M. Karplus, *Proc. Nat. Acad. Sci. USA* **86**, 152 (1989).
15. N. Qian and T. J. Sejnowski, *J. Mol. Biol.* **202**, 865 (1988).
16. D. G. Kneller, F. E. Cohen and R. Langridge, *J. Mol. Biol.* **214**, 171 (1991).
17. G. L. Wilcox, M. Poliac and M. N. Liebman, *Tetra. Comp. Meth.* **3**, 191 (1990).
18. J. Hopfield, *Proc. Nat. Acad. Sci. USA* **79**, 2554 (1982).
19. M. Sasai and P. G. Wolynes, *Phys. Rev. Lett.* **65**, 2740 (1990).
20. R. A. Goldstein, Z. A. Luthey-Schulten, and P. G. Wolynes, *Proc. Nat. Acad. Sci. USA* **89**, 4918 (1992).
21. R. A. Goldstein, Z. A. Luthey-Schulten and P. G. Wolynes, *Proc. Nat. Acad. Sci. USA* **89**, 9029 (1992).

22. R. A. Goldstein, Z. A. Luthey-Schulten and P. G. Wolynes, in *Proc. 26th Annual Hawaii Int. Conf. on System Sciences*, Vol. 1, eds. T. N. Mudge, V. Milutinovic and L. Hunter, (IEEE Computer Society Press, pp. 699–707, 1993).
23. A. V. Finkelstein and B. Reva, *Nature* (London) **351**, 497 (1991).
24. J. U. Bowie, R. Luthy and D. Eisenberg, *Science*, **25** 164 (1991).
25. D. T. Jones, W. R. Taylor and J. M. Thornton, *Nature* (London) **358**, 86 (1992).
26. A. Godzik and J. Skolnick, *Proc. Nat. Acad. Sci. USA* **89**, 12098 (1992).
27. A. Godzik, A. Kolinski and J. Skolnick, *J. Mol. Biol.* **227**, 227 (1992).
28. E. I. Shakhnovich and A. Gutin, *Europhys. Lett.* **8**, 327 (1988).
29. E. I. Shakhnovich and A. Gutin, *J. Phys.* **A22**, 1647 (1989).
30. T. Garel and H. Orland, *Europhys. Lett.* **6**, 307 (1988).
31. T. Garel and H. Orland, *Europhys. Lett.* **6**, 597 (1988).
32. W. M. Fitch and T. F. Smith, *Proc. Nat. Acad. Sci. USA* **80**, 1382 (1983).
33. D. F. Feng, M. S. Johnson and R. F. Doolittle, *J. Mol. Evol.* **21**, 112 (1985).
34. E. J. Demchuk, G. N. Esipova and V. G. Tumanyan. *Studia Biophysica* **2–3**, 193 (1989).
35. S. A. Benner, M. A. Cohen and G. H. Gonnet, *J. Mol. Biol.* **229**, 1065 (1993).
36. R. A. Goldstein, J. Katzenellenbogen, Z. A. Luthey-Schulten, D. Seielstad and P. G. Wolynes, *Proc. Nat. Acad. Sci. USA* **90**, 9949 (1993).
37. S. Miyazawa and R. L. Jernigan, *Macromolecules* **18**, 534 (1985).
38. M. J. Sippl, *J. Mol. Biol.* **213**, 859 (1990).
39. M. J. Sippl and S. Weitckus, *Proteins* **13**, 258 (1992).
40. J. D. Bryngelson and P. G. Wolynes, *Biopolymers* **30**, 171, (1990).
41. M. Levitt, *J. Mol. Biol.* **104**, 59 (1976).
42. J. Richardson, *Advan. Protein Chem.* **34**, 167 (1981).
43. A. V. Finkelstein and O. B. Ptitsyn, *Prog. Biophys. Mol. Biol.* **50**, 171 (1987).
44. A. G. Murzin and A. V. Finkelstein, *J. Mol. Biol.* **204**, 749 (1988).
45. C. Chothia, *Nature* (London) **357**, 543 (1992).
46. W. Kabsch and C. Sander, *Biopolymers* **22**, 2577 (1983).
47. A. Shrake and J. A. Rupley, *J. Mol. Biol.* **79**, 351 (1973).
48. F. M. Richards, *Ann. Rev. Biophys. Bioeng.* **6**, 151 (1977).
49. F. C. Bernstein, T. F. Koetzle, G. J. B. Williams, E. F. Meyer, Jr., M. D. Brice, J. R. Rodgers, O. Kennard, T. Shimanouchi and M. Tasumi, *J. Mol. Biol.* **112**, 535 (1977).
50. E. E. Abola, F. C. Bernstein, S. H. Bryant, T. F. Koetzle and J. Weng, in *Crystallographic Databases—Information Content, Software Systems, Scientific Applications*, eds. F. H. Allen, G. Bergerhoff, and R. Sievers, (Data Commision of the Int. Union of Crystallography, Bonn, 1987) pp. 107-132.
51. G. M. Crippen, *Biochemistry* **30**, 4232 (1991).
52. T. F. Smith and M. S. Waterman, *J. Mol. Biol.* **147**, 195 (1981).
53. K. K. Koretke, Z. Luthey-Schulten and P. G. Wolynes, *Prot. Science*, June 1996 (in press).

INDEX